T0372007

Quantum Mechanics

Foundations and Applications

Quantum Mechanics

Foundations and Applications

D G Swanson
Auburn University, Alabama, USA

Taylor & Francis is an imprint of the
Taylor & Francis Group, an informa business

CRC Press
Taylor & Francis Group
6000 Broken Sound Parkway NW, Suite 300
Boca Raton, FL 33487-2742

Printed in the United States of America on acid-free paper
10 9 8 7 6 5 4 3 2 1

International Standard Book Number-10: 1-58488-752-4 (Hardcover)
International Standard Book Number-13: 978-1-58488-752-2 (Hardcover)

Library of Congress Cataloging-in-Publication Data

Swanson, D. G. (Donald Gary), 1935-
Quantum mechanics : foundations and applications / D.G. Swanson.
p. cm.
Includes bibliographical references and index.
ISBN 1-58488-752-4 (alk. paper)
1. Quantum theory. I. Title.

QC174.12.S93 2006
530.12--dc22 2006045501

Visit the Taylor & Francis Web site at
http://www.taylorandfrancis.com

and the CRC Press Web site at
http://www.crcpress.com

Preface

This book grew out of a compilation of lecture notes from a variety of other texts in Modern Physics and Quantum Mechanics. Except for the part about digital frequency counters related to the uncertainty principle, none of the material is truly original, although virtually everything has been rewritten in order to unify the presentation. Over the course of twenty years, I taught the Quantum Mechanics off and on for physics majors in their junior or senior year, and this same course filled the gap for incoming graduate students who were not properly prepared in Quantum Mechanics to take our graduate course. Originally, it was a full year course over three quarters, but when we converted to semesters, the first half was required and the second half was optional, but recommended for majors intending to go on to graduate work in physics. In our present curriculum, the first five chapters comprise the first semester and is designed for juniors, while the remainder is designed for seniors. This leaves many of the later chapters optional for the instructor, and some sections are marked as advanced by an asterisk and can be easily deleted without affecting the continuity. I like to use the generating functions in Appendix B to establish normalization constants and recursion formulas, but they can be omitted completely. The 1-D scattering in Chapter 10 is marked as advanced, and may be skipped, but the unique connection between the 1-D Schrödinger equation and the Korteweg–deVries equation for solitons, which is solved using the inverse scattering method, where a nonlinear classical problem is solved by the linear techniques of Quantum Mechanics I found compelling.

My own first course in Quantum Mechanics was taught by Robert Leighton the first year his text came out, and his use of the Fourier transform pairs for the wave functions in coordinate and momentum space made the operator formalism transparent to me, and this feature, more than any other single factor, provided the impetus for me to write this text, since this formalism is uncommon in most of the modern texts. Some of the text follows Leighton, but there have been so many other primary texts for our course over the years (with my notes as an adjunct to the text) that the topics and problems come from many sources, most of which are listed in the bibliography. In recent years, I have expanded the notes to form a primary text which has been used successfully in preparing our students for the graduate level courses in Quantum Mechanics and Quantum Statistics.

This text may be especially appealing to Electrical Engineering students (I taught the Quantum Mechanics for Engineers course at the University of

Texas at Austin for many years) because they are already familiar with the properties of Fourier transforms.

I would like to thank Leslie Lamport and Donald Knuth for their development of LaTeX and TeX respectively, without which I would never have attemped to write this book. The figures have been set with TeXniques by Michael J. Wichura.

If a typographical or other error is discovered in the text, please report it to me at swanson@physics.auburn.edu and I will keep a downloadable errata page on my webpage at www.physics.auburn.edu/~swanson.

D. Gary Swanson

Contents

1

The Foundations of Quantum Physics

1.1 The Prelude to Quantum Mechanics

Quantum mechanics is essentially a 20th century development that is based on a number of observations that defied classical explanations. While some of these experiments have semiclassical explanations, the triumph of quantum mechanics is that it gives precise verification for an overwhelming number of experimental observations. Its extensions into relativistic quantum mechanics through quantum electrodynamics (QED), electro-weak theory, and quantum chromodynamics (QCD) have led us to the Standard Model of today. While the number of unanswered questions remains approximately constant at each stage of development, the number of answered questions that relate theory with experiment continues to grow rapidly.

In this chapter, some of the historical landmarks in the development of the theory are noted. The resolution of the dilemmas presented by classical theory will be dealt with in later chapters, but these are listed to motivate the break from classical mechanics. Because of the apparently unphysical nature of the postulates upon which quantum mechanics is founded, we supply motivation and justification for this break. In the first part of this chapter, we will review some of the experiments that confounded classical theory, and then introduce a formalism that provides some rationale for the postulates upon which quantum mechanics is based. In the end, these postulates will stand on their own.

1.1.1 The Zeeman Effect

Looking back to the last few years of the 19th century, the discovery by J.J. Thomson of the electron in 1897 was followed almost immediately by the announcement from Zeeman and Lorentz that it participated in electromagnetic radiation from atoms. Assuming that electrons in a uniform magnetic field would radiate as dipoles, they discovered that the emission from an electron in an atom, whose natural frequency is ω_0, would be split into three distinct frequencies by the magnetic field, whose magnitudes are given by

$$\omega_1 = \omega_0 + \frac{eB}{2m}, \qquad \omega_2 = \omega_0, \qquad \omega_3 = \omega_0 - \frac{eB}{2m}, \tag{1.1}$$

where B is the magnetic induction, e is the electronic charge, and m is the electron mass. While many atoms exhibited this behavior, notably the sodium D-lines that were studied by Zeeman, numerous examples were discovered that did not follow any simple pattern. Some had many additional frequencies, some had no unshifted frequency, and sometimes the splitting was larger than expected. These unusual cases were called the **anomalous Zeeman effect** while the simple cases were simply called examples of the **normal Zeeman effect**.

Problem 1.1 *Normal Zeeman effect.*

(a) Consider a particle with charge e and mass m moving in a magnetic field $\boldsymbol{B} = B\hat{\boldsymbol{e}}_z$ under an external dipole force $\boldsymbol{F} = -m\omega_0^2\boldsymbol{r}$. Write the differential equations in rectangular coordinates.

(b) Solve the differential equations and show that the three frequencies given in Equation (1.1) are the natural frequencies provided $\omega_0 \gg eB/m$. *Hint:* The change of variables to $u = x + iy$ and $w = x - iy$ may be helpful in converting from two *coupled* equations in x and y to two *uncoupled* equations in u and w.

(c) Show that the three frequencies may be identified with a linear oscillation parallel to the magnetic field and two circular motions in a plane perpendicular to the magnetic field, one where the magnetic force is inward and the other outward.

1.1.2 Black-Body Radiation

At the turn of the century, Planck first introduced the notion of **quantization**, which was revolutionary, and not altogether satisfactory even to him. That one should be led to such an outrageous notion requires severe provocation, and this serious failure of classical theory provided such an impetus. The experimental distribution of radiation from a black body was well enough measured at that time, and an empirical formula was discovered by Planck, but no theory that matched it was available. On the contrary, existing theory contradicted it.

While the radiation from different materials differs from black-body radiation to some extent, due both to the type of material and on the surface preparation, it is found that emission from a small hole in a cavity is independent of the material and has a universal character that we call black-body radiation (by definition a black body absorbs everything and reflects nothing, and a small hole lets all radiation in and virtually nothing reflects back out the hole if it is small enough). This universal distribution function depends only on temperature, and its character is shown in Figure 1.1. It should be noted from the figure that there is a peak in the distribution that depends on temperature, that the area under each curve increases with temperature, and that for most common temperatures, the peak radiation is in the red or infrared (visible radiation ranges from 0.4 microns to 0.7 microns).

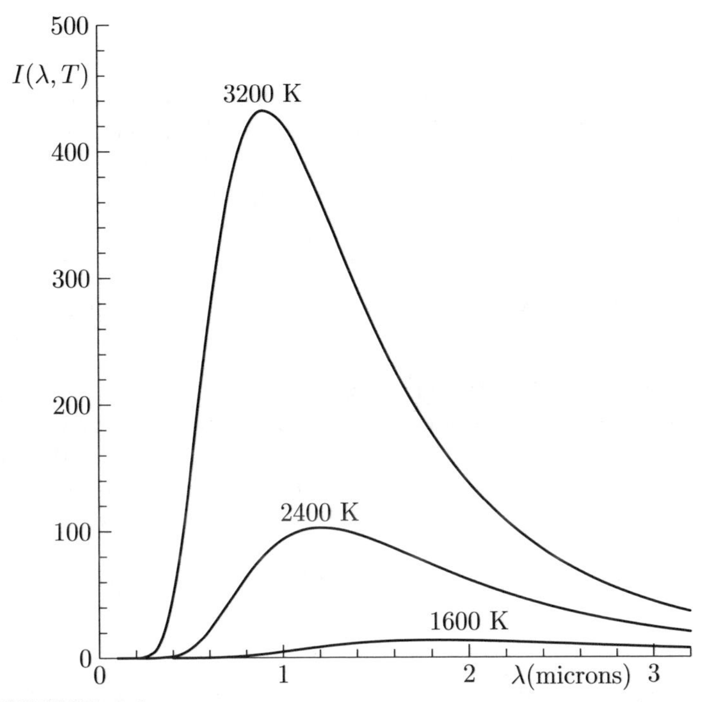

FIGURE 1.1
Spectral distribution of black-body radiation for several temperatures. $I(\lambda, T)$ is total emissive power in watts per square centimeter per micron.

The most successful theories describing this distribution function were obtained from thermodynamics and accounted for some, but not all, of the features evident in the figure. From this classical theory, the following features of the radiation were established:

1. **Kirchhoff's Law** states that for any partially absorbing media exposed to isotropic radiation of wavelength λ, the fraction absorbed, $A(\lambda, T)$, is proportional to the emissivity $E(\lambda, T)$, such that $E/A = E_{\text{BB}}$ where $E_{\text{BB}}(\lambda, T)$ is the emissivity of a black body at that wavelength and temperature.

2. The classical expression relating energy density and momentum for plane waves relate the pressure P and the energy density U for isotropic radiation:
$$P = \tfrac{1}{3}U\,.$$

3. From classical electromagnetic theory, it then follows that the total power, e_{B}, emitted per unit surface area of a black body is given simply by

$$e_{\mathrm{B}}(T) = \tfrac{1}{4}cU$$

where c is the speed of light.

4. The **Stefan–Boltzmann law** is then established by considering a Carnot engine with electromagnetic radiation as a working fluid in a cavity. This leads to the conclusion that the total power emitted per unit area is proportional to the fourth power of the temperature, or

$$e_{\mathrm{B}}(T) = \sigma T^4$$

where σ is Stefan's constant whose value could not be determined from classical theory.

5. During a reversible adiabatic expansion of black-body radiation, the radiation must remain of the same character as the temperature changes or else there would be a violation of the second law. By considering the Doppler effect from a perfectly reflecting moving piston, **Wien's displacement law** was established: If cavity radiation is slowly expanded or compressed to a new volume and temperature, the radiation that was originally present at wavelength λ will be shifted to wavelength λ' where

$$\lambda T = \lambda' T'$$

and the energy density $\mathrm{d}U = (4/c)I(\lambda, T)\,\mathrm{d}\lambda$ where $I(\lambda, T)$ is the **intensity** will in the process be changed to $\mathrm{d}U' = (4/c)I(\lambda', T')\,\mathrm{d}\lambda'$, that is given by

$$\frac{\mathrm{d}U}{\mathrm{d}U'} = \left(\frac{T}{T'}\right)^4 .$$

This reduces the problem of describing black-body radiation to finding a single function of the product λT by writing

$$\begin{aligned} I(\lambda, T) &= \left(\frac{T}{T'}\right)^4 I(\lambda', T')\frac{\mathrm{d}\lambda'}{\mathrm{d}\lambda} \\ &= \left(\frac{T}{T'}\right)^5 I\left(\frac{\lambda T}{T'}, T'\right) \\ &= T^5 f(\lambda T). \end{aligned} \tag{1.2}$$

This was found to be in excellent agreement with experiment both in relation to the shift of the peak of the distribution noted in Figure 1.1, where $\lambda_{\max} T$ =const. (which is sometimes referred to as Wien's displacement law rather than the more general statement above) but also in the overall distribution.

After these successes, the attempt to find this function $f(\lambda T)$ was an embarrassing failure. Actually, it was recognized that more than thermodynamics was needed, since although thermodynamics can give the *average* energy and velocity of a system of molecules, it cannot give the *distribution* of velocities. For this, statistical mechanics is necessary, so the attempt to find the appropriate distribution was based on classical statistical mechanics.

One treatment was based on very general arguments — that of Rayleigh and Jeans (1900) — and assumed the equipartition of energy must hold for the various electromagnetic modes of vibration of the cavity in which the radiation is situated. There should thus be κT per normal mode of oscillation times the number of normal modes per unit wavelength interval. This is obviously an *impossible* result, since the medium has a *continuum* and hence an infinite number of normal modes and hence an infinite amount of energy and an infinite specific heat. In the long wavelength limit, however, this does give both the proper form and magnitude, so that the Rayleigh-Jeans law is frequently given as

$$I(\lambda, T) = \frac{2\pi c \kappa T}{\lambda^4}.$$

This expression and its integral over the entire spectrum is obviously unbounded as $\lambda \to 0$, and this result was known as the **ultraviolet catastrophe**. Planck found an alternate approach by considering the radiation field to interact with charged, one-dimensional **harmonic oscillators**. He argued that since the nature of the wall does not affect the distribution function, a simple model should suffice. He had already discovered an empirical formula for the distribution function of the form

$$I(\lambda, T) = \frac{c_1}{\lambda^5(e^{c_2/\lambda \kappa T} - 1)}$$

that agreed in form with the Rayleigh–Jeans result at long wavelengths. He made the following hypotheses:

1. Each oscillator absorbs energy in a continuous way from the radiation field.

2. An oscillator can *radiate* energy only when its total energy in an *exact integral multiple* of a certain unit of energy for that oscillator, and when it radiates, it radiates *all* of its energy.

3. The actual radiation or nonradiation of energy of a given oscillator as its energy passes through the critical value above is determined by statistical chance. The ratio of the probability of nonemission to the probability of emission is proportional to the intensity of the radiation that excites the oscillator.

Planck assumed that the possible radiating energies were given by

$$E_n = nh\nu$$

where $\nu(= f)$ is the frequency of the oscillator, n is an integer, and h is a constant. Thus he arrived at the expression

$$I(\lambda, T) = \frac{2\pi c^2 h}{\lambda^5 (e^{ch/\lambda\kappa T} - 1)}. \tag{1.3}$$

The constant h is called Planck's constant, and has the value $h = 6.626 \times 10^{-34}$ J-s.

Problem 1.2 Show from Equation (1.2) that the wavelength λ_{max} at which $I(\lambda, T)$ is maximum satisfies the relation $\lambda_{\text{max}} T$ =const. Evaluate this constant, using Equation (1.3).

Problem 1.3 The wavelength of maximum intensity in the solar spectrum is approximately 500 nm. Assuming the sun radiates as a black body, find the surface temperature of the sun.

Problem 1.4 Over what range of wavelengths is the Rayleigh–Jeans law within 10% of the black-body (Planck) law?

1.1.3 Photoelectric Effect

In Planck's hypotheses that led to the black-body spectrum, he assumed that only the oscillators were quantized, while the radiation field was not. This was generally regarded as an *ad hoc* hypothesis, and many tried to reproduce his result without any quantum hypothesis, but failed. Soon after Planck's first publication, Einstein put forward the notion that the *radiation field itself might also be quantized.* This was put forward in his proposed explanation of the **photoelectric effect** (1905). This effect, first observed in 1887 by Hertz, exhibited the following experimentally discovered facts:

1. *Negative* particles were emitted (Hallwachs, 1889).

2. The emitted particles are forcibly ejected by the light (Hallwachs, Elster, and Geitel, 1889).

3. There is a close relationship between the **contact potential** of a metal and its *photosensitivity* (Elster and Geitel, 1889).

4. The photocurrent is *proportional to the intensity of the light* (Elster and Geitel, 1891).

5. The emitted particles are electrons (Lenard and J. J. Thomson, 1899).

6. The *kinetic energies* of the emitted particles are *independent* of the intensity of the light, while the *number* of electrons is proportional to the intensity (Lenard, 1902).

7. The emitted electrons possess a maximum kinetic energy that is greater, the shorter the wavelength of the light, and no electrons whatever are emitted if the wavelength of the light exceeds a threshold value (Lenard, 1902).

8. Photoelectrons are often emitted without measurable time delay after the illumination is turned on. For weak illumination, the mean delay is consistent with that expected for random ejection at an average rate proportional to the illumination intensity.

While the full quantitative verification of Einstein's photoelectric effect equation,

$$h(\nu - \nu_0) = \tfrac{1}{2}mv^2,$$

where ν_0 represents the threshold frequency, was delayed, the results are extremely difficult to explain with the wave theory of light. A major difficulty is to explain how the radiant energy that falls on $\sim 10^8$ atoms can be absorbed by a single electron. Another difficulty is the sharp threshold, and finally, delays as short as 3×10^{-9}s are observed, whereas classical theory would require microseconds to days to accumulate the amount of energy required, depending on the illumination. We note that this was the result that led to the Nobel prize for Einstein, but it took over 15 years before they were confident enough of the result to make the award.

Problem 1.5 In a photoelectric experiment, electrons are emitted when illuminated by light with wavelength 440 nm unless the stopping voltage is greater than (or equal to) 0.5 V. If the stopping voltage were set to zero, at what wavelength would the electrons stop emitting?

1.1.4 Atomic Structure, Bohr Theory

After the discovery of the electron, various models were proposed for the structure of atoms. These models had to deal with the evidence known at that time:

1. Electrons are present in all atoms and are the source of spectral radiation.

2. Since atoms are neutral, there must be some other source of positive charge.

3. Most of the mass must be in the positive charge, since the electron mass was known to account for only about 1/2000 of the mass of hydrogen.

4. The absence of "harmonic overtones" meant the electrons executed simple harmonic motion in the radiation process.

5. Classical electromagnetic theory requires that an accelerated charge radiate energy in proportion to the square of the acceleration, so electrons must be at rest in an atom.

This last point seemed to doom any planetary model, where the positive charge would be stationary with one or more electrons orbiting around it, since the energy loss due to radiation would lead to a collapse of the atom in less than a microsecond. As an alternative, J. J. Thomson invented the "plum pudding" model that suggested that the positive charge was uniformly distributed in a jelly-like medium ("Jellium"), and that the electrons were located like raisins in the pudding and could oscillate about their equilibrium position in simple harmonic motion. This model died when Rutherford scattered α particles from gold atoms in a thin foil, and found that some electrons were deflected by more than 90°, an impossibly large deflection for the plum pudding model. Rutherford went on to calculate the distribution of scattering angles for a central force and concluded that the positive charge occupied a very tiny fraction of the volume of an atom. Precise measurements by Geiger and Marsden (1913) verified the calculated distribution. This led to a nuclear model for the atom.

This left a deep problem, since one could now conceive of a planetary model with an electron circulating about the nucleus, but the accelerated electron would necessarily radiate away all of its energy in 10^{-8}s or less, so no stable model could be found. Without such a rotation, there was nothing to prevent the collapse of the atom.

This quandary was resolved by Bohr, who postulated that the planetary model was right, but that *the electron would not radiate if its angular momentum was an integral multiple of* $h/2\pi \equiv \hbar$. He further postulated that when radiating, the energy of the radiated photon was given by the Einstein formula, $E = h\nu$ where $E = E_1 - E_2$ is the energy difference between the two orbits. Here Planck's constant appeared in two separate ways, both in the angular momentum and in the radiation. Taking the angular velocity to be ω, the nuclear mass to be M, the electron mass to be m, and the nuclear charge Ze, we can write the equations of motion about the common center of mass, where the electron is a distance r from the center of mass and the nucleus is a distance R on the opposite side, where $mr = MR$, as:

1. The angular momentum is

$$(mr^2 + MR^2)\omega = mr^2\omega\left(1 + \frac{m}{M}\right) = n\hbar, \qquad n = 1, 2, 3, \ldots \tag{1.4}$$

2. The centripetal acceleration is produced by the Coulomb force

$$m\omega^2 r = \frac{Ze^2}{4\pi\epsilon_0(r+R)^2} = \frac{Ze^2}{4\pi\epsilon_0 r^2(1+m/M)^2}. \tag{1.5}$$

3. The total energy is the sum of the kinetic and potential energies

$$\begin{aligned} E_n &= \tfrac{1}{2}mr^2\omega^2 + \tfrac{1}{2}MR^2\omega^2 - \frac{Ze^2}{4\pi\epsilon_0(r+R)} \\ &= \tfrac{1}{2}mr^2\omega^2\left(1+\frac{m}{M}\right) - \frac{Ze^2}{4\pi\epsilon_0 r(1+m/M)} \\ &= -\tfrac{1}{2}mr^2\omega^2\left(1+\frac{m}{M}\right), \end{aligned} \tag{1.6}$$

where the last result has used Equation (1.5).

4. Each of these may be solved for $\omega^2 r^4$, such that

$$\begin{aligned} \omega^2 r^4 &= \frac{n^2\hbar^2}{m^2(1+m/M)^2} = A \\ &= \frac{Ze^2 r}{4\pi\epsilon_0 m(1+m/M)^2} = rB \\ &= -\frac{2E_n r^2}{m(1+m/M)} = r^2 E_n C. \end{aligned}$$

5. By eliminating r from the right-hand sides of the above equalities, one finds

$$E_n = \frac{B^2}{AC} = -\frac{mZ^2e^4}{32\pi^2\epsilon_0^2 n^2\hbar^2(1+m/M)} \tag{1.7}$$

for the energy in joules. The corresponding radius is

$$r_n = \frac{A}{B} = \frac{4\pi\epsilon_0 n^2\hbar^2}{mZe^2}. \tag{1.8}$$

The state with $n = 1$ is called the **ground state** and represents an unexcited atom. For hydrogen ($Z = 1$), the smallest radius is called the **first Bohr orbit** and is given by $a_0 = 4\pi\epsilon_0\hbar^2/me^2 = 5.29 \times 10^{-11}$ m.

The frequency corresponding to a transition from a state n to a state n' is given by

$$\nu = \frac{E_n - E_{n'}}{h} = \frac{mZ^2e^4}{8\epsilon_0^2h^3(1+m/M)}\left(\frac{1}{n'^2} - \frac{1}{n^2}\right) \tag{1.9}$$

and the **wave number** is given by

$$\tilde{\nu} = \frac{\nu}{c} = \frac{1}{\lambda} = Z^2R\left(\frac{1}{n'^2} - \frac{1}{n^2}\right), \tag{1.10}$$

where R is the **Rydberg constant**, given by

$$R = \frac{me^4}{8\epsilon_0^2ch^3(1+m/M)}. \tag{1.11}$$

Its value for hydrogen is $R_H = 10973731.5\ \mathrm{m}^{-1}$.

An empirical formula of this kind was discovered in 1885 by Balmer for $n' = 2$, and all of the spectral lines for $n' = 2$, $n \geq 3$ in hydrogen form the **Balmer series**.

Problem 1.6 Compare the frequency of radiation with the frequency of revolution of the electron in its orbit. If the electron makes a jump from state n to $n-1$, show that the frequency of the radiation is intermediate between the orbital frequencies of the upper and lower states. This is an example of the **Bohr's correspondence principle**, which states that in the limit of large quantum numbers, the quantum mechanical and classical results are the same.

Problem 1.7 There are four visible lines in the hydrogen spectrum, all with final quantum number $n' = 2$. Find the wavelength of each to four significant figures.

Problem 1.8 *Lyman-α emission.* Lyman-α emission is the radiation from hydrogen from the $n = 2$ to the $n = 1$ state. There are three isotopes of hydrogen, all with the same charge, but for deuterium, there is an extra neutron in the nucleus, and for tritium, there are two extra neutrons in the nucleus. Taking the mass of the nuclei to be $M = m_p$, $2m_p$, $3m_p$, calculate the *differences* between the wavelengths for the Lyman-α lines for the three isotopes, $\lambda_{H,2\to1} - \lambda_{D,2\to1}$ and $\lambda_{H,2\to1} - \lambda_{T,2\to1}$.

Unfortunately, although many extensions of the theory taking into account elliptical orbits, described by the Bohr-Sommerfeld theory, which accounted for some of the variations observed in one-electron atoms, no theory was found to account for helium or many-electron atoms in general except those that had only one outer-shell electron. This failure prompted the transition to modern quantum mechanics, and this early quantum theory came to be called the **Old Quantum Theory**.

1.2 Wave–Particle Duality and the Uncertainty Relation

The difficulty with the old quantum theory led many to search for some other type of theory, and some of the crucial steps along the way were due to de Broglie and Heisenberg who deepened the gulf between classical theory and modern theory.

1.2.1 The Wave Properties of Particles

In 1923, Louis de Broglie wrote a dissertation based on special relativistic considerations that suggested that particles have wave-like properties. After

Einstein had shown that electromagnetic waves had particle-like properties, this indicated the flip side by arguing that particles had wave-like properties. His principal result was that particle momentum was related to the particle wavelength through the relation

$$\lambda = h/p.$$

Through the use of Planck's constant, it also became a quantum relationship, and led to the interpretation that the stability of the Bohr orbits was due to the fact that each orbit was an integral number of wavelengths in circumference. The **de Broglie relation** was verified in 1928 by Davisson and Germer who measured electron diffraction and showed that the wavelengths of the electrons corresponded to that given by de Broglie.

Problem 1.9 Show that the various circular orbits of the Bohr theory correspond to the condition that the circumference of each orbit is an integral number of particle wavelengths. The stability of the orbits is then interpreted as the condition where the electrons interfere constructively with themselves. This notion has been found to have very deep significance in modern quantum theory.

1.2.2 The Uncertainty Principle

It is often imagined that the uncertainty principle is uniquely connected with quantum mechanics and that there is no classical analog. This is demonstrably false, as we will show in the following discussion. Since the uncertainty relation was one of the crucial starting points in the development of the Heisenberg formulation of quantum mechanics, it is worthwhile to examine its role in the formulation by Schrödinger. Although Schrödinger did not use the following arguments in the development of his method, they provide a rationale for the postulates of quantum mechanics that may appear more natural.

1.2.2.1 Digital Frequency Counters

We begin by discussing how a digital frequency counter works. The device is used for measuring how many cycles per second (whether the voltage signal is sinusoidal, square, or triangular is immaterial) of an electric signal. The fundamentals of the device include:

1. A trigger circuit that senses when the voltage switches from negative to positive (or from positive to negative, but generally not both), and sends a pulse to the counter each time the trigger fires,

2. A digital counter that simply adds the number of pulses it receives, and

3. A timed gate that determines how long a period of time the counter will accept pulses. This gate sets the count to zero when it starts, and stops

the counter from accepting any more counts when the predetermined time is over.

We now consider the accuracy inherent in this kind of counter. We will first presume that it never makes any mistakes in counting and that the gate time is exact. This would seem to be a perfect counter. There are uncertainties, however, that prevent it from being perfect. If we consider that the gate time is one second, for example, and imagine a case where there are thousands or millions of counts per second, it is highly unlikely that the counter will record the first count immediately after the gate opens. What fraction of a cycle will pass after the gate opens and before the first trigger is essentially random, so the counting process is intrinsically uncertain by up to one count. Now we could arrange the gate to start precisely at a trigger, and eliminate this error, although this is not always done. We want to be as accurate as possible, however, so we will presume there is no uncertainty in the starting process.

At the end of the gate time, however, we cannot avoid the error, since it is highly unlikely that there will be precisely an integral number of cycles during the gate period. This still leaves us with the uncertainty of one cycle. For the gate time of one second, then, the frequency will be $f = n + \delta$ where n is the number of counts in the counter and δ is a number between 0 and 1. In a typical counter, where there is uncertainty at both the start and finish, the specifications would typically list this as $f = n \pm 1$. If one wanted the frequency more accurate than this (the fractional uncertainty for this case would be $1/n$), one can increase the gate time to 10 seconds to reduce the uncertainty by a factor of 10, or to T seconds to reduce the uncertainty by a factor of T. We then would denote the uncertainty in frequency by $\Delta f = 1/T$. If we wanted to take the probable uncertainty, we might estimate that $\langle\delta\rangle = \frac{1}{2}$, and take $\Delta f \sim 1/2T$.

We see here the complementarity of the relation, since higher accuracy in frequency means the gate is open for a longer time, and the shorter the gate time used to make the measurement, the worse the accuracy in the frequency. If we designate the gate period as Δt, then the symmetry in the accuracy relation is apparent by writing the relation as

$$(\Delta f)(\Delta t) \geq \frac{1}{2}. \tag{1.12}$$

If one were able to trigger on both positive and negative zero-crossings, one could reduce the error a little more, but fundamentally, this is an intrinsic limit on the accuracy of the measurement of frequency.

Consider now a different type of counter that starts counting when the signal crosses from negative to positive and ends the counting the next time the signal crosses from negative to positive. During this cycle, it counts the number of cycles of an incredibly accurate atomic clock, apparently reducing the error or uncertainty to virtually zero. This counter presumes, however, that each cycle is precisely the same as the last, and for the case when the system is precisely periodic, there is no uncertainty, and the system is said to be in a stationary state. If the system is not precisely periodic, however, as in an FM radio signal where the information is transmitted by varying the frequency, not the amplitude, of the signal, there will be uncertainty since only the length of one cycle is known and they are not all the same. If we were to take hundreds or thousands of measurements of the signal, we could determine the *average* frequency (the carrier frequency) to high accuracy, and also calculate the *spread* in frequency (the bandwidth). The rms deviation from the carrier frequency we designate the *width*, Δf, although the maximum deviations may be several times wider. In order to determine $\langle f \rangle$ and Δf with any reliability, we must make many individual measurements, so that Δt increases as the accuracy in f increases, and again we have the same basic uncertainty relation.

1.2.2.2 Relation to Quantum Mechanics

If we now use Einstein's relation between frequency and energy of a photon, where $E = hf$ or $E = \hbar\omega$, multiplying Equation (1.12) by h leads to

$$(\Delta E)(\Delta t) \geq \frac{h}{2} \,. \tag{1.13}$$

The mathematical limit is actually found to be smaller than indicated by a factor of 2π. The limit in Equation (1.12) is found by considering arbitrarily shaped periodic functions of time and their associated functions in frequency that are related by $F(\omega) = \mathcal{F}[f(t)]$ where $\mathcal{F}$ indicates the Fourier transform. In the following sections, we investigate the properties and implications of Fourier transforms in describing systems that have intrinsic uncertainty.

1.2.2.3 The Gamma-Ray Microscope

Consider the problem of trying to measure the *position* of an electron very accurately, say with a microscope. For high resolution, we will of course use a large aperture microscope with very short wavelength light. In order to disturb the electron as little as possible, we will use only *one photon.* We thus use very low intensity illumination with very high energy photons. The general configuration of the microscope is shown in Figure 1.2.

It is immediately apparent that we can know the location of the electron *only approximately*, since the location is determined by the diffraction pattern and we can assume the photon *probably* landed in the first diffraction ring. This leaves us with an uncertainty of position, from physical optics, of

$$\Delta x \approx \frac{f\lambda}{D}$$

where f is the focal length of the lens, D is the diameter of the lens, and λ is the wavelength of the photon. In the reflection of the photon from the

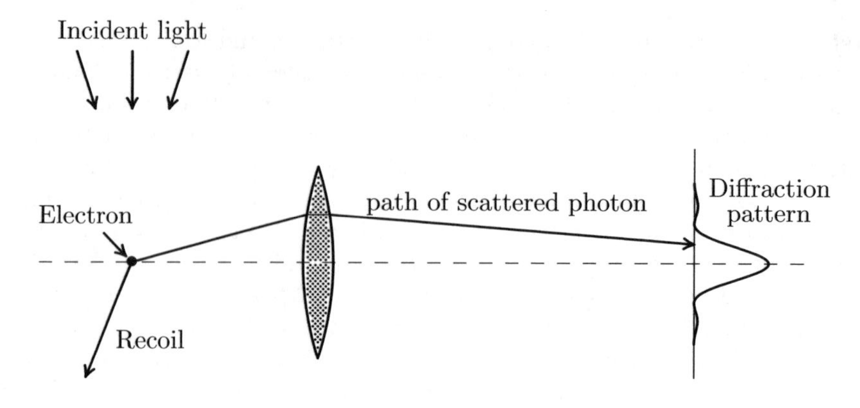

FIGURE 1.2
Sketch of a "gamma-ray microscope."

electron, the electron has recoiled *by an unknown amount.* Knowing only that the photon went through the lens, but not where, we estimate the x-component of the momentum to be

$$\Delta p_x \approx \frac{h}{\lambda}\frac{D}{2f}$$

where the first factor gives the photon momentum and the second factor approximates the tangent of the angle ($\tan\theta \approx \theta$). It is now seen that to make Δx small, it is necessary to make Δp_x large, since $f\lambda/D$ are reciprocal in the two relationships. This factor can be eliminated by taking the product of the two, so that

$$\Delta x\, \Delta p_x \geq \tfrac{1}{2}h.$$

This result can be extended to any pair of **conjugate quantities** (in the sense of Hamilton's canonical equations) P and Q so that one may write quite generally $\Delta P\, \Delta Q \geq h$.

1.2.3 Fourier Transforms

The wave-particle duality and the uncertainty principle suggest that the mathematics of conjugate quantities may be useful in describing quantum phenomena, and one example of such a canonical relationship is embodied in Fourier transforms, where typically one begins with a function of time $f(t)$ and taking the Fourier transform, one obtains the corresponding function of frequency, $F(\omega)$. This can just as well be cast into a function of space through some $f(x)$ and its transformed quantity $F(k)$, where eventually we will identify k with the x-component of momentum.

1.2.3.1 Fourier Transform Pairs

Definition of the Fourier Transform:

$$\mathcal{F}[f(x)] \equiv F(k) = \frac{1}{\sqrt{2\pi}} \int_{-\infty}^{\infty} f(x) \mathrm{e}^{-\mathrm{i}kx} \mathrm{d}x\,.$$

Definition of the inverse Fourier Transform:

$$\mathcal{F}^{-1}[F(k)] \equiv f(x) = \frac{1}{\sqrt{2\pi}} \int_{-\infty}^{\infty} F(k) \mathrm{e}^{\mathrm{i}kx} \mathrm{d}k\,.$$

A few examples of Fourier transform pairs are given in Table 1.1

TABLE 1.1
Fourier Transform Pair Examples.

$f(x)$	Plot of $f(x)$	$F(k)$	Plot of $F(k)$
$f(x) = \begin{cases} \cos\left(\frac{\pi x}{2a}\right) \\ 0 \end{cases}$	1, $-a$, a, x	$F(k) = \frac{2a\sqrt{2\pi}\cos ka}{\pi^2 - 4k^2a^2}$	$a(\frac{2}{\pi})^{3/2}$, $-\frac{3\pi}{2a}$, $\frac{3\pi}{2a}$, k
$f(x) = \dfrac{A}{a^2 + x^2}$	$\frac{A}{a^2}$, $-a$, a, x	$F(k) = A\sqrt{\frac{\pi}{2}}\frac{\mathrm{e}^{-\lvert k \rvert a}}{a}$	$\sqrt{\frac{\pi}{2}}\frac{A}{a}$, $-\frac{1}{a}$, $\frac{1}{a}$, k
$f(x) = \begin{cases} 1 + \frac{x}{a} \\ 1 - \frac{x}{a} \\ 0 \end{cases}$	1, $-a$, a, x	$F(k) = \frac{4\sin^2(ka/2)}{\sqrt{2\pi}ak^2}$	$\frac{a}{\sqrt{2\pi}}$, $-\frac{4\pi}{a}$, $-\frac{2\pi}{a}$, $\frac{2\pi}{a}$, $\frac{4\pi}{a}$, k

1.2.3.2 Probability Densities

We wish to construct from the transform pairs a probability density, but neither $f(x)$ nor $F(k)$ are suitable as they stand, since they are complex functions in general. We choose rather to represent the probability densities by the magnitude squared of the amplitudes, so that

$$P(x) = |A\, f(x)|^2$$
$$P(k) = |A\, F(k)|^2$$

where A is a normalization constant determined from the condition that the probability of finding something *somewhere* is unity, so that we require

$$\int_{-\infty}^{\infty} P(x)\, \mathrm{d}x = |A|^2 \int_{-\infty}^{\infty} |f(x)|^2\, \mathrm{d}x = 1\,,$$

which guarantees that

$$\int_{-\infty}^{\infty} P(k)\,\mathrm{d}k = |A|^2 \int_{-\infty}^{\infty} |F(k)|^2\,\mathrm{d}k = 1\,.$$

The mean or expected value of any function of x or k may then be obtained by

$$\begin{aligned}\langle x^n \rangle &= \int_{-\infty}^{\infty} x^n P(x)\,\mathrm{d}x = |A|^2 \int_{-\infty}^{\infty} x^n |f(x)|^2\,\mathrm{d}x \\ \langle k^n \rangle &= \int_{-\infty}^{\infty} k^n P(k)\,\mathrm{d}k = |A|^2 \int_{-\infty}^{\infty} k^n |F(k)|^2\,\mathrm{d}k\,.\end{aligned}$$

In order to estimate the spread or uncertainty in either x or k, we use the rms deviation from average, and following literally the recipe "root mean squared deviation from average," we obtain

$$\begin{aligned}(\Delta x) &= \sqrt{\langle (x - \langle x \rangle)^2 \rangle} \\ (\Delta k) &= \sqrt{\langle (k - \langle k \rangle)^2 \rangle}\,,\end{aligned}$$

which can be simplified by expanding the squared term and taking the mean value of each term so that

$$(\Delta x) = \sqrt{\langle x^2 \rangle - 2\langle x \rangle \langle x \rangle + \langle x \rangle^2} = \sqrt{\langle x^2 \rangle - \langle x \rangle^2}\,,$$

and similarly for (Δk).

Example 1.1

Fourier pairs. As an example of the kinds of calculations involved in these procedures, we shall examine the second example from Table 1.1 in detail.

Beginning first with the normalization of $P(x)$,

$$\begin{aligned}|A|^2 \int_{-\infty}^{\infty} \frac{\mathrm{d}x}{(a^2 + x^2)^2} &= |A|^2 \left[\frac{x}{2a^2(a^2 + x^2)} + \frac{1}{2a^3} \tan^{-1}\left(\frac{x}{a}\right) \right]_{-\infty}^{\infty} \\ &= \frac{\pi |A|^2}{2a^3} = 1\end{aligned}$$

so

$$|A| = \sqrt{\frac{2a^3}{\pi}}\,.$$

We then find $F(k)$ either from a table of integrals or from

$$F(k) = \frac{1}{\sqrt{2\pi}} \sqrt{\frac{2a^3}{\pi}} \int_{-\infty}^{\infty} \frac{\mathrm{e}^{\mathrm{i}kx}\mathrm{d}x}{a^2 + x^2} = \frac{a^{3/2}}{\pi} \int_{-\infty}^{\infty} \frac{\mathrm{e}^{\mathrm{i}kx}\,\mathrm{d}x}{(x + \mathrm{i}a)(x - \mathrm{i}a)}\,.$$

If $k > 0$ (and we must consider both cases since we want $-\infty \leq k \leq \infty$), then we may close the contour in the complex x-plane (illustrated to the right of the equation above) below and pick up the enclosed pole at $x = -\mathrm{i}a$ with the result

$$F(k) = \frac{a^{3/2}}{\pi}\left(\frac{-2\pi\mathrm{i}}{-2\mathrm{i}a}\right)\mathrm{e}^{-ka} = \sqrt{a}\mathrm{e}^{-ka},$$

where the upper minus sign comes because the contour was taken clockwise, and the lower minus sign comes from evaluating the residue at $x = -\mathrm{i}a$. If $k < 0$, then we may close the contour above and pick up the residue from the enclosed pole at $x = \mathrm{i}a$, with the result

$$F(k) = \frac{a^{3/2}}{\pi}\left(\frac{2\pi\mathrm{i}}{2\mathrm{i}a}\right)\mathrm{e}^{ka} = \sqrt{a}\mathrm{e}^{ka},$$

and these two results can be combined to give

$$F(k) = \sqrt{a}\mathrm{e}^{-|k|a},$$

and this function is already normalized (since $f(x)$ is).

For the various averages, we find

$$\langle x\rangle = |A|^2\int_{-\infty}^{\infty}\frac{x\,\mathrm{d}x}{(a^2+x^2)^2} = 0 \quad \text{(odd integrand between symmetric limits)}$$

$$\langle x^2\rangle = |A|^2\int_{-\infty}^{\infty}\frac{x^2\,\mathrm{d}x}{(a^2+x^2)^2}$$

$$= \frac{2a^3}{\pi}\left[\frac{-x}{2(a^2+x^2)} + \frac{1}{2a}\tan^{-1}\left(\frac{x}{a}\right)\right]_{-\infty}^{\infty} = \frac{2a^3}{\pi}\frac{\pi}{2a} = a^2$$

$$(\Delta x) = \sqrt{a^2 - 0} = a$$

$$\langle k\rangle = a\int_{-\infty}^{\infty} k\mathrm{e}^{-2|k|a}\mathrm{d}k = 0 \quad \text{(odd integrand again)}$$

$$\langle k^2\rangle = a\int_{-\infty}^{\infty} k^2\mathrm{e}^{-2|k|a}\mathrm{d}k = 2a\int_0^{\infty} k^2\mathrm{e}^{-2ka}\mathrm{d}k$$

$$= 2a\mathrm{e}^{-2ka}\left[\frac{k^2}{-2a} - \frac{2k}{4a^2} - \frac{2}{8a^3}\right]_0^{\infty}$$

$$= 2a\left(\frac{2}{8a^3}\right) = \frac{1}{2a^2}$$

$$(\Delta k) = \sqrt{\frac{1}{2a^2} - 0} = \frac{1}{\sqrt{2}a}.$$

From these results, we find that the uncertainty product is given by

$$(\Delta x)(\Delta k) = \frac{a}{\sqrt{2}a} = \frac{1}{\sqrt{2}}.$$

The general nature of this result is illustrated graphically in Table 1.1, where the complementarity of the function and its transform pair is apparent from the fact that if either one is broadened by changing a, the other is narrowed, leaving the product unchanged. □

1.2.3.3 Two Generalizations

1. Many of the wave functions we encounter are symmetric or antisymmetric about the origin. If the function is symmetric about the origin and the limits are likewise symmetric, the mean value of x or k is always zero because multiplying a symmetric (even) function by x or k (both odd) renders the integrand odd, so the integral vanishes by symmetry.

2. For calculating $\langle x^2 \rangle$ or $\langle k^2 \rangle$ with a symmetric probability density function, the integrand is even, so it may be simpler to evaluate the integral from the origin to one limit and multiply by two rather than evaluate the result at both limits.

1.2.3.4 Interpretation

From the example above, we may observe several features of this problem. First, we note that the "width" of the $f(x)$ function is given by a, so that as a gets larger, the probability distribution function gets wider and any measure of x becomes more uncertain. On the other hand, however, the complementary function, $F(k)$, becomes narrower as a is increased, as the "width" varies inversely with a. This means that the product of the two uncertainties is independent of the width of either, in this example given by the value $1/\sqrt{2}$, a pure number. This complementary nature of Fourier transform pairs is intrinsic, so that spreading one function invariably shrinks the other, and the uncertainty product is independent of the width of either, and depends only on the *shape* of the two functions. Because this product is independent of the details of either distribution, we use Fourier transforms to model the quantum mechanical functions, and we shall thereby be guaranteed that the uncertainty product will always satisfy the Heisenberg Uncertainty Principle.

Problem 1.10 *Mean values.*

(a) Find the normalization constant for the last case of Table 1.1 by setting

$$|A|^2 \int_{-a}^{a} f^2(x)\, dx = 1\,.$$

(b) Evaluate $\langle x \rangle$ using the first generalization.

(c) Evaluate $\langle x^2 \rangle$ by doing the indicated integrals.

(d) Evaluate $\langle k \rangle$ using the first generalization.

(e) Noting that the normalization constant is the same for both $f(x)$ and $F(k)$, evaluate $\langle k^2 \rangle$ using

$$\int_{-\infty}^{\infty} \frac{\sin^4 u}{u^2} \mathrm{d}u = \frac{\pi}{2}.$$

(f) By what percentage does the uncertainty product $(\Delta x)(\Delta k)$ exceed the minimum of $1/2$?

1.3 Fourier Transforms in Quantum Mechanics

1.3.1 The Quantum Mechanical Transform Pair

In order to ensure that we obtain the proper form for the Uncertainty Principle, we may use x as a length, but must choose $k = p/\hbar$ (the de Broglie relation) to ensure that $kx = px/\hbar$ is dimensionless. The choice of $\hbar$ as the constant comes from the desired form of the Uncertainty Principle, since we want $(\Delta k)(\Delta x) \to (\Delta p)(\Delta x)/\hbar \sim 1$ so that $(\Delta p)(\Delta x) \sim \hbar$. Inserting $k = p/\hbar$ uniformly leaves the transform pair asymmetric*, so we choose to put $\sqrt{\hbar}$ in each, with the result written as

$$\phi(p) = \frac{1}{\sqrt{2\pi\hbar}} \int_{-\infty}^{\infty} \psi(x) \mathrm{e}^{-\mathrm{i}px/\hbar} \mathrm{d}x$$

$$\psi(x) = \frac{1}{\sqrt{2\pi\hbar}} \int_{-\infty}^{\infty} \phi(p) \mathrm{e}^{\mathrm{i}px/\hbar} \mathrm{d}p.$$

We could let $2\pi\hbar \to h$, but we choose to show explicitly the 2π that comes from the Fourier transforms. The corresponding expressions for the normalizations and the expected values are

$$1 = \int_{-\infty}^{\infty} |\psi(x)|^2 \, \mathrm{d}x$$

$$1 = \int_{-\infty}^{\infty} |\phi(p)|^2 \, \mathrm{d}p$$

$$\langle x^n \rangle = \int_{-\infty}^{\infty} x^n |\psi(x)|^2 \, \mathrm{d}x$$

$$\langle p^n \rangle = \int_{-\infty}^{\infty} p^n |\phi(p)|^2 \, \mathrm{d}p$$

*Fourier transforms traditionally omit the factor of 2π in calculating $F(k)$, but insert the $1/2\pi$ when calculating the $f(x)$ via the inverse Fourier transform. We choose to keep the transform and the inverse transform completely symmetric except for the exponent, so we use $1/\sqrt{2\pi}$ in each.

and

$$(\Delta x) = \sqrt{\langle x^2 \rangle - \langle x \rangle^2}\,, \tag{1.14}$$

$$(\Delta p) = \sqrt{\langle p^2 \rangle - \langle p \rangle^2}\,. \tag{1.15}$$

1.3.2 Operators

It is possible for every $\psi(x)$ that is continuous and has a continuous first derivative to calculate the corresponding $\phi(p)$ (although it may be difficult), and use $\psi(x)$ to calculate the expected value of every function of x and use $\phi(p)$ to calculate the expected value of every function of p, but this is never done in practice. It is much easier to deal only with $\psi(x)$ *or* $\phi(p)$ by using *operators* to represent functions of p in coordinate space. To see this, we start with the definition

$$\begin{aligned}
\langle p \rangle &\equiv \int_{-\infty}^{\infty} \mathrm{d}p\, \phi^*(p)\phi(p)p \\
&= \int_{-\infty}^{\infty} \mathrm{d}p\, \phi^*(p) \frac{1}{\sqrt{2\pi\hbar}} \int_{-\infty}^{\infty} \mathrm{d}x\, \psi(x) p \mathrm{e}^{-\mathrm{i}px/\hbar} \\
&= \int_{-\infty}^{\infty} \mathrm{d}p\, \phi^*(p) \frac{1}{\sqrt{2\pi\hbar}} \int_{-\infty}^{\infty} \mathrm{d}x\, \frac{\hbar}{\mathrm{i}} \frac{\mathrm{d}\psi(x)}{\mathrm{d}x} \mathrm{e}^{-\mathrm{i}px/\hbar} \\
&= \frac{1}{h} \int_{-\infty}^{\infty} \mathrm{d}p \int_{-\infty}^{\infty} \mathrm{d}x'\, \psi^*(x') \mathrm{e}^{\mathrm{i}px'/\hbar} \int_{-\infty}^{\infty} \mathrm{d}x\, \frac{\hbar}{\mathrm{i}} \frac{\mathrm{d}\psi(x)}{\mathrm{d}x} \mathrm{e}^{-\mathrm{i}px/\hbar} \\
&= \frac{1}{h} \int_{-\infty}^{\infty} \mathrm{d}x \int_{-\infty}^{\infty} \mathrm{d}x' \psi^*(x') h\delta(x'-x) \frac{\hbar}{\mathrm{i}} \frac{\mathrm{d}\psi(x)}{\mathrm{d}x} \\
&= \int_{-\infty}^{\infty} \mathrm{d}x\, \psi^*(x) \frac{\hbar}{\mathrm{i}} \frac{\mathrm{d}\psi(x)}{\mathrm{d}x}\,,
\end{aligned} \tag{1.16}$$

where the first line is the definition, the second line has replaced $\phi(p)$ by its integral expression, the third line represents the result of integrating by parts in x (noting that $|\psi(x)| \to 0$ as $|x| \to \infty$ in order for $\psi(x)$ to be normalizable), the fourth line represents the replacement of $\phi^*(p)$ by its integral expression, and the fifth line results from integrating over p, where

$$\int_{-\infty}^{\infty} \mathrm{e}^{\mathrm{i}(x'-x)p/\hbar} \mathrm{d}p = \hbar \int_{-\infty}^{\infty} \mathrm{e}^{\mathrm{i}(x'-x)u} \mathrm{d}u \equiv 2\pi\hbar\delta(x'-x) = h\delta(x'-x)\,,$$

and the last line represents the integral over x' using the Dirac δ-function.

1.3.2.1 The Dirac δ-Function.

The Dirac δ-function may be defined for any function $f(x)$, by the properties

$$\int_a^b f(x)\delta(x-x_0)\,\mathrm{d}x \equiv \begin{cases} 0\,, & x_0 < a\,, \\ f(x_0)\,, & a \le x_0 \le b\,, \\ 0\,, & x_0 > b\,, \end{cases} \tag{1.17}$$

so that every integration with a Dirac δ-function is trivial, simply evaluating the integrand at a fixed point. Since we generally deal with the infinite range $[-\infty, \infty]$, the fixed point is almost never beyond the range of the integral.

The δ-function may be represented by the integral

$$\delta(x) = \frac{1}{2\pi} \int_{-\infty}^{\infty} \mathrm{e}^{\mathrm{i}kx} \, \mathrm{d}k \,. \tag{1.18}$$

The properties of the δ-function from this representation appear paradoxical, since if $x = 0$, the result is infinite, and if $x \neq 0$, the result appears undefined. There are two ways to gain more insight into the properties and use of the δ-function by considering alternative expressions that approach the definition in Equation (1.17) above.

1. For our first case, we assume there is some "tapering," so that the integrand vanishes at the end points, and then let the tapering vanish. This is represented by the limit

$$\begin{aligned}
\delta(x) &= \lim_{\epsilon \to 0} \frac{1}{2\pi} \int_{-\infty}^{\infty} \mathrm{e}^{\mathrm{i}kx - \epsilon^2 k^2} \, \mathrm{d}k \\
&= \lim_{\epsilon \to 0} \frac{1}{2\pi} \mathrm{e}^{-x^2/4\epsilon^2} \int_{-\infty}^{\infty} \exp\left[-\left(\epsilon k - \frac{\mathrm{i}x}{2\epsilon} \right)^2 \right] \mathrm{d}k \\
&= \lim_{\epsilon \to 0} \frac{1}{2\sqrt{\pi}\epsilon} \mathrm{e}^{-x^2/4\epsilon^2} \\
&= \begin{cases} \infty \,, & x = 0 \,, \\ 0 \,, & x \neq 0 \,, \end{cases}
\end{aligned}$$

where ϵ is real. The integral is bounded, however, since

$$\begin{aligned}
\int_{-\infty}^{\infty} \delta(x) \, \mathrm{d}x &= \lim_{\epsilon \to 0} \frac{1}{2\sqrt{\pi}\epsilon} \int_{-\infty}^{\infty} \mathrm{e}^{-x^2/4\epsilon^2} \, \mathrm{d}x \\
&= 1 \,.
\end{aligned}$$

We thus discover that the δ-function has no finite width, but has unit area at $x = 0$.

2. Another similar representation also uses a deferred limit method, where

$$\begin{aligned}
\delta(x) &= \lim_{a \to \infty} \frac{1}{2\pi} \int_{-a}^{a} \mathrm{e}^{\mathrm{i}kx} \, \mathrm{d}k \\
&= \lim_{a \to \infty} \frac{\sin ax}{\pi x} \,, \\
&= \lim_{a \to \infty} \frac{a}{\pi} \frac{\sin ax}{ax} \\
&= \begin{cases} \infty \,, & x = 0 \,, \\ \text{oscillatory with zero period} \,, & x \neq 0 \,. \end{cases}
\end{aligned}$$

The integral is again bounded, however, since

$$\begin{aligned}\int_{-\infty}^{\infty} \delta(x)\,\mathrm{d}x &= \lim_{a\to\infty} \frac{a}{\pi}\int_{-\infty}^{\infty} \frac{\sin ax}{ax}\,\mathrm{d}x \\ &= 1\,.\end{aligned}$$

This result is essentially the same as the other case.

The importance of the result from Equation (1.16) is that *all* quantities may be calculated from the knowledge of $\psi(x)$ alone. In **coordinate space**, then, we write the mean value expressions as

$$\langle x^n\rangle = \int_{-\infty}^{\infty} \psi^*(x)x^n\psi(x)\,\mathrm{d}x \tag{1.19}$$

$$\langle p^n\rangle = \int_{-\infty}^{\infty} \psi^*(x)\left(\frac{\hbar}{\mathrm{i}}\frac{\mathrm{d}}{\mathrm{d}x}\right)^n \psi(x)\,\mathrm{d}x\,, \tag{1.20}$$

and the same recipe may be used for any function of momentum if we let

$$p \to \hat{p} = \frac{\hbar}{\mathrm{i}}\frac{\mathrm{d}}{\mathrm{d}x}\,.$$

Note that the operator for p lies *between* $\psi^*(x)$ and $\psi(x)$ so that it operates *only* on $\psi(x)$.

This representation of the variable p by the **operator** $\hat{p}$ in coordinate space is possible because of the Fourier transform relationship between the coordinate and momentum wave amplitudes. As long as we stay in coordinate space, $\hat{x} = x$ since the coordinate simply multiplies whatever follows. We could also work in **momentum space** where the coordinate operator becomes

$$x \to \hat{x} = \mathrm{i}\hbar\frac{\mathrm{d}}{\mathrm{d}p}\,,$$

and $\hat{p} = p$ since again in this space, the operation is simply a multiplication. In two or three dimensions, of course, the derivatives become partial derivatives.

Example 1.2

Operator method. As an example of using the operator method, we revisit Example 1.1 and calculate $\langle k^2\rangle$ from $f(x)$ instead of from $F(k)$ using $k \to \hat{k} = (1/\mathrm{i})\mathrm{d}/\mathrm{d}x$. The integral becomes

$$\langle k^2\rangle = -|A|^2\int_{-\infty}^{\infty} \frac{1}{a^2+x^2}\frac{\mathrm{d}^2}{\mathrm{d}x^2}\frac{1}{a^2+x^2}\,\mathrm{d}x$$

and the simplest way to evaluate this is to integrate by parts once, noting that there is no contribution from the end points, so the result is

$$\langle k^2\rangle = \frac{2a^3}{\pi}\int_{-\infty}^{\infty}\left(\frac{\mathrm{d}}{\mathrm{d}x}\frac{1}{a^2+x^2}\right)^2\mathrm{d}x$$

$$= \frac{2a^3}{\pi} \int_{-\infty}^{\infty} \left(\frac{-2x}{(a^2+x^2)^2} \right)^2 \mathrm{d}x$$
$$= \frac{8a^3}{\pi} \int_{-\infty}^{\infty} \frac{x^2}{(a^2+x^2)^4} \mathrm{d}x$$
$$= \frac{8a^3}{\pi} \frac{\pi}{16a^5} = \frac{1}{2a^2},$$

which is the same result as before. ▯

Problem 1.11 Show that

$$\int_{-\infty}^{\infty} \mathrm{e}^{\mathrm{i}px/\hbar} \mathrm{d}p = h\delta(x).$$

Problem 1.12 For the wave function

$$\psi(x) = A\mathrm{e}^{-x^2/2a^2 + \mathrm{i}p_0x/\hbar},$$

(a) Find $|A|$, $\phi(p)$, $\langle x \rangle$, $\langle x^2 \rangle$, and (Δx) from $\psi(x)$.

(b) Find $\langle p \rangle$, $\langle p^2 \rangle$, and (Δp) from $\phi(p)$.

(c) Find $(\Delta x)(\Delta p)$.

(d) Show that $\langle p^2 \rangle$ calculated from $\psi(x)$ using the momentum operator is the same as from part (b).

Problem 1.13 Show that the following two functions are also valid representation for the Dirac δ-function where ϵ is real and positive:

(a)

$$\delta(x) = \frac{1}{\sqrt{\pi}} \lim_{\epsilon \to 0} \frac{1}{\sqrt{\epsilon}} \mathrm{e}^{-x^2/\epsilon}$$

(b)

$$\delta(x) = \frac{1}{\pi} \lim_{\epsilon \to 0} \frac{\epsilon}{x^2 + \epsilon^2}$$

1.4 The Postulatory Basis of Quantum Mechanics

Having established that Fourier transform pairs provide a useful mathematical model for dealing with canonically conjugate quantities, we now take some of these relationships to establish a set of **postulates** upon which quantum mechanics is to be founded. These will provide recipes for the interpretation of data, and *however surprising the results are*, the ultimate test of the validity of these postulates is the comparisons of the results with actual observations. The basis stated here is only for nonrelativistic quantum mechanics, but some relativistic extensions will be included in later chapters.

Since many of the expressions that relate to experiment are derived from classical expressions, which are then translated to quantum expressions via the recipes in the postulates, it is presumed that the classical system has a Lagrangian function given by

$$L(q_j, \dot{q}_j, t) = T - V\,,$$

where T and V are the kinetic and potential energies expressed in terms of the coordinates, q_j, and their time derivatives, $\dot{q}_j$, respectively, from which a set of momenta p_j that are canonically conjugate to the q_j may be constructed by the Hamiltonian treatment by the recipe

$$p_j = \frac{\partial L}{\partial \dot{q}_j},$$

and the momenta are then used to eliminate the velocities $\dot{q}_j$ in constructing the **Hamiltonian function**

$$H(q_j, p_j, t) = \sum_{j=1}^{N} p_j \dot{q}_j - L(q_j, \dot{q}_j, t).$$

For such a classical system, the equations of motion for the system are given by Hamilton's pair of equations,

$$\dot{p}_j = -\frac{\partial H}{\partial q_j}, \qquad \text{and} \qquad \dot{q}_j = \frac{\partial H}{\partial p_j}. \tag{1.21}$$

These equations define trajectories in (p_j, q_j), which we call **phase space**, and for given initial conditions in classical mechanics, give *exact trajectories* that lead to *exact predictions* about the system at a later time.

In quantum mechanics, the uncertainty principle limits the exactness of our knowledge. It should be understood that although the equations of quantum mechanics are *deterministic in nature* (so that for a particular initial condition, the system at a later time is completely determined), the initial conditions cannot be determined exactly, so the outcome can only be stated as a probability. In view of this, the postulates describe the **state** of a system in terms of probabilities and probability amplitudes. The first postulate puts this notion into quantitative form.

1.4.1 Postulate 1

There exist two complex **probability amplitudes** (also called **wave functions**), $\Psi(q_j, t)$ and $\Phi(p_j, t)$ that completely define the state of a quantum-mechanical system in the following way: If at time t, the **coordinates** of the system are measured, the **probability** that these will be found to lie within the ranges q_j to $q_j + \mathrm{d}q_j$ is

$$W_q(q_j, t)\,\mathrm{d}q_1 \cdots \mathrm{d}q_N = \Psi^*(q_j, t)\Psi(q_j, t)\,\mathrm{d}q_1 \cdots \mathrm{d}q_N, \tag{1.22}$$

while if, instead, the **momenta** were measured at time t, the **probability** that they would be found to lie within the interval between p_j and $p_j + \mathrm{d}p_j$ is

$$W_p(p_j, t)\,\mathrm{d}p_1 \cdots \mathrm{d}p_N = \Phi^*(p_j, t)\Phi(p_j, t)\,\mathrm{d}p_1 \cdots \mathrm{d}p_N.$$

Since each coordinate must have *some* probability of being found *somewhere*, it is also usually required that the total integrated probability be unity, or that

$$\int \Psi^*\Psi\,\mathrm{d}^N q = 1, \qquad \text{and} \qquad \int \Phi^*\Phi\,\mathrm{d}^N p = 1,$$

where $\mathrm{d}^N q = \mathrm{d}q_1 \cdots \mathrm{d}q_N$ and a single $\int$ sign represents an integration over *all* of the variables over their entire range. Occasionally, the integral will be set to some arbitrary value, or only relative probabilities may be given where the ratio of two unbounded integrals is finite. While we represent the wave function of space and time by $\Psi(q_j, t)$, the function of space *only* shall be denoted by $\psi(q_j)$ and similarly for $\Phi(p_j, t)$, the function of momentum *only* shall be denoted by $\phi(p_j)$.

The wave functions must be *continuous* so that the probability of finding a particle at a particular place or with a particular momentum will not change by a finite amount an infinitesimal distance away in either coordinate or momentum space. This guarantees that the wave functions are differentiable over the range defined. Because the derivatives of $\psi(q_j)$ are related to the momentum and the derivatives of $\phi(p_j)$ are related to the coordinate, we also require that the first derivatives must be continuous.

Problem 1.14 *Probability integrals for locating a particle.* A wave function $\psi(x) = A \sin \pi x/a$ is defined for $0 \leq x \leq a$ (it vanishes outside this range).

(a) Find $|A|$.

(b) What is the probability that a particle will be found in the range $0 \leq x \leq a/3$?

(c) What is the probability that the particle will be found in the range $a/3 \leq x \leq 2a/3$?

(d) Can you think of a way to evaluate part (c) *without* evaluating another integral?

Problem 1.15 *Probability integrals for estimating the momentum of a particle.* A momentum wave function $\phi(p) = A_p \cos \pi a p/2\hbar$ is defined for $-\hbar/a \leq p \leq \hbar/a$ (it vanishes outside this range).

(a) Find $|A_p|$.

(b) What is the probability that the momentum of the particle will be found in the range $-\hbar/2a \leq p \leq \hbar/2a$?

(c) What is the probability that the momentum of the particle will be found in the range $\hbar/2a \leq p \leq \hbar/a$?

1.4.2 Postulate 2

The probability amplitudes $\Psi(q_j, t)$ and $\Phi(p_j, t)$ are connected by the relations

$$\Phi(p_j, t) = (2\pi\hbar)^{-1/2N} \int \Psi(q_j, t) \exp\left(-\mathrm{i}\sum_{j=1}^{N} \frac{p_j q_j}{\hbar}\right) \mathrm{d}^N q\,, \tag{1.23}$$

$$\Psi(q_j, t) = (2\pi\hbar)^{-1/2N} \int \Phi(p_j, t) \exp\left(+\mathrm{i}\sum_{j=1}^{N} \frac{p_j q_j}{\hbar}\right) \mathrm{d}^N p\,. \tag{1.24}$$

These are obviously Fourier transform pairs, where each of the two functions are inverse transforms of one another.

Problem 1.16 For the wave function

$$\psi(x) = A \exp[-(x - x_0)^2/2a^2] \exp(\mathrm{i}p_0 x/\hbar),$$

(a) Find $\langle x \rangle$.

(b) Even though $\langle x \rangle \neq 0$, show that (Δx) is the same as in Problem 1.12a.

Problem 1.17 Find the expectation value of the kinetic energy of a particle of mass m whose motion is instantaneously described by the wave function in Problem 1.16.

Problem 1.18 At $t = 0$ the wave function for the electron in a hydrogen atom is $\psi(x, y, z) = A \exp[-(x^2 + y^2 + z^2)^{1/2}/a_0]$. Find A, $\langle x \rangle$, $(\Delta x)^2$, and $\langle 1/r \rangle$. (*Hint:* Convert to spherical coordinates and integrate over the volume.)

1.4.3 Postulate 3

The expectation value of any dynamical quantity $F(q_j, p_j)$ may be evaluated by using either of the relations

$$\langle F \rangle = \int \psi^* \hat{F}_q \psi \,\mathrm{d}^N q\,, \tag{1.25}$$

$$\langle F \rangle = \int \phi^* \hat{F}_p \phi \,\mathrm{d}^N p\,, \tag{1.26}$$

where $\hat{F}_q$ is a **linear Hermitian operator** obtained by replacing p_j by $(\hbar/\mathrm{i})\partial/\partial q_j$ in $F(q_j, p_j)$, and $\hat{F}_p$ is a **linear Hermitian operator** obtained by replacing q_j by $\mathrm{i}\hbar\,\partial/\partial p_j$ in $F(q_j, p_j)$.

Problem 1.19 *Evaluation of* $\langle p^2 \rangle$. Show that $\langle p^2 \rangle$ can always be evaluated by the integral (in one dimension)

$$\langle p^2 \rangle = \int_{-\infty}^{\infty} \left| \frac{\hbar}{\mathrm{i}} \frac{\mathrm{d}\psi}{\mathrm{d}x} \right|^2 \mathrm{d}x \,. \tag{1.27}$$

1.4.4 Postulate 4

The wave functions Ψ and Φ evolve in time according to the equations

$$\hat{\mathcal{H}}_q \Psi = -\frac{\hbar}{\mathrm{i}} \frac{\partial \Psi}{\partial t}, \tag{1.28}$$

$$\hat{\mathcal{H}}_p \Phi = -\frac{\hbar}{\mathrm{i}} \frac{\partial \Phi}{\partial t}. \tag{1.29}$$

These two equations are each the **time-dependent Schrödinger equation**. Only one is needed, of course, and we will generally stay in coordinate space from this point on, but the symmetry between the coordinate and momentum representations are now fully evident.

In addition to the apparent symmetry between coordinates and momentum apparent in Equations (1.28) and (1.29), there is another kind of symmetry from the 4-vector representation of special relativity. Just as $p_x \to (\hbar/\mathrm{i})(\partial/\partial x)$, we now have from the fourth component of the 4-vector momentum $p_4 = \mathrm{i}W/c \to (\hbar/\mathrm{i})(\partial/\mathrm{i}c\,\partial t)$, where in nonrelativistic theory W is simply the sum of the kinetic and potential energy, so that $W \to H$.

Although we will find that the Schrödinger equation is at the heart of quantum mechanics, we will defer further comment about it until we have occasion to solve it for some special cases, at which time its central role will become more apparent.

1.5 Operators and the Mathematics of Quantum Mechanics

The requirement that valid quantum-mechanical operators must be linear and Hermitian comes from practical considerations related to what we observe.

1.5.1 Linearity

Let $\hat{F}$ denote an operator, and let ψ and χ be operands that satisfy the conditions of continuity, finiteness, and integrability which are required of all valid wave functions. Then the requirement that an operator $\hat{F}$ be linear means that

$$\hat{F}(\psi + \chi) = \hat{F}\psi + \hat{F}\chi \tag{1.30}$$

$$\hat{F}(C\psi) = C\hat{F}\psi \tag{1.31}$$

for any real or complex constant C. Linearity essentially guarantees that amplitudes may be added either before or after some operation is performed. This is fundamental to the wave character of the theory, and the fact that interference and diffraction effects, both of which are fundamentally linear (since amplitudes, not intensities, add), characterize both light *and* particles. Another way of stating the requirement of linearity is to postulate the **principle of superposition** which states that the sum of two independent solutions of the Schrödinger equation (for example) is also a solution.

1.5.2 Hermitian Operators

A **Hermitian** operator is required to guarantee that quantum mechanics will lead to *real* observable results even though wave functions are complex in general. This follows from the definition and properties of this type of operator. An operator is Hermitian if it satisfies the equation

$$\int \psi^* \hat{F}\chi \, \mathrm{d}^N q = \int \chi (\hat{F}\psi)^* \, \mathrm{d}^N q \tag{1.32}$$

for arbitrary wave functions ψ and χ. This may be related to the **adjoint** of an operator, which is defined by

$$\int \psi^* \hat{F}\chi \, \mathrm{d}^N q = \int \chi (\hat{F}^\dagger \psi)^* \, \mathrm{d}^N q \, , \tag{1.33}$$

where $\hat{F}^\dagger$ is the adjoint of $\hat{F}$. From this property, it is apparent that a Hermitian operator is **self-adjoint**, or that the adjoint of a Hermitian operator is identical to the operator, so that $\hat{F}^\dagger = \hat{F}$ is a sufficient condition that the operator is Hermitian. If $\hat{F}$, $\hat{G}$, and $\hat{H}$ are all Hermitian operators, then it follows that

$$(\hat{F} + \hat{G})\psi = \hat{F}\psi + \hat{G}\psi \qquad \text{definition of } \hat{F} + \hat{G}$$

$$(\hat{F}\hat{G})\psi = \hat{F}(\hat{G}\psi) \qquad \text{definition of } \hat{F}\hat{G}$$

$$[\hat{F}(\hat{G} + \hat{H})]\psi = \hat{F}(\hat{G}\psi) + \hat{F}(\hat{H}\psi) \qquad \text{distributive law.}$$

If $\hat{F}$ and $\hat{G}$ are Hermitian operators, then the operators

$$\hat{S} = \tfrac{1}{2}(\hat{F}\hat{G} + \hat{G}\hat{F}) \tag{1.34}$$

$$\hat{A} = \tfrac{1}{2\mathrm{i}}(\hat{F}\hat{G} - \hat{G}\hat{F}) \tag{1.35}$$

are also Hermitian operators that are respectively symmetric and antisymmetric with respect to an interchange of $\hat{F}$ and $\hat{G}$.

The guarantee that **expectation values** of dynamical quantities will be *real* is apparent from the relations

$$\langle F \rangle = \int \psi^* \hat{F} \psi \, \mathrm{d}^N q \qquad \text{and} \qquad \langle F \rangle^* = \int \psi (\hat{F} \psi)^* \, \mathrm{d}^N q \,,$$

which by Equation (1.32) with $\chi = \psi$ gives $\langle F \rangle = \langle F \rangle^*$ and is therefore real. While the mathematical structure of the Hermitian property may seem arcane, it is absolutely mandatory if quantum mechanics is to describe the real world.

1.5.3 Commutators

It should be noted that linear Hermitian operators do not, in general, commute, that is $\hat{F}\hat{G} \neq \hat{G}\hat{F}$ in general, or the **commutator** defined by $[\hat{F}, \hat{G}] \equiv \hat{F}\hat{G} - \hat{G}\hat{F} \neq 0$, except in special cases. Since commutators are written without a wave function on the right, it is sometimes useful to imagine a "phantom" ψ to the right.

Example 1.3
Evaluating the commutator. As an example, we evaluate the commutator $[\hat{\mathcal{H}}, \hat{x}]$, where

$$\hat{\mathcal{H}} = -\frac{\hbar^2}{2m} \frac{\partial^2}{\partial x^2}$$

is a free-particle Hamiltonian, in a series of steps as

$$\begin{aligned}
(\hat{\mathcal{H}}\hat{x} - \hat{x}\hat{\mathcal{H}})\psi &= -\frac{\hbar^2}{2m} \left[\frac{\partial^2 (x\psi)}{\partial x^2} - x \frac{\partial^2 \psi}{\partial x^2} \right] \\
&= -\frac{\hbar^2}{2m} \left[\frac{\partial}{\partial x} \left(x \frac{\partial \psi}{\partial x} + \psi \right) - x \frac{\partial^2 \psi}{\partial x^2} \right] \\
&= -\frac{\hbar^2}{2m} \left[x \frac{\partial^2 \psi}{\partial x^2} + \frac{\partial \psi}{\partial x} + \frac{\partial \psi}{\partial x} - x \frac{\partial^2 \psi}{\partial x^2} \right] \\
&= -2 \frac{\hbar^2}{2m} \frac{\partial \psi}{\partial x} \\
&= -\frac{\mathrm{i}\hbar}{m} \hat{p}_x \psi \,,
\end{aligned}$$

but it is customary to *not* write this wave function on the right. In executing expressions of this kind, however, it is essential that one always remember that one is often taking the derivative of a product even though the product is not evident because we have suppressed this "phantom" ψ on the right. □

Problem 1.20 Find the commutator operator (a) of $\hat{q}_s$ and $\hat{p}_s$, (b) of $\hat{q}_s$ and $\hat{p}_r$ $(s \neq r)$, (c) of $\hat{q}_s$ and $\hat{q}_r$, and (d) of $\hat{p}_s$ and $\hat{p}_r$.

Problem 1.21 Find the operator in coordinate language that corresponds to the angular momentum about the z-axis.

$$\textit{Ans.} \qquad \hat{L}_z = \frac{\hbar}{\mathrm{i}} \left(x\frac{\partial}{\partial y} - y\frac{\partial}{\partial x} \right).$$

Problem 1.22 Show that the operator above is also equal to $(\hbar/\mathrm{i})(\partial/\partial\phi)$, where ϕ is the azimuthal angle in spherical coordinates about the z-axis. (*Hint:* This problem is easier in one direction than the other.)

Problem 1.23 Find the commutator operator of the x- and y- components of angular momentum, $[\hat{L}_x, \hat{L}_y]$.

1.5.4 Matrices as Operators

All of the operators described in this section are related to functions of a continuous variable and integrals are used to describe the scalar product. We will also encounter operators that are represented by matrices and wave functions that are either row or column vectors. The original wave mechanics developed by Heisenberg was based entirely on matrix operators, and was sometimes referred to as Matrix Mechanics. The properties of Hermitian matrices is discussed in Appendix A.

1.5.5 Dirac Notation

Instead of writing out integrals each time an expectation value is required, Dirac developed a shorthand notation that has become common. An example of this notation is the writing of the Hermitian property, given in Equation (1.32) as

$$\langle\psi|\hat{F}|\chi\rangle = \langle\hat{F}\psi|\chi\rangle, \tag{1.36}$$

where the term on the left is the **bra** and the term on the right is the **ket**, the pair forming a **bra-ket**. It is possible to deal with each piece separately, so that the bra for the first expression is $\langle\psi|$ and the ket for the first expression is $|\chi\rangle$, and the operator in between always operates on the ket. It is always understood that the bra involves a complex conjugate, while the ket does not. It is also understood that the integration is to be carried out over the entire range of each variable. If the range is to be restricted for some reason, then the integral expression is required. When there is no operator between the bra and the ket, one of the vertical bars is deleted, so that a normalization integral would be represented by $\langle\psi|\psi\rangle = 1$. Sometimes, bras and kets are separated and appear to be in backward order, such as

$$\hat{O}_{nm} = |\psi_n\rangle\langle\chi_m|.$$

In such a case, $\hat{O}_{nm}$ is called a dyad or a dyadic operator. When evaluated, it lies between a bra and a ket, so that, for example,

$$\langle \psi_j | \hat{O}_{nm} | \psi_k \rangle = \langle \psi_j | \psi_n \rangle \langle \chi_m | \psi_k \rangle \,.$$

1.6 Properties of Quantum Mechanical Systems

Having discovered a number of unusual properties of the real world, quantum mechanics was formulated on the basis of a series of postulates to model those properties. It now behooves us to see how well these postulates meet the requirements. Detailed calculations giving specific numerical predictions of the results of experiments will be treated in subsequent chapters, but several general features may be noted at this point.

1.6.1 Wave–Particle Duality

Since the apparently contradictory properties of matter that have both wave-like and particle-like characteristics was one of the motivations for the structure of quantum mechanics, it is now worth noting that the wave function carries the wave-like properties and the probability density exhibits the particle-like properties. For a free particle, the wave function has the form $\psi(x) = A \exp(ip_0 x/\hbar)$, which has the form of a wave, and if we were to imagine an electron incident on a double slit arrangement, which is a familiar example of interference in optics, the electron wave function will likewise have a similar diffraction pattern beyond the slits. We are not permitted to ask which slit the electron passed through any more than we could ask which slit the photon went through in optics. The diffraction pattern gives the probability of finding the electron at a particular location, where the detection of the electron at some specific location again emphasizes the particle character of the electron.

It is interesting to note that when one is describing a particle, the description includes "vector" notions such as "where is it?" or "where is it going?" while the wave description includes "scalar" properties such as the wavelength or energy. This vector-scalar duality is inherent in the 4-vectors of relativity, composed of 3-component vectors plus a scalar fourth component. In this way of understanding the wave-particle duality, it is not a uniquely quantum mechanical phenomenon. We will find other phenomena in quantum mechanics that are rooted in the special theory of relativity and may not be understood classically or quantum mechanically without relativity, such as spin.

1.6.2 The Uncertainty Principle and Schwartz's Inequality

Since the Fourier transform pairs of Postulate 2 were shown to have the uncertainty principle built in, it is obvious that our postulates embody the uncertainty principle. The result is more general than we can simply infer from our postulates, however. Using the formulations above, we now wish to establish the formal uncertainty principle in a general way by proving that for *any* two noncommuting Hermitian operators $\hat{P}$ and $\hat{Q}$, that

$$(\Delta P)(\Delta Q) \geq \tfrac{1}{2}|\langle \mathrm{i}[\hat{P},\hat{Q}]\rangle| , \tag{1.37}$$

where (ΔP) and (ΔQ) are the root-mean-square (rms) deviations from average. To prove this, we begin by proving **Schwartz's Inequality**, which states that for any two square integrable functions (i.e., functions whose absolute squares integrated over their domain are bounded) $F(x)$ and $G(x)$ over any interval a to b, that (in Dirac notation)

$$\langle F|F\rangle\langle G|G\rangle \geq |\langle F|G\rangle|^2. \qquad \text{Schwartz's Inequality} \tag{1.38}$$

To prove this, we start by noting that we may write the relation

$$\int \left|F(x)\langle G|G\rangle - G(x)\langle G|F\rangle\right|^2 \mathrm{d}x \geq 0 \tag{1.39}$$

since the integrand is an absolute magnitude squared and hence positive definite. Expanding this expression leads to Equation (1.38). The equality can only occur if F is proportional to G, or

$$G(x) = cF(x) , \tag{1.40}$$

where c is a constant. Expanding Equation (1.39), the inequality can be written out as

$$\langle F\langle G|G\rangle - G\langle F|G\rangle|F\langle G|G\rangle - G\langle G|F\rangle\rangle \geq 0 ,$$

or finally as

$$\langle F|F\rangle\langle G|G\rangle^2 - \langle F|G\rangle\langle G|F\rangle\langle G|G\rangle - \langle G|F\rangle\langle F|G\rangle\langle G|G\rangle \\ +\langle G|G\rangle\langle F|G\rangle\langle G|F\rangle \geq 0 .$$

The last two terms cancel, and when we factor out the $\langle G|G\rangle$ term, then the remaining terms may be rearranged into the form of Equation (1.38).

This inequality is sometimes called the **triangle inequality**. In vector form, the result for any two real vectors $\boldsymbol{A}$ and $\boldsymbol{B}$ may be written as

$$A^2B^2 \geq |\boldsymbol{A}\cdot\boldsymbol{B}|^2 ,$$

where the equality occurs only when $\boldsymbol{A} = c\boldsymbol{B}$. This inequality may be proved by noting that if $\boldsymbol{C} = \boldsymbol{A}+\boldsymbol{B}$, then the triangle with sides A, B, and C always has $|C| \leq |A| + |B|$ and the equality requires $\boldsymbol{A}$ parallel to $\boldsymbol{B}$. Squaring both this scalar result and the vector relation, we have

$$|C|^2 \leq |A|^2 + |B|^2 + 2|A||B|\,,$$
$$|C|^2 = |A|^2 + |B|^2 + 2|\boldsymbol{A} \cdot \boldsymbol{B}|\,,$$

and the difference between these two gives the triangle inequality.

In order to relate these results to the uncertainty relation, we now define two Hermitian operators

$$\hat{\mathcal{P}} \equiv \hat{P} - \langle P \rangle\,,$$
$$\hat{\mathcal{Q}} \equiv \hat{Q} - \langle Q \rangle\,,$$

and apply the Schwartz inequality of Equation (1.38) by letting

$$F = \hat{\mathcal{P}}\psi\,,$$
$$G = \hat{\mathcal{Q}}\psi\,,$$

so that again using Dirac notation, we have

$$\begin{aligned}\langle F|F\rangle &= \langle \hat{\mathcal{P}}\psi|\hat{\mathcal{P}}\psi\rangle = \langle \psi|\hat{\mathcal{P}}^2|\psi\rangle \\ &= \langle \psi|(\hat{P} - \langle P\rangle)^2|\psi\rangle \equiv (\Delta P)^2\,,\end{aligned}$$

where we have used the Hermitian property of $\hat{\mathcal{P}}$. In the same manner, we have for G the result

$$\langle G|G\rangle = (\Delta Q)^2\,.$$

From these and Equation (1.38), it follows that

$$(\Delta P)^2(\Delta Q)^2 \geq \left|\langle \hat{\mathcal{Q}}\psi|\hat{\mathcal{P}}\psi\rangle\right|^2 = \left|\langle \psi|\hat{\mathcal{Q}}\hat{\mathcal{P}}|\psi\rangle\right|^2\,, \tag{1.41}$$

where we have used the Hermitian property of $\hat{\mathcal{Q}}$. By adding and subtracting terms, we may write the product of the operators as

$$\hat{\mathcal{Q}}\hat{\mathcal{P}} = \tfrac{1}{2}(\hat{\mathcal{Q}}\hat{\mathcal{P}} + \hat{\mathcal{P}}\hat{\mathcal{Q}}) + \tfrac{1}{2}(\hat{\mathcal{Q}}\hat{\mathcal{P}} - \hat{\mathcal{P}}\hat{\mathcal{Q}})\,,$$

so the right-hand side of Equation (1.41) is

$$|\langle \hat{\mathcal{P}}\hat{\mathcal{Q}}\rangle|^2 = \left|\tfrac{1}{2}\langle \hat{\mathcal{Q}}\hat{\mathcal{P}} + \hat{\mathcal{P}}\hat{\mathcal{Q}}\rangle + \tfrac{1}{2\mathrm{i}}\langle \mathrm{i}(\hat{\mathcal{Q}}\hat{\mathcal{P}} - \hat{\mathcal{P}}\hat{\mathcal{Q}})\rangle\right|^2\,.$$

Both operators, $\hat{\mathcal{Q}}\hat{\mathcal{P}} + \hat{\mathcal{P}}\hat{\mathcal{Q}}$ and $\mathrm{i}(\hat{\mathcal{Q}}\hat{\mathcal{P}} - \hat{\mathcal{P}}\hat{\mathcal{Q}}) = \mathrm{i}[\hat{\mathcal{Q}}, \hat{\mathcal{P}}]$, are Hermitian, from Equations (1.34) and (1.35), and hence have real expectation values. Therefore, Equation (1.41) becomes

$$(\Delta P)^2(\Delta Q)^2 \geq \tfrac{1}{4}\left[\langle \hat{\mathcal{Q}}\hat{\mathcal{P}} + \hat{\mathcal{P}}\hat{\mathcal{Q}}\rangle^2 + \langle \mathrm{i}[\hat{\mathcal{Q}}, \hat{\mathcal{P}}]\rangle^2\right]. \tag{1.42}$$

Now from the definitions of $\hat{\mathcal{P}}$ and $\hat{\mathcal{Q}}$, their commutator may be written

$$[\hat{\mathcal{Q}}, \hat{\mathcal{P}}] = (\hat{Q} - \langle Q \rangle)(\hat{P} - \langle P \rangle) - (\hat{P} - \langle P \rangle)(\hat{Q} - \langle Q \rangle) = [\hat{Q}, \hat{P}] .$$

Then, since $\langle \hat{\mathcal{Q}}\hat{\mathcal{P}} + \hat{\mathcal{P}}\hat{\mathcal{Q}} \rangle^2$ is nonnegative, the original theorem of Equation (1.37) is proved.

We further note from Equation (1.40) and the definitions of F and G that equality occurs only when

$$(\hat{Q} - \langle Q \rangle)\psi = c(\hat{P} - \langle P \rangle)\psi \tag{1.43}$$

everywhere, *and* when the anticommutator vanishes, or

$$\langle \hat{\mathcal{Q}}\hat{\mathcal{P}} + \hat{\mathcal{P}}\hat{\mathcal{Q}} \rangle = 0. \tag{1.44}$$

As an example, consider $\hat{P} = \hat{p}_x$ and $\hat{Q} = \hat{x}$. Since $\mathrm{i}[\hat{p}_x, \hat{x}] = \hbar$, we immediately have that

$$(\Delta p_x)(\Delta x) \geq \tfrac{1}{2}\hbar ,$$

which is the most common form of Heisenberg's uncertainty principle.

Problem 1.24 For the special case with $\hat{P} = \hat{p}_x$ and $\hat{Q} = x$, find the wave function that will minimize the uncertainty product.

(a) First assume $\langle p_x \rangle = \langle x \rangle = 0$ and let $\hat{p}_x\psi = cx\psi$ and solve the differential equation for ψ.

(b) For what values of the constant c is ψ bounded?

(c) Show that for these choices of c, $\langle \hat{\mathcal{Q}}\hat{\mathcal{P}} + \hat{\mathcal{P}}\hat{\mathcal{Q}} \rangle = 0$ so that equality is assured.

Problem 1.25 Show that the commutator $[\hat{p}_x, \hat{x}]$ is the same in both the coordinate and the momentum representations.

1.6.3 The Correspondence Principle

While some (or all) of the postulates of quantum mechanics seem to defy our intuition and appear to have little correspondence to the world in which we live, where classical mechanics appears to work quite well, we need assurance that whenever the quantity h is small, there should be a correspondence between classical mechanics and quantum mechanics. Called the **Correspondence Principle**, this principle states that whenever the finite size of h can be ignored, classical mechanics should be valid. A few examples will give the sense of this principle, but it is possible to prove it quite generally.

Consider a system described by the Hamiltonian function

$$\mathcal{H} = p^2/2m + V(x, y, z) ,$$

where $p^2 = p_x^2 + p_y^2 + p_z^2$. It is desired to calculate the time rate of change of the expectation value of one of the coordinates, $\mathrm{d}\langle x\rangle/\mathrm{d}t$, for example. From

$$\langle x\rangle = \int \Psi^*(x,y,z,t)x\Psi(x,y,z,t)\,\mathrm{d}x\,\mathrm{d}y\,\mathrm{d}z\,,$$

this is given by

$$\begin{aligned}\frac{\mathrm{d}\langle x\rangle}{\mathrm{d}t} &= \int \left[\frac{\partial\Psi^*}{\partial t}x\Psi + \Psi^* x\frac{\partial\Psi}{\partial t}\right]\mathrm{d}x\,\mathrm{d}y\,\mathrm{d}z \\ &= \int \left[\frac{\mathrm{i}}{\hbar}(\hat{\mathcal{H}}\Psi)^* x\Psi + \Psi^* x\left(-\frac{\mathrm{i}}{\hbar}\hat{\mathcal{H}}\Psi\right)\right]\mathrm{d}x\,\mathrm{d}y\,\mathrm{d}z \\ &= \int \Psi^*\frac{\mathrm{i}}{\hbar}(\hat{\mathcal{H}}x - x\hat{\mathcal{H}})\Psi\,\mathrm{d}x\,\mathrm{d}y\,\mathrm{d}z \\ &= \left\langle\Psi\left|\frac{\mathrm{i}}{\hbar}[\hat{\mathcal{H}},\hat{x}]\right|\Psi\right\rangle,\end{aligned} \tag{1.45}$$

where we used the time-dependent Schrödinger equation at the second step. This tells us that the *x-velocity* is given by the expectation value of $\mathrm{i}/\hbar$ times the commutator of the Hamiltonian with x. Since the Hamiltonian is given by

$$\hat{\mathcal{H}} = -\frac{\hbar^2}{2m}\left(\frac{\partial^2}{\partial x^2} + \frac{\partial^2}{\partial y^2} + \frac{\partial^2}{\partial z^2}\right) + V(x,y,z)\,,$$

the commutator is given by

$$\begin{aligned}\hat{\mathcal{H}}\hat{x} - \hat{x}\hat{\mathcal{H}} &= -\frac{\hbar^2}{2m}\left(\frac{\partial^2}{\partial x^2}x - x\frac{\partial^2}{\partial x^2}\right) \\ &= -2\frac{\hbar^2}{2m}\frac{\partial}{\partial x} \\ &= -\frac{\mathrm{i}\hbar}{m}\hat{p}_x\,.\end{aligned} \tag{1.46}$$

The final result is that

$$\frac{\mathrm{d}\langle x\rangle}{\mathrm{d}t} = \frac{\langle p_x\rangle}{m}\,, \tag{1.47}$$

which states that the time rate of change of the expectation value of x is the expectation value of the x-component of momentum divided by the mass. In this example, we did not invoke the smallness of h to obtain the classical result, but the correspondence is seen in that the expectation value and the measured value only coincide when the uncertainty is negligible, or when h is small.

If we now wish to know the time rate of change of the x-component of momentum, we find

$$\frac{\mathrm{d}\langle p_x\rangle}{\mathrm{d}t} = \int \Psi^*\frac{\mathrm{i}}{\hbar}(\hat{\mathcal{H}}\hat{p}_x - \hat{p}_x\hat{\mathcal{H}})\Psi\,\mathrm{d}x\,\mathrm{d}y\,\mathrm{d}z$$

$$
\begin{aligned}
&= \int \Psi^* \left(\hat{\mathcal{H}} \frac{\partial}{\partial x} - \frac{\partial}{\partial x} \hat{\mathcal{H}} \right) \Psi \, \mathrm{d}x \, \mathrm{d}y \, \mathrm{d}z \\
&= \int \Psi^* \left(- \frac{\partial V}{\partial x} \right) \Psi \, \mathrm{d}x \, \mathrm{d}y \, \mathrm{d}z \\
&= \left\langle - \frac{\partial V}{\partial x} \right\rangle . \qquad (1.48)
\end{aligned}
$$

Equations (1.47) and (1.48) are known as **Ehrenfest's Theorem**, which states that the classical equations of motion can be replaced by the quantum mechanical expectation values. In the classical system, we recognize the negative gradient of the potential as the force, and thus recover Newton's law.

We may generalize Equation (1.45) such that for *any* dynamical quantity $F(x, p_x)$, we may write

$$
\frac{\mathrm{d}}{\mathrm{d}t} \langle F \rangle = \left\langle \left| \frac{\mathrm{i}}{\hbar} [\hat{\mathcal{H}}, \hat{F}] \right| \right\rangle . \qquad (1.49)
$$

By comparing with Equation (1.37), letting $P \to F$ and $Q \to \mathcal{H}$, we find

$$
(\Delta F)(\Delta \mathcal{H}) \geq \tfrac{1}{2} |\langle \mathrm{i}[\hat{F}, \hat{\mathcal{H}}] \rangle|
$$

or, using Equation (1.49),

$$
(\Delta F)(\Delta \mathcal{H}) \geq \frac{\hbar}{2} \left| \frac{\mathrm{d}\langle F \rangle}{\mathrm{d}t} \right| .
$$

If we define the quantity,

$$
\left| \frac{(\Delta F)}{\mathrm{d}\langle F \rangle / \mathrm{d}t} \right| \equiv \Delta t_F
$$

then the quantity Δt_f corresponds to the time for $\langle F \rangle$ to change by the amount (ΔF). This corresponds to the minimum time it would take to determine that $\langle F \rangle$ has changed since it must change by at least the width to be measurable. If we consider all such functions that vary in time and choose the shortest of these times as Δt, then (letting $(\Delta \mathcal{H}) \to (\Delta E)$ as it is conventionally written),

$$
(\Delta E)(\Delta t) \geq \tfrac{1}{2} \hbar . \qquad (1.50)
$$

Problem 1.26 Fill in the steps leading to Equations (1.46) and (1.48).

1.6.4 Eigenvalues and Eigenfunctions

As observed in spectroscopic measurements long before the advent of even the old quantum theory, atoms appeared to have specific quantized energy states rather than a continuum that was more in line with the expectations of classical mechanics. We have already seen that quite generally one expects things *not* to be precisely defined in quantum mechanics, since a "spread" appears to be the norm due to the uncertainty principle, but there are cases where precisely defined quantities exist. If such states do exist, we want to know (a) under what conditions they exist, and if they exist at some instant in time, (b) what happens to them as time evolves?

Question (a) is easy enough to answer, since we can show that if for some dynamical quantity $F(q_j, p_j)$ there exists a wave function such that

$$\hat{F}\psi = F_0\psi\,, \tag{1.51}$$

where F_0 is a real constant, then that dynamical quantity is precisely defined. This follows from evaluating

$$\begin{aligned}\langle F\rangle &= \langle\psi|\hat{F}|\psi\rangle \\ &= \langle\psi|F_0\psi\rangle \\ &= F_0, \end{aligned} \tag{1.52}$$

$$\begin{aligned}\langle F^2\rangle &= \langle\psi|\hat{F}|\hat{F}\psi\rangle \\ &= \langle\psi|\hat{F}|F_0\psi\rangle \\ &= F_0\langle\psi|\hat{F}|\psi\rangle \\ &= F_0^2. \end{aligned} \tag{1.53}$$

From these if follows immediately that

$$(\Delta F)^2 = \langle F^2\rangle - \langle F\rangle^2 = F_0^2 - F_0^2 = 0\,,$$

so there is no uncertainty in this quantity and F has the precise value of F_0. Equation (1.51) is called an **eigenvalue equation**, the value F_0 is called an **eigenvalue**, and the wave function ψ is called an **eigenfunction of $\hat{F}$ corresponding to the eigenvalue F_0**.

Example 1.4

Eigenvalues. A simple yet important example of an eigenvalue problem is related to the z-component of angular momentum. In Problem 1.21, the operator leads to the eigenvalue equation

$$\hat{L}_z\psi = \frac{\hbar}{\mathrm{i}}\frac{\partial\psi}{\partial\phi} = L_{z0}\psi\,,$$

which may be integrated simply to yield

$$\psi(\phi) = A\exp(\mathrm{i}L_{z0}\phi/\hbar). \tag{1.54}$$

For this example, the wave function $\psi(\phi)$ is not single-valued, since for any integer n, $\phi + 2\pi n$ represents the same point is space, but $\psi(\phi)$ from Equation (1.54) is not necessarily the same. We cannot permit a wave function to have more than one value at a single point in space, or else our interpretation of probability density makes no physical sense. In order to recover some physical sense, we demand that $\psi(\phi+2\pi)=\psi(\phi)$. This in turn requires that $2\pi L_{z0}/\hbar = 2\pi m$ where m is any integer. We conclude that the eigenvalue L_{z0} can have any of the values

$$L_{z0} = m\hbar, \qquad m = 0, \pm 1, \pm 2, \ldots$$

This result is in agreement with Bohr theory, where angular momentum was quantized, but, as will be demonstrated later, *the energy eigenvalues of the Bohr theory can occur with zero angular momentum*, so there is no necessary connection between n, which relates to the energy, and m, which relates to the angular momentum, except that we will find that $|m| < n$. □

1.6.4.1 Simultaneous Eigenfunctions

It sometimes happens that an eigenfunction of some dynamical quantity is also an eigenfunction of some other dynamical quantity. In fact, let us assume that there exist several such eigenfunctions (perhaps with different eigenvalues), so that a more general eigenfunction that is a linear combination of these also is a simultaneous eigenfunction. Then we have for each individual eigenfunction

$$\hat{F}\psi_s = F_s\psi_s \qquad \text{and} \qquad \hat{G}\psi_s = G_s\psi_s\,,$$

for the two dynamical quantities $F(q_j, p_j)$ and $G(q_j, p_j)$. We may then write

$$\begin{aligned}\hat{F}\hat{G}\psi_s &= \hat{F}(G_s\psi_s) = G_s\hat{F}\psi_s = G_sF_s\psi_s \\ \hat{G}\hat{F}\psi_s &= \hat{G}(F_s\psi_s) = F_s\hat{G}\psi_s = F_sG_s\psi_s\end{aligned}$$

and since $F_sG_s = G_sF_s$ as these are simply constants, it follows that

$$\hat{F}\hat{G}\psi_s = \hat{G}\hat{F}\psi_s, \qquad \text{or} \qquad [\hat{F},\hat{G}]\psi_s = 0\,, \tag{1.55}$$

so that $\hat{F}$ and $\hat{G}$ commute, provided that $F_s \neq 0$ and $G_s \neq 0$. This means therefore that for two quantities to be simultaneously precisely defined, *their operators must commute* (it may be shown that this is both necessary as well as sufficient).

Question (b) raised above had to do with the time evolution of expectation values. From Equation (1.45), it is apparent that the time evolution of the expectation value of any quantity is related to its commutator with the Hamiltonian. Quite generally, *any quantity will vary with time unless its operator commutes with the Hamiltonian.* Thus, even if it is precisely defined at some instant, it may not remain precisely defined in time. On the other hand, if its operator does commute with the Hamiltonian, $\langle F\rangle$ remains F_0 and $\langle F^2\rangle$ remains F_0^2, so it remains precisely defined. A special example is the energy, since the evolution of the energy is of particular interest. The corresponding operator is the Hamiltonian itself, and for this case the actual time dependence can be found explicitly. The eigenvalue equation is

$$\hat{\mathcal{H}}\Psi(q_j,t) = E_0\Psi(q_j,t),$$

but from the time-dependent Schrödinger equation, we have also

$$\hat{\mathcal{H}}\Psi(q_j,t) = -\frac{\hbar}{\mathrm{i}}\frac{\partial\Psi}{\partial t} = E_0\Psi(q_j,t).$$

From the latter equality, we immediately obtain

$$\Psi(q_j,t) = \psi(q_j)\exp(-\mathrm{i}E_0 t/\hbar)\,,$$

where $\psi(q_j)$ represents the *spatial* portion of the wave function and the temporal evolution is given by the exponential phase factor. Using this representation, the equation for the spatial part of the wave function is given by

$$-\frac{\hbar^2}{2m}\nabla^2\psi + (V - E_0)\psi = 0\,, \tag{1.56}$$

which is known as the **time-independent Schrödinger equation**.

1.6.4.2 Orthogonality of Eigenfunctions

Another useful property of eigenfunctions and eigenvalues of operators is the fact that different eigenfunctions of a given operator that have different eigenvalues are orthogonal to one another, which means that the integral of their product vanishes. This is easily demonstrated by considering two eigenfunctions, ψ_n and ψ_m, that satisfy $\hat{F}\psi_n = F_n\psi_n$ and $\hat{F}\psi_m = F_m\psi_m$. Then, we multiply the first by ψ_m^* and the second by ψ_n^* and subtract the complex conjugate of the second product from the first and integrate to obtain

$$\int [\psi_m^*\hat{F}\psi_n - \psi_n(\hat{F}\psi_m)^*]\,\mathrm{d}^N q = (F_n - F_m^*)\langle\psi_m|\psi_n\rangle\,.$$

Using the Hermitian property, the integral on the left vanishes, and the eigenvalues are real, so we have

$$(F_n - F_m)\langle\psi_m|\psi_n\rangle = 0\,, \tag{1.57}$$

which requires either $F_n = F_m$ or $\langle\psi_m|\psi_n\rangle = 0$, so that if the eigenvalues are distinct, the eigenfunctions are orthogonal. When the eigenvalues are not distinct, the eigenfunctions are said to be **degenerate**, and we may still construct orthogonal eigenfunctions from a linear combination of the degenerate eigenfunctions.

Problem 1.27 Justify the steps in the derivation of Equations (1.52) and (1.53).

Problem 1.28 Show that if $\hat{F}$ and $\hat{\mathcal{H}}$ commute, then both $\mathrm{d}\langle F\rangle/\mathrm{d}t$ and $\mathrm{d}\langle F^2\rangle/\mathrm{d}t$ vanish, so that $F(q_j,p_j)$ remains precisely defined.

Problem 1.29 It was shown in Equation (1.55) that if a function is simultaneously an eigenfunction of two operators then the operators commute. Show the converse, namely, that if two operators commute, they have simultaneous eigenfunctions.

1.6.5 Conservation of Probability

Postulate 1 required that the total integrated probability density be unity so that the particle exists somewhere. Having introduced the fact that the wave function evolves in time, it is necessary to ensure that the total probability does not change, even though the local probability density may change. Denoting the total probability density as $P(t) = \int \Psi^*(q_n,t)\Psi(q_n,t)\,\mathrm{d}^N q$, then we investigate

$$\begin{aligned}\frac{\mathrm{d}P}{\mathrm{d}t} &= \int \frac{\partial}{\partial t}(\Psi^*\Psi)\,\mathrm{d}^N q \\ &= \int \left(\frac{\partial \Psi^*}{\partial t}\Psi + \Psi^*\frac{\partial \Psi}{\partial t}\right)\mathrm{d}^N q\,.\end{aligned}$$

By using the time-dependent Schrödinger equation, the partial derivatives may be eliminated, with the result

$$\frac{\mathrm{d}P}{\mathrm{d}t} = \int \frac{\mathrm{i}}{\hbar}[(\hat{\mathcal{H}}^*\Psi^*)\Psi - \Psi^*(\hat{\mathcal{H}}\Psi)]\,\mathrm{d}^N q\,.$$

The Hermitian character of $\hat{H}$ then ensures that the right-hand side vanishes and the total probability is conserved.

Having established that the total probability is conserved, it is worthwhile to observe that it is possible to write a conservation law for the probability *density* also, just as there is an equation of continuity for charge density or fluid flow. Taking a simple example where the Hamiltonian is given by

$$\hat{\mathcal{H}} = -\frac{\hbar^2}{2m}\nabla^2 + V(x,y,z)\,, \tag{1.58}$$

we may write the rate of change of the probability density as

$$\frac{\partial W}{\partial t} = \frac{\partial \Psi^*}{\partial t}\Psi + \Psi^*\frac{\partial \Psi}{\partial t} = \frac{\mathrm{i}}{\hbar}[(\hat{\mathcal{H}}^*\Psi^*)\Psi - \Psi^*(\hat{\mathcal{H}}\Psi)].$$

Using the explicit form of Equation (1.58), however, this may be written as

$$\frac{\partial W}{\partial t} + \frac{\hbar}{2m\mathrm{i}}(\Psi^*\nabla^2\Psi - \Psi\nabla^2\Psi^*) = 0\,,$$

or, through the use of a vector identity,

$$\frac{\partial W}{\partial t} + \nabla\cdot\frac{\hbar}{2m\mathrm{i}}(\Psi^*\nabla\Psi - \Psi\nabla\Psi^*) = 0\,. \tag{1.59}$$

From this, we may identify the vector $\boldsymbol{S} = \frac{1}{2}(\hbar/m\mathrm{i})(\Psi^*\nabla\Psi - \Psi\nabla\Psi^*)$ as a **probability current density**, whose integral over a closed surface represents the change of probability inside the surface. Using this notation, the conservation law may be written as

$$\frac{\partial W}{\partial t} + \nabla\cdot\boldsymbol{S} = 0.$$

Problem 1.30 Supply the missing steps in the derivation of Equation (1.59).

2

The Schrödinger Equation in One Dimension

In the previous chapter, the postulates helped us to use the wave function to find the expected results of experimental measurements by letting a dynamical quantity $F(x, p, t) \to F(\hat{x}, \hat{p}, t)$ and integrating over space. In the examples, however, the wave functions were typically given and not derived from first principles. We now endeavor to find the appropriate wave functions for a variety of physical systems in one dimension by solving the Schrödinger equation where for time-independent cases, the total energy is an eigenvalue. In subsequent chapters, we will extend the method to three dimensions and develop the appropriate operators, eigenvalues, and eigenfunctions for angular momentum. We will also develop methods for examining time-dependent systems. The latter chapters address special systems of interest that may be referred to as applications of quantum mechanics.

2.1 The Free Particle

As our first example of solving the Schrödinger equation, we choose the **free particle** in one dimension, which means that the particle is free of any forces. It thus has a constant potential, so the time-independent Schrödinger equation in one dimension is written as

$$\hat{\mathcal{H}}\psi = -\frac{\hbar^2}{2m}\frac{\mathrm{d}^2\psi}{\mathrm{d}x^2} + V\psi = E\psi\,, \tag{2.1}$$

where $\psi = \psi(x)$ is the spatial part of the wave function, and the complete wave function is $\Psi(x, t) = \psi(x)\exp(-\mathrm{i}Et/\hbar)$. Just as the time-dependent equation was easily solved in the previous chapter, so is the solution of Equation (2.1) easily solved, with solution

$$\psi(x) = A\exp\left[\pm\mathrm{i}\sqrt{2m(E-V)}x/\hbar\right]\,. \tag{2.2}$$

Problem 2.1 Show that the solution of the 1-dimensional Schrödinger equation for a free particle can be written as $\psi(x) = A\exp(\pm\mathrm{i}kx)$ where $\hbar k$ is the classical momentum of a free particle of energy E.

Problem 2.2 Show that the eigenfunctions of Equation (2.2) are simultaneously eigenfunctions of the momentum operator, and that the positive sign corresponds to a particle traveling with momentum $\hbar k$ in the positive x-direction.

2.1.1 Boundary Conditions and Normalization

Using the postulates of the previous chapter, which require that the probability density vanish at $\pm\infty$, then we have the rather surprising result that this requires $A = 0$, which appears trivial. The difficulty has to do with the interpretation of what we wanted the normalization constant to accomplish. Usually, a particle exists in some bounded volume, or at least the probability of finding it at large $|x|$ decreases exponentially outside some volume. In this case, however, the simplest case of the free particle gives uniform probability *everywhere*, and the only relevant questions are (1) what is the *relative* probability of finding the particle in one place compared to another, or (2) what is the flux of particles. For these questions, it is only required to consider the ratio of the probability of finding the particle in one place compared to the probability of finding it anyplace else, and the constant factor cancels out, leaving a valid answer for the **relative probability**.

This is still trivial unless there is a change in the potential somewhere, and we will consider a variety of examples where the potential is not uniform in space, but is *constant* in each separate region, changing in discrete steps. With this restriction, the solution is of the same *form* everywhere, but the constants will vary as $V(x)$ changes. We can construct stair-step barriers with different levels on opposite sides, and we can construct various kinds of wells with rectangular bottoms.

Whenever the potential changes abruptly, the solutions on either side must match somehow, since they are different from one another, but somehow connected. From simple considerations of the interpretation of the wave function, we can determine that:

1. *The wave function must be continuous at the boundary*, since the probability of finding the particle on one side of the boundary must match the probability of finding it on the other side of the boundary, since our concept of a real particle is that it must cross the boundary as a unit.

2. *The derivative of the wave function must be continuous at the boundary.* This is required since *momentum* must be conserved across boundaries.

Quite generally, when considering a free particle, it is easier to consider a *beam* of particles rather than a single particle, so that the probability density represents the number of beam particles to be found per meter per second (or the fraction of the beam intensity).

2.1.2 Wave Packets

If one were to consider a beam that is not monoenergetic, then the momenta of the beam components will vary, but if the momenta are nearly uniform, one can imagine a situation where all are in phase at one particular point in space so that all of the individual components add in phase. Since there is some variation, if one goes far enough away, they are eventually going to be out of phase, and some cancellation will occur, so that the probability density is localized. For any finite number of beam components, they will eventually come back in phase, and the amplitude will oscillate. As the number of components is increased, however, it is possible to make the period in space very long, so that over a large region in space, the wave function is localized. We call such a superposition of waves a **wave packet**, and in the limit as the number of components becomes infinite (a continuous variation in E), there may be a single localized region where the probability is finite. We may alternately wish to consider a beam made up of individual electrons that are localized, and the wave packet description of a superposition of waves provides the localization in both configuration space and momentum space.

Example 2.1
Gaussian wave packet. A particular example is the wave function

$$\psi(x) = A\exp\left[-(x-x_0)^2/2a^2\right]\,\exp(\mathrm{i}p_0x/\hbar)\,, \tag{2.3}$$

where A is a constant. We encountered this wave function in Problem 1.16, but now we can interpret it as a wave function centered about momentum p_0, but composed of a continuum of components with varying amplitudes that decrease rapidly as the momentum differs from p_0. In order to see that this interpretation is valid, we first examine the free-particle wave function with time,

$$\Psi(x,t) = A\exp\left(\mathrm{i}px/\hbar\right)\,\exp\left(-\mathrm{i}p^2t/2m\hbar\right)\,, \tag{2.4}$$

where $p^2/2m$ is the energy of a free particle with $V=0$. Since the eigenvalues for energy are continuously distributed, we can construct a wave function from a superposition of individual states that is an integral over the momentum, of the form

$$\Psi(x,t) = \frac{1}{2\pi\sqrt{\hbar}}\int_{-\infty}^{\infty} A(p)\exp\left(\mathrm{i}px/\hbar\right)\,\exp\left(-\mathrm{i}p^2t/2m\hbar\right)\mathrm{d}p\,. \tag{2.5}$$

The introduction of the factor $A(p)$ is necessary, since the uncertainty principle indicates that if a particle is localized in space, it must be localized in momentum, and $A(p)$ represents an amplitude centered about p_0 with finite width, just as the coordinate of a single electron is centered about $x_0(t)$ with finite width. We recognize that Equation (2.5) is simply the expression for $\Psi(x,t)$ in terms of the momentum wave function $\Phi(p,t)$, and from Problem

1.16, we choose $A(p)$ to have the form

$$A(p) = \left(\frac{a}{\hbar\sqrt{\pi}}\right)^{1/2} \mathrm{e}^{-a^2(p-p_0)^2/2\hbar^2}\,\mathrm{e}^{-\mathrm{i}(p-p_0)x_0/\hbar}\,,$$

so that at $t = 0$, $\Psi(x,0)$ will be just the expression in Equation (2.3) with $A(p_0) = (a\sqrt{\pi})^{-1/2}$. With this choice, $\Psi(x,t)$ becomes

$$\Psi(x,t) = \left(\frac{a}{\hbar\sqrt{\pi}}\right)^{1/2} \mathrm{e}^{\mathrm{i}p_0x_0/\hbar} \int_{-\infty}^{\infty} \mathrm{e}^{-a^2(p-p_0)^2/2\hbar^2} \exp\left[\mathrm{i}p(x-x_0)/\hbar\right] \times \exp\left[-\mathrm{i}p^2t/2m\hbar\right] \mathrm{d}p\,.$$

Evaluating this integral, after some rearrangement it may be written as

$$\Psi(x,t) = \pi^{-1/4}\left(a + \frac{\mathrm{i}\hbar t}{ma}\right)^{-1/2} \exp\left[\mathrm{i}\left(\frac{p_0x}{\hbar} - \frac{p_0^2t}{2m\hbar}\right)\right] \times \exp\left[-\frac{(x-x_0-p_0t/m)^2(1-\mathrm{i}\hbar t/ma^2)}{2(a^2+\hbar^2t^2/m^2a^2)}\right]. \tag{2.6}$$

At $t = 0$, it is clear that this is the same as Equation (2.3), but as time evolves, we note the following properties, most of which are evident in Figure 2.1:

1. The probability distribution function is

$$W_q(x,t) = \pi^{-1/2}\left(a^2 + \frac{\hbar^2t^2}{m^2a^2}\right)^{-1/2} \exp\left[-\frac{(x-x_0-p_0t/m)^2}{(a^2+\hbar^2t^2/m^2a^2)}\right].$$

   This indicates that the distribution is always gaussian and normalized, but the position of the maximum moves and the width changes with time as shown in Figure 2.1.

2. The position of the maximum of $W_q(x,t)$ moves with speed p_0/m, which corresponds to the average momentum of the particle. The peak of the envelope in Figure 2.1 moves at this speed.

3. The width of the distribution $W_q(x,t)$ increases with time in a way that depends on the two uncertainties associated with the coordinate and the momentum. The quantity $(\hbar/ma)t$ is equal to the distance traversed in time t at a speed $\sqrt{2}(\Delta p/m)$, where $\Delta p = \hbar/\sqrt{2}a = \hbar/2\Delta x$, which is the initial uncertainty in p. The greater the uncertainty in p, the more rapid the spread. The width has more than doubled in Figure 2.1, growing as $w(t) = \sqrt{a^2 + \hbar^2t^2/m^2a^2}$ or linearly in time for long times.

4. The phase factor $\exp\left[\mathrm{i}(p_0x/\hbar - p_0^2t/2m\hbar)\right]$ is the space and time factor for a particle whose momentum is precisely p_0. The actual uncertainty in momentum causes this to be only approximately true in the full wave function.

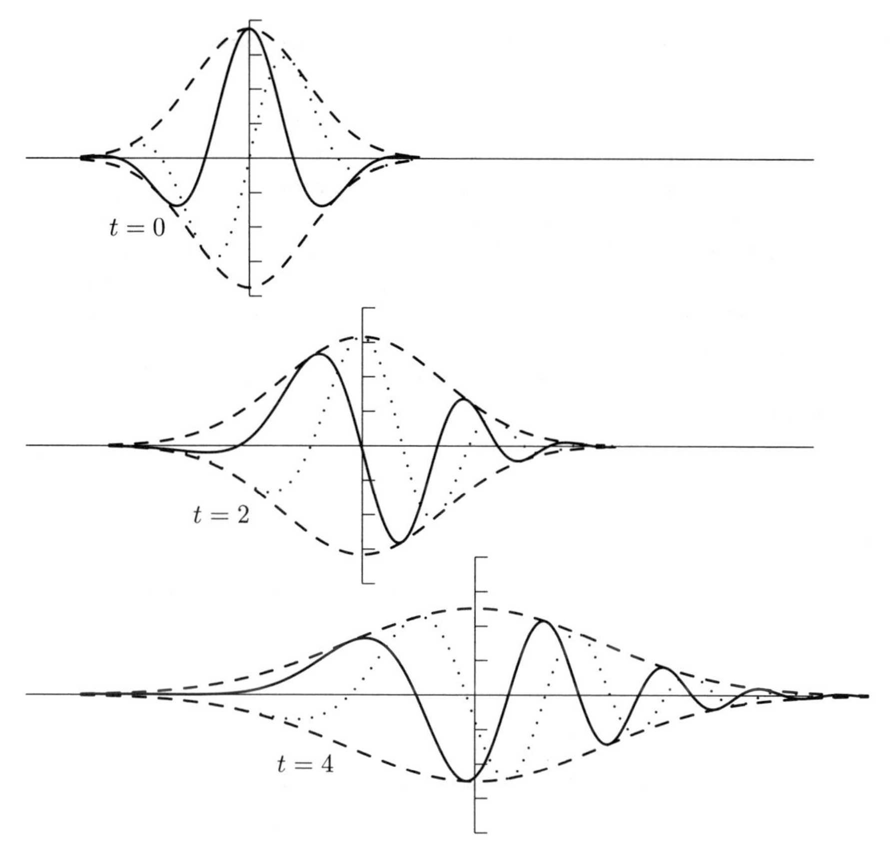

FIGURE 2.1
Spreading of a wave packet as it propagates to the right. The real part of $\Psi(x,t)$ is solid, the imaginary part is dotted, and the envelope (absolute magnitude) is dashed. For this example, the phase velocity is about half the group velocity.

5. The more complicated phase factor leads to the effect of the wave function oscillating more rapidly in the front of the wave packet than in the rear. This is due to the spread in momentum so that the "faster" components run on ahead of the packet. In the example of Figure 2.1, this is evident only at the leading edge. If the figure were a movie, it would appear that the wave is being generated at the front of the packet and disappearing into the tail.

$\square$

2.1.3 Transmission and Reflection at a Barrier

There are a great many problems in quantum mechanics where the exact solution is either unattainable or too difficult to find. Before expending a great effort, it is useful in many cases to approximate the general features of the problem by an easily solved problem, from which the related problem may be understood, even if only qualitatively. As an example of this kind of approximation, we will consider a particle impinging on a barrier, and approximate the shape of the barrier to be a step barrier, so the potential makes a discrete jump at one point in space. This is not entirely realistic, since any real barrier will have some finite extent, and the edge will vary smoothly, but the principal feature is the jump in energy, and that is well described by our model.

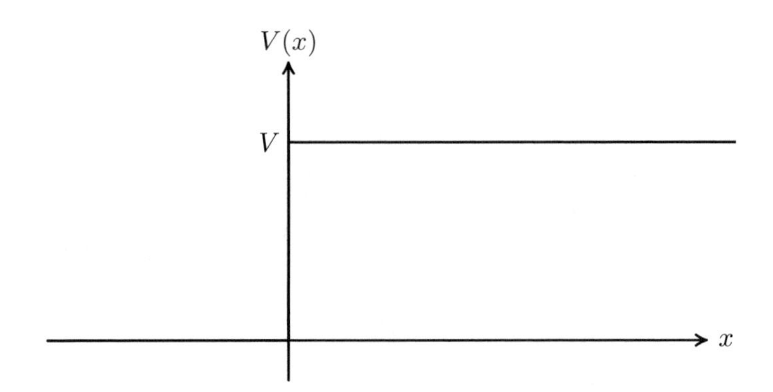

FIGURE 2.2
A semiinfinite potential barrier of height V.

This example is illustrated in Figure 2.2, where a particle is presumed to be incident from the left, and if it has sufficient energy to mount the potential barrier of height V ($E > V$), then it will be at least partially transmitted, and if it lacks sufficient energy ($E < V$), it is expected to be totally reflected.

For this type of problem we must solve the Schrödinger equation in two separate regions of space, one on the left where the potential vanishes, and one on the right where the potential energy is V. The form of the wave function is given by Equation (2.2) with $V = 0$ on the left, and $V \neq 0$ on the right. We will have to establish connections between the solutions in the two regions through the matching of boundary conditions. The solutions in the two regions are each of the form

$$\begin{aligned} \psi_\ell(x) &= A\mathrm{e}^{ikx} + B\mathrm{e}^{-ikx}, \qquad x \leq 0, \\ \psi_r(x) &= C\mathrm{e}^{i\kappa x} + D\mathrm{e}^{-i\kappa x}, \qquad x \geq 0, \end{aligned} \tag{2.7}$$

where

$$k = \sqrt{2mE}/\hbar\,, \qquad \kappa = \sqrt{2m(E-V)}/\hbar\,.$$

We note first that if $E < 0$ the only solutions are trivial, since both k and κ are imaginary, and if the wave functions are bounded, one coefficient in each region must vanish. The remaining terms cannot satisfy both the continuity of the wave function and the continuity of its derivative unless the other two constants vanish, so the solution is trivial (by trivial, we mean that $\psi_\ell = \psi_r = 0$ everywhere).

Next, if $0 < E < V$, then k is real but κ is imaginary. Changing to a real $\kappa = \sqrt{2m(V-E)}/\hbar$, then we have

$$\begin{aligned} \psi_\ell(x) &= A\mathrm{e}^{\mathrm{i}kx} + B\mathrm{e}^{-\mathrm{i}kx}\,, \qquad & x \le 0\,, \\ \psi_r(x) &= C\mathrm{e}^{\kappa x} + D\mathrm{e}^{-\kappa x}\,, \qquad & x \ge 0\,, \end{aligned} \tag{2.8}$$

and we must immediately set $C = 0$ for a bounded solution since $\mathrm{e}^{\kappa x} \to \infty$ as $x \to \infty$. The boundary conditions at $x = 0$ are then $\psi_\ell(0) = \psi_r(0)$, and $\psi'_\ell(0) = \psi'_r(0)$, so we have $A + B = D$ and $\mathrm{i}k(A - B) = -\kappa D$. Solving for the amplitude ratios, we find

$$\frac{B}{A} = \frac{\mathrm{i}k + \kappa}{\mathrm{i}k - \kappa} \qquad \text{and} \qquad \frac{D}{A} = \frac{2\mathrm{i}k}{\mathrm{i}k - \kappa}\,. \tag{2.9}$$

We interpret this result by noting that A represents the amplitude for the incident wave from the left and B represents the amplitude of the reflected wave, so for the probability density for the reflected wave, we have $|B/A|^2 = 1$, which tells us that there is total reflection in this case. This was expected, since the particle did not have enough energy to get over the barrier. We note, however, that $D \neq 0$, which means that there is a finite probability of finding the particle *on the right*, where classically it cannot go. This means that it penetrates some finite distance beyond the barrier, but for macroscopic parameters, the distance is extremely short. The standing wave character of the solution on the left that leads to nodes and peaks is easily understood in analogy with waves on a stretched string when one recalls the wave nature of particles. The barrier penetration is a purely quantum mechanical phenomenon, however, corresponding to evanescent waves in electricity and magnetism.

When $E > V > 0$, then both k and κ are real, but now we must set $D = 0$ since that term would represent particles incident *from the right*, and by postulate, we are considering a beam of particles incident from the left. The boundary conditions then give $A + B = C$ and $\mathrm{i}k(A - B) = \mathrm{i}\kappa C$. These lead to the ratios

$$\frac{B}{A} = \frac{k - \kappa}{k + \kappa} \qquad \text{and} \qquad \frac{C}{A} = \frac{2k}{k + \kappa}\,. \tag{2.10}$$

For this case, we see that there will be some transmission ($|C/A|^2 \neq 0$) of the beam where particles move past the barrier with reduced kinetic energy, and there will also be some reflection ($|B/A|^2 > 0$). Classically, this latter result is unexpected, but it is very typical of what happens when a light wave encounters a dielectric.

Problem 2.3 Fill in the missing steps leading to Equations (2.9) and (2.10).

Problem 2.4 Show that quite generally, when $E > V > 0$, $|C/A|^2 > 1$, which at first glance seems to say that the transmitted beam is greater than the incident beam. Show that the result does make sense, as the solution satisfies Equation (1.59), the continuity equation, indicating that the number of particles striking the barrier per unit time is equal to the number leaving the barrier per unit time. Discuss the interpretation of $|C/A|^2 > 1$.

Problem 2.5 Solve the corresponding problem if the beam is incident from the *right*. Show that *not all of the particles are transmitted, even though they gain energy crossing the barrier.* This result is again very typical of what happens when a light wave exits from a dielectric.

Problem 2.6 Show that it is possible to make a **perfectly transmitting barrier** by providing a thin intermediate layer before the step of Figure 2.2 with $V_i < V < E$ if the value of the intermediate potential and the thickness of the intermediate layer are chosen properly. Find the intermediate potential V_i in terms of E and V and the thickness of the layer. This is analogous to an anti-reflection coating on a good camera lens. (*Hint:* It is easier to set $R = 0$ than to set $T = 1$.)

2.1.4 The Infinite Potential Well

If the barrier of the previous section were infinite, there would of course be no transmission and 100% reflection. If there are two infinite barriers with a particle trapped in a one-dimensional box, then presumably it will bounce back and forth. If the waves representing the bouncing particle do not have the proper phase, however, they may have destructive interference. Such states do not represent real particles, so real particles must have only certain wavelengths or energies. This will be true for either an infinitely deep well or a finite well. Looking at the infinite well first, the wave function is nonvanishing only on the inside of the well, since $V\psi$ is unbounded for finite ψ and infinite V. The general solution of the Schrödinger equation is a free particle inside, with $\psi = A\mathrm{e}^{\mathrm{i}kx} + B\mathrm{e}^{-\mathrm{i}kx}$ but with $\psi = 0$ at each side of the box. If the box extends from 0 to a, then the boundary conditions at $x = 0$ and $x = a$ are

$$A + B = 0\,, \qquad A\mathrm{e}^{\mathrm{i}ka} + B\mathrm{e}^{-\mathrm{i}ka} = 0\,,$$

so that the first requires $B = -A$, and this result inserted into the second gives

$$A(\mathrm{e}^{\mathrm{i}ka} - \mathrm{e}^{-\mathrm{i}ka}) = 2\mathrm{i}A\sin ka = 0\,.$$

This requires that $ka = n\pi$ with $n = \pm 1, \pm 2, \ldots$ so that only special values of k and hence of $E = \hbar^2 k^2/2m$ are allowed ($n = 0$ is forbidden because that solution is trivial). The resulting allowed energies are given by

$$E_n = \frac{\hbar^2}{2m}\left(\frac{n\pi}{a}\right)^2 . \tag{2.11}$$

These discrete energies correspond to waves bouncing back and forth in the well where there is constructive interference, and hence standing waves.

The wave functions may be written as $\psi_n(x) = A \sin k_n x = A \sin(n\pi x/a)$. The normalization constant is determined by

$$\begin{aligned}
\int_0^a |\psi_n|^2 \, \mathrm{d}x &= |A|^2 \int_0^a \sin^2(n\pi x/a) \, \mathrm{d}x \\
&= \frac{|A|^2 a}{n\pi} \int_0^{n\pi} \sin^2 u \, \mathrm{d}u \\
&= \frac{|A|^2 a}{n\pi} \left(\frac{u}{2} - \frac{\sin 2u}{4} \right)_0^{n\pi} \\
&= \frac{|A|^2 a}{2} = 1
\end{aligned} \tag{2.12}$$

so $A = \sqrt{2/a}$ (where we let $u = n\pi x/a$ on the second line).

The wave functions for different n are mutually orthogonal to one another since

$$\begin{aligned}
\int_0^a \psi_n^*(x)\psi_m(x) \, \mathrm{d}x &= \frac{2}{a} \int_0^a \sin\left(\frac{n\pi x}{a}\right) \sin\left(\frac{m\pi x}{a}\right) \mathrm{d}x \\
&= \frac{2}{\pi} \int_0^{\pi} \sin nu \sin mu \, \mathrm{d}u \\
&= \frac{1}{\pi} \left[\frac{\sin(n-m)u}{n-m} - \frac{\sin(n+m)u}{n+m} \right]_0^{\pi} \\
&= \begin{cases} 0 & n \neq m \\ 1 & n = m \end{cases}
\end{aligned} \tag{2.13}$$

where we again changed variables on the second line by letting $u = \pi x/a$. When a set of functions are both orthogonal and normalized, we say the set is **orthonormal**. We write this as

$$\int_0^a \psi_n^*(x)\psi_m(x) \, \mathrm{d}x = \delta_{n,m} , \tag{2.14}$$

where $\delta_{n,m}$ is the **Kronecker delta**, which is unity for $n = m$ and zero otherwise.

This orthonormal set is a **complete set** in that we may represent *any* function, $f(x)$, in terms of these functions such that

$$f(x) = \sum_{n=1}^{\infty} c_n \psi_n(x) = \sqrt{\frac{2}{a}} \sum_{n=1}^{\infty} c_n \sin\left(\frac{n\pi x}{a}\right) . \tag{2.15}$$

The c_n may be found by multiplying both sides by $\psi_m^*(x)$ and integrating to find

$$\begin{aligned}\int_0^a \psi_m^*(x)f(x)\,\mathrm{d}x &= \sum_{n=1}^{\infty} c_n \int_0^a \psi_m^*\psi_n\,\mathrm{d}x \\ &= \sum_{n=1}^{\infty} c_n\delta_{m,n} \\ &= c_m \end{aligned} \tag{2.16}$$

since the Kronecker delta collapses the sum as the only term that survives the sum over n is when $n = m$.

Example 2.2

Completeness. As an example of completeness, let $f(x) = Ax(a-x)$, and although it is not necessary, we will normalize $f(x)$ so that

$$\begin{aligned}\int_0^a |f(x)|^2\,\mathrm{d}x &= |A|^2 \int_0^a x^2(a-x)^2\,\mathrm{d}x \\ &= |A|^2 \int_0^a (a^2x^2 - 2ax^3 + x^4)\,\mathrm{d}x \\ &= |A|^2\left(\frac{a^5}{3} - \frac{a^5}{2} + \frac{a^5}{5}\right) \\ &= \frac{a^5|A|^2}{30} = 1\end{aligned}$$

so $A = \sqrt{30/a^5}$. Then, from Equation (2.16), we have

$$\begin{aligned}c_m &= \int_0^a f(x)\psi_m(x)\,\mathrm{d}x \\ &= \sqrt{\frac{2}{a}}\sqrt{\frac{30}{a^5}}\int_0^a \sin\left(\frac{m\pi x}{a}\right)x(a-x)\,\mathrm{d}x \\ &= \frac{2\sqrt{15}}{a^3}\left[\frac{a^3}{(m\pi)^2}\int_0^{m\pi} u\sin u\,\mathrm{d}u - \frac{a^3}{(m\pi)^3}\int_0^{m\pi} u^2\sin u\,\mathrm{d}u\right] \\ &= \begin{cases} 8\sqrt{15}/(m\pi)^3, & m \text{ odd} \\ 0, & m \text{ even}\end{cases}.\end{aligned} \tag{2.17}$$

The series for $f(x)$ can then be expressed as

$$f(x) = \frac{8\sqrt{15}}{\pi^3}\left[\psi_1(x) + \frac{\psi_3(x)}{3^3} + \frac{\psi_5(x)}{5^3} + \cdots\right] \tag{2.18}$$

□

Problem 2.7 *Approximating a function in an infinite well.*

(a) A wave function is given in an infinite well of width a as

$$f(x) = \begin{cases} \dfrac{2}{\sqrt{a}} \sin \dfrac{2\pi x}{a}, & 0 \le x \le \dfrac{a}{2} \\ 0, & \text{otherwise} \end{cases}$$

Find N for the eigenfunction expansion so that

$$\Psi_N(x) = \sum_{k=1}^{N} a_k \psi_k(x)$$

will be within 5% of $f(x)$ at $x = a/4$.

(b) Calculate $\Psi_N(a/2)$ and $\Psi_N(3a/4)$ and sketch $\Psi_N(x)$ from 0 to a.

Problem 2.8 *Expansion coefficients.*

(a) Fill in the missing steps in the derivation of Equation (2.17) showing why the even terms vanish.

(b) Calculate c_1, c_3 and c_5 for this example.

(c) It may be shown that $\sum_{n=1}^{\infty} |c_n|^2 = 1$. Find the deviation from unity for the partial sum $c_1^2 + c_3^2 + c_5^2$.

(Note: this amounts to a proof that $1+1/3^6+1/5^6+1/7^6+\cdots = \sum_{n=0}^{\infty} 1/(2n+1)^6 = \pi^6/960$.)

Problem 2.9 *Uncertainty product.*

(a) Find the uncertainty product, $(\Delta p)(\Delta x)$, for a particle in a one-dimensional box for any n.

(b) For the $n = 1$ state, what percentage is this result *above* the minimum uncertainty product?

2.1.5 The Finite Potential Well

One may instead consider a more realistic example where a particle is trapped in a finite well that has both finite depth and finite width. This problem is set up much like the previous problem, except that now there are three regions, although we will take the potential height to be the same on both sides of the well, so that the left and right regions differ only in location. Again, in realistic potential wells, the square shape is only a model, but it has most of the basic characteristics of realistic wells, only it is not quantitative. Our model is depicted in Figure 2.3, where the central well is taken to extend from $-a$ to a.

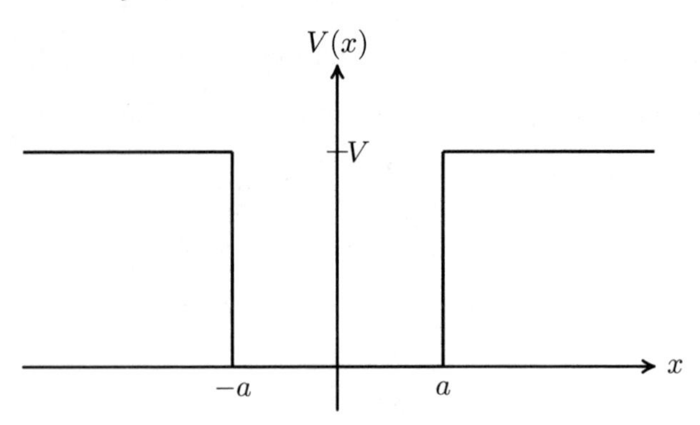

FIGURE 2.3
A rectangular potential well of depth V and width $2a$.

This case again has only trivial solutions for $E < 0$, but the interesting case is for $0 < E < V$. The wave functions are then

$$\begin{aligned} \psi_\ell(x) &= A\mathrm{e}^{\kappa x} + B\mathrm{e}^{-\kappa x}\,, & x &\le -a\,, \\ \psi_c(x) &= C\mathrm{e}^{\mathrm{i}kx} + D\mathrm{e}^{-\mathrm{i}kx}\,, & -a &\le x \le a\,, \\ \psi_r(x) &= F\mathrm{e}^{\kappa x} + G\mathrm{e}^{-\kappa x}\,, & x &\ge a\,, \end{aligned} \tag{2.19}$$

where again we define both k and κ to be real such that

$$k = \sqrt{2mE}/\hbar\,, \qquad \kappa = \sqrt{2m(V-E)}/\hbar\,.$$

In order to satisfy the boundary conditions at $\pm\infty$, we must first set $B = F = 0$ for a bounded solution. We must then satisfy boundary conditions for ψ and ψ' at $\pm a$, where we have

$$\begin{aligned} A\mathrm{e}^{-\kappa a} &= C\mathrm{e}^{-\mathrm{i}ka} + D\mathrm{e}^{\mathrm{i}ka}\,, \\ \kappa A\mathrm{e}^{-\kappa a} &= \mathrm{i}kC\mathrm{e}^{-\mathrm{i}ka} - \mathrm{i}kD\mathrm{e}^{\mathrm{i}ka}\,, \\ G\mathrm{e}^{-\kappa a} &= C\mathrm{e}^{\mathrm{i}ka} + D\mathrm{e}^{-\mathrm{i}ka}\,, \\ -\kappa G\mathrm{e}^{-\kappa a} &= \mathrm{i}kC\mathrm{e}^{\mathrm{i}ka} - \mathrm{i}kD\mathrm{e}^{-\mathrm{i}ka}\,. \end{aligned}$$

By eliminating A from the first pair of equations and G from the second pair, we easily find that $C = \pm D$. From this we find that if $C = D$, then $A = G$, and if $C = -D$, then $A = -G$. Thus the wave functions exhibit either *even or odd symmetry*, or **parity**, which is characteristic of any Hamiltonian that is even in x. Had we chosen the well to range from 0 to $2a$, this symmetry would not have been so apparent. This property of wave functions is so pervasive in quantum mechanics, that it even has conservation laws associated with it, such that in almost all interactions, parity is conserved. The most general form of the conservation law is that CPT is conserved, where $C = \pm 1$ stands for charge conjugation ($C = -1$ indicates a change in sign of the charge), $P = \pm 1$ stands for the parity ($P = -1$ if the parity changes), and $T = \pm 1$ stands for time-reversal invariance ($T = -1$ if time reverses). In weak interactions, changes in both C and P have been observed, but no violations of CP have been observed, so that T appears to be conserved.

Since the solutions clearly fall into two separate groups, those that are symmetric in space and those that are antisymmetric, it is convenient to deal with each case separately. We now observe that for the equations to be compatible, only certain values of k and κ are possible, and these lead to the **eigenvalues** for this problem. Looking first at the symmetric case ($P = +1$) with $C = D$, we find that we have the relation

$$e^{ika} + e^{-ika} = \frac{k}{i\kappa}(e^{ika} - e^{-ika}), \quad \text{or} \quad \cos ka = \frac{k}{\kappa} \sin ka .$$

In a similar fashion, the antisymmetric case ($P = -1$) with $C = -D$, leads to

$$\sin ka = -\frac{k}{\kappa} \cos ka .$$

These two results may be cast into the form

$$\cot k' = \frac{k'}{\sqrt{b^2 - k'^2}}, \tag{2.20}$$

$$-\tan k' = \frac{k'}{\sqrt{b^2 - k'^2}}, \tag{2.21}$$

where

$$k' \equiv ka = \frac{a}{\hbar}\sqrt{2mE} \quad \text{and} \quad b \equiv \frac{a}{\hbar}\sqrt{2mV} . \tag{2.22}$$

While the solutions that give $k'(b)$ are transcendental, we may approximate for large b that

$$\frac{k'}{\sqrt{b^2 - k'^2}} \simeq \frac{k'}{b}\left(1 + \frac{k'^2}{2b^2} - \cdots\right),$$

and $\cot[(n + \frac{1}{2})\pi - \epsilon] = \tan \epsilon \simeq \epsilon$ and $-\tan(n\pi - \epsilon) = \tan \epsilon \simeq \epsilon$ so that

$$k' \simeq \frac{n\pi}{2(1 + 1/b)}, \tag{2.23}$$

where the two expressions have been combined. This indicates that as $b \to \infty$, we have a well of infinite depth (a well the particle *can't* get out of, or a box), we have equally spaced values of k' every $\pi/2$, and the energy eigenvalues from Equation (2.23) in the limiting case are

$$E_n = \frac{n^2 h^2}{32ma^2}, \qquad n = 1, 2, 3, \ldots \tag{2.24}$$

As an illustration, graphical solutions of Equations (2.20) and (2.21) are shown in Figure 2.4 where the well depth is given by $V = 5h^2/16ma^2$. It is evident that there is one bound state solution in each successive range of $\pi/2$ up to b, so that since in this case $3\pi/2 < b < 2\pi$, there are four states. The graphical method indicates a solution whenever the curves representing either $\cot k'$ or $-\tan k'$ cross the plot of $k'/\sqrt{b^2 - k'^2}$.

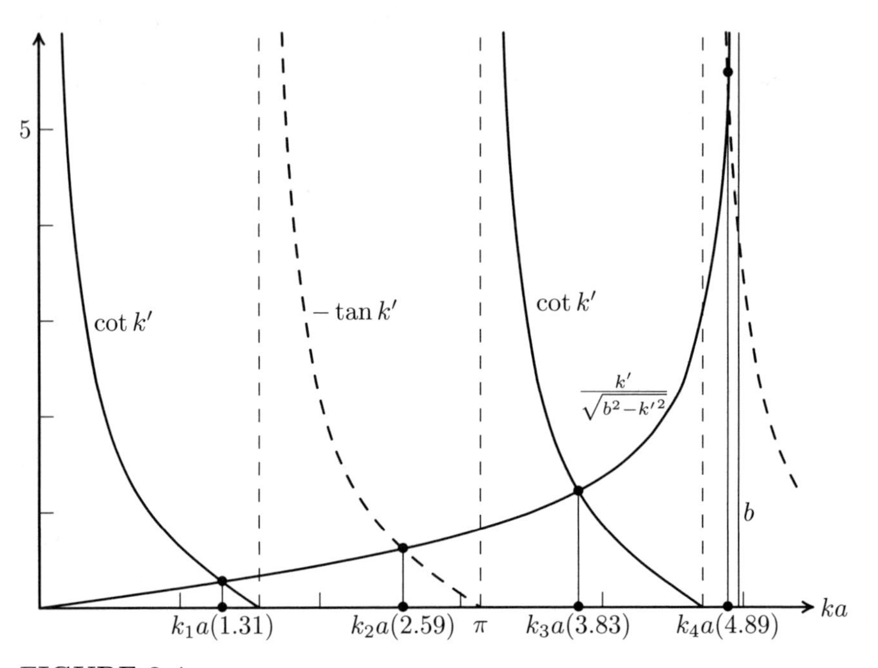

FIGURE 2.4
Illustration of the solutions of $\cot k' = k'/(b^2 - k'^2)^{1/2}$ and $-\tan k' = k'/(b^2 - k'^2)^{1/2}$ for $V = 5h^2/4mc^2$ where $c = 2a$ is the width.

Problem 2.10 Supply the missing steps leading to the result that $C = \pm D$ and $A = \pm G$.

Problem 2.11 Supply the missing steps leading to Equation (2.23).

Problem 2.12 *Bound state energy levels in the finite well.* A finite well as depicted in Figure 2.3 has a depth given by $V = \pi^2\hbar^2/ma^2$. Find the energies of the ground state, the first and second excited states to three significant figures by estimating a first value and then improving it by successive approximation through the use of Equations (2.20) and (2.21).

2.1.6 Transmission through a Barrier

While it is obviously possible in a classical problem for a beam of electrons to pass *over* a barrier where the kinetic energy exceeds the height of a potential barrier and the barrier is only of finite thickness, it is also possible in quantum mechanics for a particle to pass *through* a barrier where it has insufficient energy to pass over it, or $E < V$. This process is called **tunneling**. This configuration is illustrated in Figure 2.5, and again we divide space into three regions. For $E < V$, we have

$$\begin{aligned} \psi_\ell(x) &= A\mathrm{e}^{\mathrm{i}kx} + B\mathrm{e}^{-\mathrm{i}kx}, & x &\leq -a\,, \\ \psi_c(x) &= C\mathrm{e}^{\kappa x} + D\mathrm{e}^{-\kappa x}, & -a \leq\ & x \leq a\,, \\ \psi_r(x) &= F\mathrm{e}^{\mathrm{i}kx}, & x &\geq a\,, \end{aligned} \tag{2.25}$$

with $\kappa = \sqrt{2m(V-E)}/\hbar$, while for $E > V$, we have

$$\begin{aligned} \psi_\ell(x) &= A\mathrm{e}^{\mathrm{i}kx} + B\mathrm{e}^{-\mathrm{i}kx}, & x &\leq -a\,, \\ \psi_c(x) &= C\mathrm{e}^{\mathrm{i}\kappa x} + D\mathrm{e}^{-\mathrm{i}\kappa x}, & -a \leq\ & x \leq a\,, \\ \psi_r(x) &= F\mathrm{e}^{\mathrm{i}kx}, & x &\geq a\,, \end{aligned} \tag{2.26}$$

with $\kappa = \sqrt{2m(E-V)}/\hbar$.

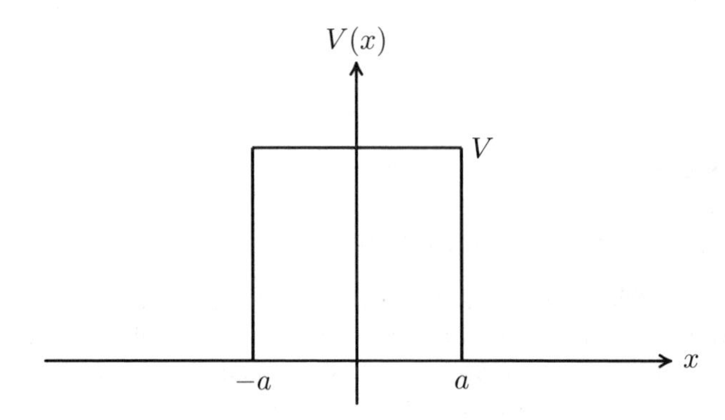

FIGURE 2.5
A rectangular potential barrier of height V and width $2a$.

In this case, the exponentially decaying solution inside the barrier may become small, but is nonzero on the other side. This connects to a transmitted wave and a "reflected" exponential solution back on the incident side that must be used to match with the incident and reflected waves. When $E > V$, there exist special values where the transmission reaches 100%, which is the classical expectation, but the transmission is generally less than 100% except in the limit $E/V \to \infty$.

Problem 2.13 *Transmission and tunneling.*

(a) Solve for F in terms of A and prove that for $E > V$,

$$\left|\frac{F}{A}\right| = \left[1 + \frac{V^2 \sin^2 \sqrt{8m(E-V)}a/\hbar}{4E(E-V)}\right]^{-1/2} . \tag{2.27}$$

(b) Show that for $E < V$, one need simply interchange E and V and let $\sin \to \sinh$.

(c) Find the values of $E - V$ that lead to 100% transmission.

Problem 2.14 *Total transmission.* An electron encounters a barrier of height 10 eV and width 1 Å (=10^{-10} m). What is the lowest energy $E > V$ (answer in electron volts, or eV) that the electron must have to pass the barrier without reflection?

Problem 2.15 *Partial transmission.* Using part (b) from Problem 2.13, find what percentage of a beam of 5 eV electrons incident on a barrier of 25 volts in height and width $a = 5.2 \times 10^{-11}$ m will be transmitted.

2.2 One-Dimensional Harmonic Oscillator

While potential barriers and wells that are rectangular are much easier to solve than more complicated shapes, the real world is rarely well-approximated by these simple shapes. Another potential distribution, that of a parabolic potential well, which is characteristic of the harmonic oscillator problem, is an excellent fit to many real problems, especially for their lowest energy states. Furthermore, for a great many real potentials, the very bottom of the potential well may be approximated by a quadratic, since if we expand a general potential $V(x)$ about the lowest point, x_0, then

$$\begin{aligned} V(x) &= V(x_0) + V'(x_0)(x - x_0) + \tfrac{1}{2}V''(x_0)(x - x_0)^2 + \cdots \\ &\simeq V(x_0) + \tfrac{1}{2}V''(x_0)(x - x_0)^2 , \end{aligned}$$

since at the bottom (by definition), $V'(x_0) = 0$. For the lowest energy states in a broad variety of problems, the harmonic oscillator problem provides a good first approximation. When additional terms in the series are required, perturbation theory, treated in Chapter 5, allows us to extend the accuracy.

2.2.1 The Schrödinger Equation

For a force of the type $F(x) = -kx$, the potential energy is given by $V(x) = -\int^x F(x')\,\mathrm{d}x' = \frac{1}{2}kx^2$ so the Hamiltonian is

$$\hat{\mathcal{H}} = -\frac{\hbar^2}{2m}\frac{\mathrm{d}^2}{\mathrm{d}x^2} + \tfrac{1}{2}kx^2\,. \tag{2.28}$$

For the time-independent Schrödinger equation, then, we may write

$$-\frac{\hbar^2}{2m}\frac{\mathrm{d}^2\psi}{\mathrm{d}x^2} + \tfrac{1}{2}kx^2\psi = E\psi\,. \tag{2.29}$$

In order to solve this equation, we shall use the general procedure:

1. *Express the equation in dimensionless form.* This is for convenience, so that there will only be one constant in the equation.

2. *Determine the asymptotic behavior of $\psi(x)$ as $|x| \to \infty$.* Since the wave function must be normalizable, the solution must vanish as $|x|$ tends toward infinity. This procedure generally allows one to eliminate one possible solution which is unbounded.

3. *Change to a new dependent variable that has better asymptotic behavior by factoring out the asymptotic behavior.* Without this step, a power series solution typically has a three- or four-term recursion formula which makes the analysis of the convergence of the power series almost impossible. Factoring out this asymptotic behavior usually results in a simpler differential equation that can be solved by a power series with a two-term recursion formula, permitting the analysis of the convergence of the series.

4. *Solve the equation for this new dependent variable by the power series method.* This is usually straightforward, leading to a two-term recursion formula if the previous steps are properly carried out.

5. *Determine what values of energy will lead to well-behaved solutions.* Typically, the power series will diverge for arbitrary values of the remaining dimensionless constant. The constant must be chosen so that the overall solution is convergent, typically by truncating the series to a polynomial.

6. *Write the complete solution in terms of the original constants.* Returning to the original constants in the wave function and the energy completes the solution of the differential equation and prepares one for the interpretation of the results.

2.2.2 Changing to Dimensionless Variables

The first step is to change variables to

$$\xi = \alpha x\,,$$

(or $\xi = \alpha(x - x_0)$ if the origin is offset) where α is yet to be determined. Denoting the derivative with respect to the argument (ξ in this case) by ψ', we have

$$-\frac{\hbar^2\alpha^2}{2m}\psi'' + \left(\frac{k\xi^2}{2\alpha^2} - E\right)\psi = 0\,, \tag{2.30}$$

and dividing by the leading constant term, we obtain

$$\psi'' + \left(\frac{2mE}{\hbar^2\alpha^2} - \frac{km}{\hbar^2\alpha^4}\xi^2\right)\psi = 0\,. \tag{2.31}$$

The equation can now be simplified by choosing

$$\alpha^4 \equiv \frac{km}{\hbar^2} \qquad \text{and} \qquad \lambda = \frac{E}{\hbar\sqrt{k/m}}\,,$$

so that the equation in dimensionless form is

$$\psi'' + (2\lambda - \xi^2)\psi = 0\,. \tag{2.32}$$

In terms of λ, we find $E = \lambda\hbar\omega$ since $\omega = \sqrt{k/m}$ for the classical harmonic oscillator.

2.2.3 The Asymptotic Form

As $|\xi| \to \infty$, Equation (2.32) takes the form

$$\psi'' \approx \xi^2\psi\,, \tag{2.33}$$

which if ξ in the coefficient *were constant* would have solutions of the form

$$\psi \to \mathrm{e}^{\pm\xi\cdot\xi} = \mathrm{e}^{\pm\xi^2}\,.$$

Since ξ is not a constant, we try a solution of the form $\psi \approx \mathrm{e}^{\pm c\xi^2}$ where c is a constant to be found. Then from Equation (2.33) we find that

$$4c^2\xi^2\mathrm{e}^{\pm c\xi^2} \pm 2c\,\mathrm{e}^{\pm c\xi^2} - \xi^2\mathrm{e}^{\pm c\xi^2} \approx 0\,,$$

so that we must choose $c = 1/2$ for the *dominant terms* as $|\xi| \to \infty$ to cancel. This leaves us with the asymptotic form

$$\psi \approx \mathrm{e}^{\pm\xi^2/2}\,,$$

but we must reject the positive sign since we require the solution to be bounded.

2.2.4 Factoring out the Asymptotic Behavior

The approximate solution may have the correct asymptotic form, but we require an exact solution that has the same asymptotic form, so we assume a solution of the form

$$\psi(\xi) = N\mathrm{e}^{-\xi^2/2}H(\xi)\,, \tag{2.34}$$

where N is a normalization constant. Substituting this solution into Equation (2.32) and factoring out the common exponential factor, we find that $H(\xi)$ satisfies

$$H'' - 2\xi H' + (2\lambda - 1)H = 0\,. \tag{2.35}$$

Problem 2.16 *Hermite equation.* Show that $H(\xi)$ as defined by Equation (2.34) satisfies Equation (2.35) if $\psi(\xi)$ satisfies Equation (2.32). This equation is known as the Hermite equation.

2.2.5 Finding the Power Series Solution

We assume Equation (2.35) has a power series solution of the form

$$H(\xi) = \sum_{\ell=0}^{\infty} a_\ell \xi^\ell\,. \tag{2.36}$$

Substituting this into Equation (2.35), this leads to

$$\sum_{\ell=0}^{\infty}[\ell(\ell-1)a_\ell\xi^{\ell-2} - 2\ell a_\ell\xi^\ell + (2\lambda-1)a_\ell\xi^\ell] = 0\,. \tag{2.37}$$

Since this must be satisfied for any value of ξ, the coefficient of each power of ξ must vanish separately. (If even one coefficient did not vanish, the sum would not vanish for arbitrary values of ξ.) Allowing no negative powers (to avoid a singularity at the origin), we set $\ell = s+2$ in the first term of Equation (2.37) and $\ell = s$ in the next two terms to obtain

$$\sum_{s=0}^{\infty}[(s+2)(s+1)a_{s+2} - 2sa_s + (2\lambda-1)a_s]\xi^s = 0\,, \tag{2.38}$$

so setting each coefficient to zero gives the recursion relation,

$$\frac{a_{s+2}}{a_s} = \frac{2s - (2\lambda - 1)}{(s+2)(s+1)}\,. \tag{2.39}$$

Since for large s, $a_{s+2}/a_s \to 2/s \to 0$, the series will be convergent, but we need to examine its form. If we examine the series expression for e^{ξ^2}, we find that it also has the ratio for successive terms $a_{s+2}/a_s \to 2/s \to 0$, so it seems that $H(\xi) \to \mathrm{e}^{\xi^2}$ and hence $\psi(\xi) \to \mathrm{e}^{\xi^2/2}$, which is again unbounded as $|\xi| \to \infty$. The only way to prevent this from happening is to truncate the series at some value $s = n$ by choosing λ so that the numerator of Equation (2.39) vanishes when $s = n$. We thus arrive at the necessary relation that

$$\lambda_n = n + \tfrac{1}{2}\,, \tag{2.40}$$

where n is a nonnegative integer.

This procedure has several implications we should note. First, it leads to the result that $H_n(\xi)$ is a polynomial of order n (we introduce the subscript n to denote a polynomial of order n). Second, these polynomials must be either even or odd, since the recursion relation relates terms of every other order, and the condition on λ_n can only truncate either an even series or an odd series, but not both simultaneously.

The first few polynomials may be found by noting that the lowest value of λ_n is $\lambda_0 = \frac{1}{2}$ so the lowest order polynomial is a constant ($a_2 = 0$, and we choose $a_1 = 0$ to eliminate all of the odd terms), and we choose a_0 to be unity. Hence $H_0(\xi) = 1$. Setting $n = 1$ so that $\lambda_1 = \frac{3}{2}$, $a_3 = 0$ and we choose $a_0 = 0$ to eliminate the even terms, so only $a_1 \neq 0$ and we choose to define $a_1 = 2$ so $H_1(\xi) = 2\xi$. We choose the arbitrary first coefficient to make the polynomials agree with the conventional Hermite polynomials [the highest order term of $H_n(\xi)$ is always $(2\xi)^n$]. These polynomials may be conveniently represented by the expression

$$H_n(\xi) = (-1)^n e^{\xi^2} \frac{d^n}{d\xi^n} e^{-\xi^2}\,. \tag{2.41}$$

Problem 2.17 *Hermite polynomials.* Expand Equation (2.41) for $n = 1$ through $n = 5$ and show that the highest order term is $(2\xi)^n$ for each n.

2.2.6 The Energy Eigenvalues

Since only certain values of λ_n are allowed, it follows from the definition of λ_n that there are only certain values for E_n. From

$$\lambda_n = \frac{E_n}{\hbar\omega}\,,$$

it follows that

$$E_n = \lambda_n \hbar\omega = \hbar\omega(n + \tfrac{1}{2})\,. \tag{2.42}$$

Using the classical relation $\omega = \sqrt{k/m}$, it is useful to note that

$$\alpha^2 = \frac{m\omega}{\hbar}\,. \tag{2.43}$$

2.2.7 Normalized Wave Functions

Writing the normalized wave functions as

$$\psi_n(x) = N_n e^{-\alpha^2 x^2/2} H_n(\alpha x)\,, \tag{2.44}$$

the normalization constant is shown in Appendix B.1 to be

$$N_n = \left(\frac{\alpha}{\sqrt{\pi} 2^n n!} \right)^{1/2} . \tag{2.45}$$

The first few normalized wave functions are:

$$\psi_0(x) = \frac{\alpha^{1/2}}{\pi^{1/4}} \mathrm{e}^{-\alpha^2 x^2/2} \tag{2.46}$$

$$\psi_1(x) = \frac{\alpha^{1/2}}{\sqrt{2}\pi^{1/4}} 2\alpha x \mathrm{e}^{-\alpha^2 x^2/2} \tag{2.47}$$

$$\psi_2(x) = \frac{\alpha^{1/2}}{\sqrt{8}\pi^{1/4}} (4\alpha^2 x^2 - 2) \mathrm{e}^{-\alpha^2 x^2/2} \tag{2.48}$$

$$\psi_3(x) = \frac{\alpha^{1/2}}{4\sqrt{3}\pi^{1/4}} (8\alpha^3 x^3 - 12\alpha x) \mathrm{e}^{-\alpha^2 x^2/2} . \tag{2.49}$$

There are several features of the harmonic oscillator wave functions that can most easily be illustrated in a series of graphs. In Figure 2.6, the first six wave functions are illustrated, but each one has the zero reference shifted by E_n so they can be plotted along with the potential function (the argument is $z = \alpha x$). With this vertical shift, one can see that at the value of the argument where $E_n - V(z)$ changes sign, each wave function changes from oscillatory (inside) to exponential (outside). The alternating of even and odd functions with even and odd n is also apparent.

In Figure 2.7, the absolute magnitudes $|\psi(z)|^2$ are shown for several wave functions, this time on the same plot. For the $n = 3$ case, it begins to be apparent that the outermost amplitude peaks are largest, indicating that the particle is most likely to be found there as the classical velocity vanishes at the turning point and moves quickly through the middle of the potential well where the velocity peaks. Figure 2.8 is a plot of the probability density for $n = 15$ and is compared to the classical probability density. Here we begin to see how the quantum mechanical probability density begins to correspond to the classical probability density (shown as a dotted curve), which is proportional to the time a particle spends between x and $x + \mathrm{d}x$, so that it is maximum at the end points.

We may note that the wave functions tabulated in Equations (2.46) through (2.49) are final solutions back in ordinary variables since α is *not* dimensionless but carries a combination of the original constants through Equation (2.43).

Problem 2.18 *Nonclassical behavior.*

(a) Find the classical limit of the x-motion for an oscillator whose energy is $\frac{1}{2}\hbar\omega$.

(b) Find the probability that the particle will be found outside this limit. [*Hint*: Use ψ_0. The integral cannot be evaluated exactly, but leads to the error function, erf(x), which is a tabulated function.]

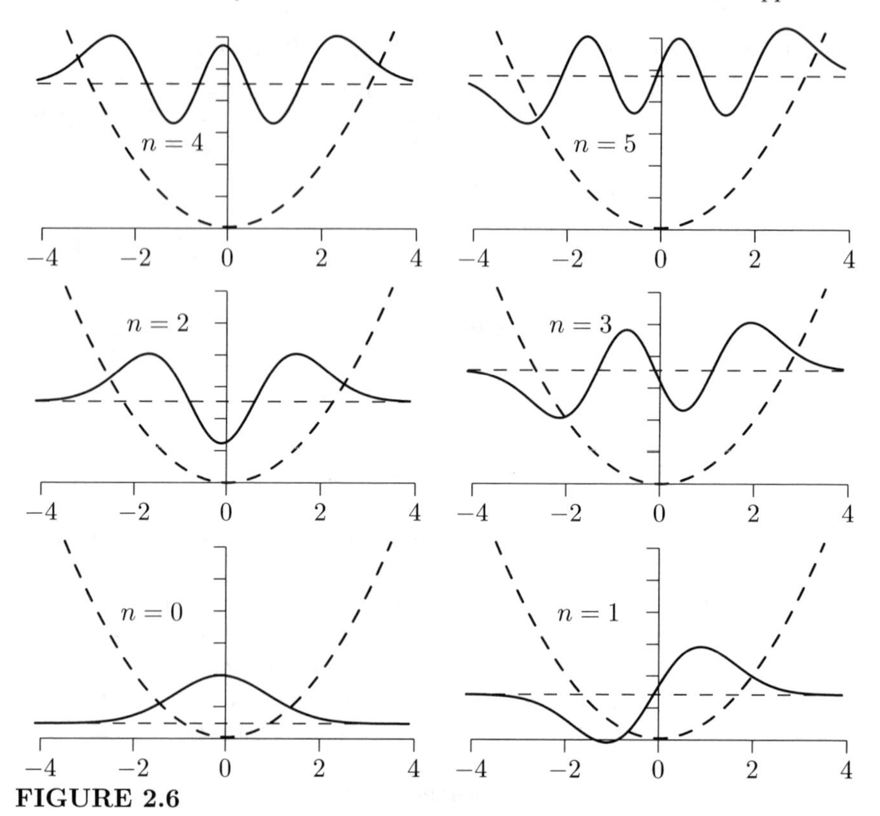

FIGURE 2.6
Wave functions (not normalized) for $n = 0$ through $n = 5$ compared to the classical density.

(c) Find the probability that the particle will be found more than *twice* this far from the origin. (See hint above.)

Problem 2.19 *Higher order wave functions.*

(a) Find $\psi_4(x)$ and $\psi_5(x)$.

(b) Show by direct integration that $\psi_4(x)$ is normalized.

Problem 2.20 *Mean speed.* A harmonic oscillator is in a state $\psi = c_0\psi_0 + c_1\psi_1$ where ψ_0 and ψ_1 are the ground state and first excited state of a harmonic oscillator. Find

$$\langle v \rangle = \frac{\mathrm{d}\langle x \rangle}{\mathrm{d}t} .$$

For what values of c_0 and c_1 will this mean speed be maximized (with $|c_0|$ and $|c_1|$ constant)?

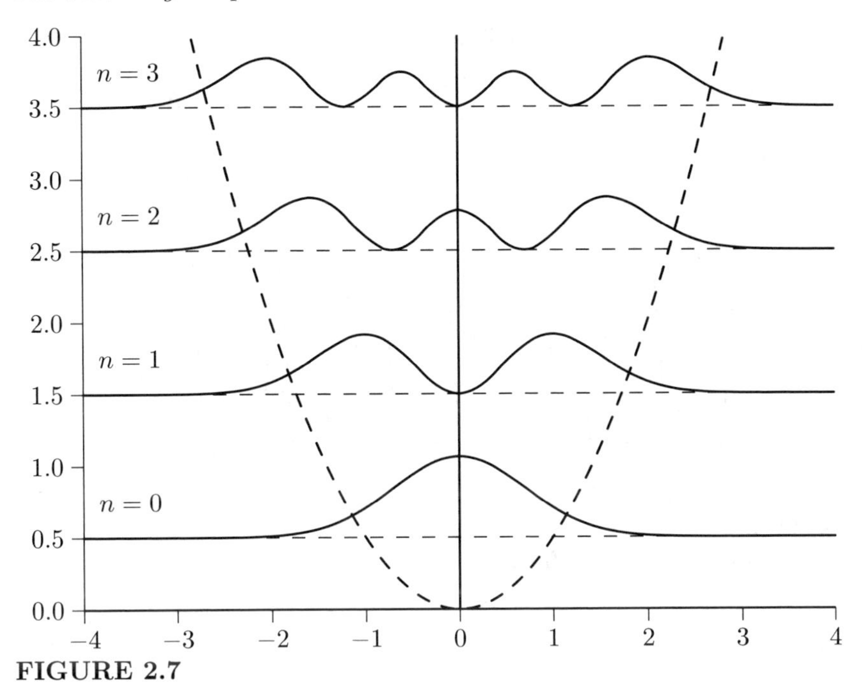

FIGURE 2.7
Probability densities for the first four harmonic oscillator wave functions.

2.3 Time Evolution and Completeness

For general cases, a harmonic oscillator system may not be in only one of the eigenstates at $t = 0$. It is of interest to be able to characterize the solution as a function of time for an arbitrary initial condition. For this task, we will use the **completeness** of the set of eigenfunctions and their **independence**, that derives from their orthogonality.

2.3.1 Completeness

If we consider an arbitrary function of ξ that can be expressed by a power series, so that

$$f(\xi) = \sum_{n=0}^{N} c_n \xi^n ,$$

where N may extend to infinity, then since each Hermite polynomial, $H_n(\xi)$, is of order n, it appears that $f(\xi)$ can be expressed also as a sum of Hermite

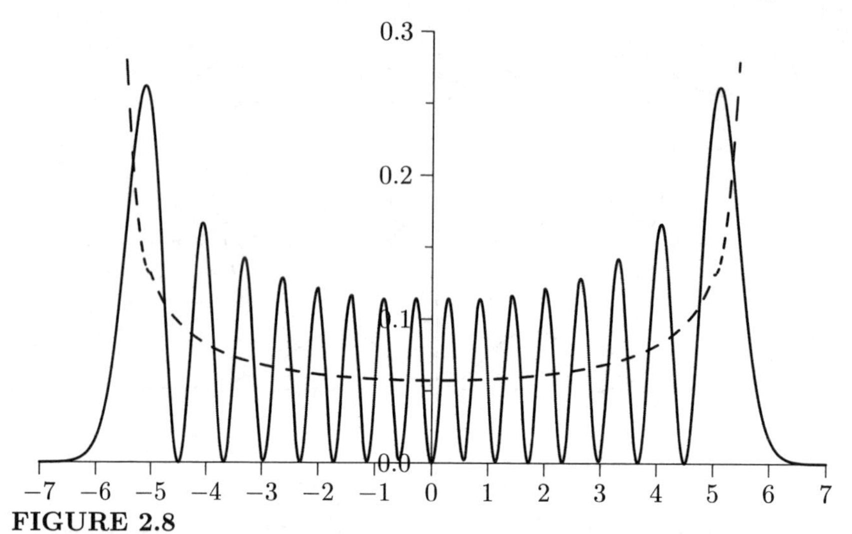

FIGURE 2.8
Probability density for $n = 15$ compared to the classical density.

polynomials, so that

$$f(\xi) = \sum_{n=0}^{N} a_n H_n(\xi)\,. \tag{2.50}$$

The statement that this can be done for *any* function, if $N \to \infty$, is an assertion that the Hermite polynomials form a **complete set**. It is relatively straightforward to find the coefficients a_n, since if we multiply both sides of Equation (2.50) by $H_m(\xi)\mathrm{e}^{-\xi^2}$ and integrate, we have

$$\begin{aligned}\int_{-\infty}^{\infty} \mathrm{e}^{-\xi^2} H_m(\xi) f(\xi)\,\mathrm{d}\xi &= \sum_{n=0}^{\infty} a_n \int_{-\infty}^{\infty} \mathrm{e}^{-\xi^2} H_m(\xi) H_n(\xi)\,\mathrm{d}\xi \\ &= \sum_{n=0}^{\infty} a_n \sqrt{\pi} 2^n n! \delta_{mn} \\ &= a_m \sqrt{\pi} 2^m m!\,.\end{aligned}$$

where we have used Equations (B.14) and (B.15) to obtain the second line so that the coefficients are given by

$$a_n = \frac{1}{\sqrt{\pi} 2^n n!} \int_{-\infty}^{\infty} \mathrm{e}^{-\xi^2} H_n(\xi) f(\xi)\,\mathrm{d}\xi\,. \tag{2.51}$$

2.3.2 Time Evolution

The advantage of using this expansion in eigenfunctions is that the time evolution of each eigenfunction is already known, and that each evolves independently of the others, so that $\Psi_n(x,t) = \psi_n(x)\exp(-\mathrm{i}E_n t/\hbar)$. Then, given

$\Psi(x,0)$, we can find $\Psi(x,t)$ by the expansion

$$\Psi(x,t) = \sum_{n=0}^{\infty} a_n \Psi_n(x,t) . \tag{2.52}$$

Since the a_n are not functions of time, we can evaluate them at $t = 0$ by multiplying Equation (2.52) by Ψ_m^* and integrating to obtain

$$\int_{-\infty}^{\infty} \Psi_m^* \Psi(x,0)\, \mathrm{d}x = \sum_{n=0}^{\infty} a_n \int_{-\infty}^{\infty} \Psi_m^* \Psi_n \, \mathrm{d}x = a_m , \tag{2.53}$$

where we have used the orthogonality of the eigenfunctions for the final result. With the coefficients thus defined, Equation (2.52) then gives the evolution of $\Psi(x,t)$ given $\Psi(x,0)$.

Problem 2.21 *Mixed states.* A harmonic oscillator is initially in the state

$$\Psi(x,0) = A\mathrm{e}^{-\alpha^2 x^2/2} \alpha x (2\alpha x + \mathrm{i}) .$$

(a) Find the wave function as a function of the time.

(b) Find the average energy of the oscillator.

2.4 Operator Method

There is an entirely independent method for solving the harmonic oscillator problem without any apparent solution of a differential equation. Using this method, all of the pertinent physical quantities can be calculated without ever knowing the particular form of the wave function. The technique of using raising and lowering operators will be useful here and again later when we study angular momentum.

2.4.1 Raising and Lowering Operators

We begin by introducing the operator notation with $\xi = q$

$$\hat{q} \equiv \xi = q \tag{2.54}$$

$$\hat{p} \equiv -\mathrm{i}\frac{\mathrm{d}}{\mathrm{d}\xi} = -\mathrm{i}\frac{\mathrm{d}}{\mathrm{d}q} . \tag{2.55}$$

Then $\hat{p}^2 = -\mathrm{d}^2/\mathrm{d}q^2$ and Equation (2.32) may be written

$$(\hat{p}^2 + \hat{q}^2)\psi = 2\lambda\psi . \tag{2.56}$$

Now $\hat{p}$ and $\hat{q}$ do not commute, since

$$[\hat{p}, \hat{q}] = -\mathrm{i}\frac{\mathrm{d}}{\mathrm{d}q}q + \mathrm{i}q\frac{\mathrm{d}}{\mathrm{d}q} = -\mathrm{i}\,, \tag{2.57}$$

so that the usual factors of the operator $\hat{p}^2 + \hat{q}^2$ become

$$(\hat{q} + \mathrm{i}\hat{p})(\hat{q} - \mathrm{i}\hat{p}) = \hat{q}^2 + \hat{p}^2 + 1 \tag{2.58}$$

$$(\hat{q} - \mathrm{i}\hat{p})(\hat{q} + \mathrm{i}\hat{p}) = \hat{q}^2 + \hat{p}^2 - 1\,. \tag{2.59}$$

If these two expressions are added, then we find

$$\hat{p}^2 + \hat{q}^2 = \tfrac{1}{2}[(\hat{q} + \mathrm{i}\hat{p})(\hat{q} - \mathrm{i}\hat{p}) + (\hat{q} - \mathrm{i}\hat{p})(\hat{q} + \mathrm{i}\hat{p})]\,.$$

If we now define two new operators based on the two factors as

$$\hat{a} \equiv \frac{1}{\sqrt{2}}(\hat{q} + \mathrm{i}\hat{p}) = \frac{1}{\sqrt{2}}\left(q + \frac{\mathrm{d}}{\mathrm{d}q}\right) \tag{2.60}$$

$$\hat{a}^\dagger \equiv \frac{1}{\sqrt{2}}(\hat{q} - \mathrm{i}\hat{p}) = \frac{1}{\sqrt{2}}\left(q - \frac{\mathrm{d}}{\mathrm{d}q}\right)\,, \tag{2.61}$$

then

$$\hat{p}^2 + \hat{q}^2 = \hat{a}\hat{a}^\dagger + \hat{a}^\dagger\hat{a}\,,$$

and Equation (2.56) may be written

$$(\hat{a}\hat{a}^\dagger + \hat{a}^\dagger\hat{a})\psi = 2\lambda\psi\,. \tag{2.62}$$

The operators satisfy the commutation relation

$$[\hat{a}, \hat{a}^\dagger] = \hat{a}\hat{a}^\dagger - \hat{a}^\dagger\hat{a} = 1\,. \tag{2.63}$$

Using Equation (2.63), we may write the Schrödinger equation, Equation (2.62), in either of the two forms:

$$\hat{a}\hat{a}^\dagger\psi = \left(\lambda + \tfrac{1}{2}\right)\psi \tag{2.64}$$

$$\hat{a}^\dagger\hat{a}\psi = \left(\lambda - \tfrac{1}{2}\right)\psi\,. \tag{2.65}$$

The task now is to construct an infinite set of eigenfunctions from any solution of Equation (2.62), Equation (2.64), or Equation (2.65). To do this, we first operate on Equation (2.64) with $\hat{a}^\dagger$ on the left so that

$$\hat{a}^\dagger\hat{a}\hat{a}^\dagger\psi = \left(\lambda + \tfrac{1}{2}\right)\hat{a}^\dagger\psi\,. \tag{2.66}$$

Then using the commutation relation $\hat{a}^\dagger\hat{a} = \hat{a}\hat{a}^\dagger - 1$, we have

$$(\hat{a}\hat{a}^\dagger - 1)\hat{a}^\dagger\psi = \left(\lambda + \tfrac{1}{2}\right)\hat{a}^\dagger\psi$$

and moving the $\hat{a}^\dagger\psi$ term on the left over to the right, we obtain

$$\hat{a}\hat{a}^\dagger(\hat{a}^\dagger\psi) = \left(\lambda + \tfrac{3}{2}\right)\hat{a}^\dagger\psi\,. \tag{2.67}$$

If we now add Equation (2.66) and Equation (2.67), we get the result

$$(\hat{a}\hat{a}^\dagger + \hat{a}^\dagger\hat{a})(\hat{a}^\dagger\psi) = 2(\lambda+1)(\hat{a}^\dagger\psi)\,. \tag{2.68}$$

We thus have the surprising result that if ψ is an eigenfunction of Equation (2.62) with eigenvalue λ, then $\hat{a}^\dagger\psi$ is also an eigenfunction but with eigenvalue $\lambda+1$.

We may do this again, operating on Equation (2.67), to obtain

$$\hat{a}^\dagger(\hat{a}\hat{a}^\dagger)(\hat{a}^\dagger\psi) = (\hat{a}^\dagger\hat{a})(\hat{a}^\dagger)^2\psi = (\hat{a}\hat{a}^\dagger - 1)(\hat{a}^\dagger)^2\psi = \left(\lambda + \tfrac{3}{2}\right)(\hat{a}^\dagger)^2\psi\,. \tag{2.69}$$

From this it follows that

$$\hat{a}\hat{a}^\dagger(\hat{a}^\dagger)^2\psi = \left(\lambda + \tfrac{5}{2}\right)(\hat{a}^\dagger)^2\psi\,, \tag{2.70}$$

and adding these two equations, the result is

$$(\hat{a}\hat{a}^\dagger + \hat{a}^\dagger\hat{a})(\hat{a}^\dagger)^2\psi = 2(\lambda+2)(\hat{a}^\dagger)^2\psi\,. \tag{2.71}$$

We may continue this process indefinitely, so the operator $\hat{a}^\dagger$ has the effect of raising any eigenfunction ψ_n with eigenvalue λ_n to the next higher eigenfunction ψ_{n+1} with eigenvalue $\lambda_{n+1} = \lambda_n + 1$.

In similar fashion, the operator $\hat{a}$ is a lowering operator. We see this by operating on the left by $\hat{a}$ on Equation (2.65), obtaining (with the use of the commutator again)

$$\hat{a}\hat{a}^\dagger\hat{a}\psi = (1 + \hat{a}^\dagger\hat{a})\hat{a}\psi = \left(\lambda - \tfrac{1}{2}\right)\hat{a}\psi\,, \tag{2.72}$$

or

$$\hat{a}^\dagger\hat{a}(\hat{a}\psi) = \left(\lambda - \tfrac{3}{2}\right)(\hat{a}\psi)\,.$$

Adding, this leads to

$$(\hat{a}\hat{a}^\dagger + \hat{a}^\dagger\hat{a})(\hat{a}\psi) = 2(\lambda-1)(\hat{a}\psi)\,. \tag{2.73}$$

Hence, $\hat{a}$ is a lowering operator just as $\hat{a}^\dagger$ is a raising operator.

2.4.2 Eigenfunctions and Eigenvalues

For the lowering operator, there is a limit, since we may not lower the eigenvalue forever or it will go negative. To demonstrate this, we calculate the expectation value of the energy (or of λ), such that

$$\begin{aligned}\langle\lambda\rangle &= \int_{-\infty}^{\infty}\psi^*\lambda\psi\,\mathrm{d}q = \tfrac{1}{2}\int_{-\infty}^{\infty}\psi^*(\hat{p}^2 + \hat{q}^2)\psi\,\mathrm{d}q \\ &= -\tfrac{1}{2}\int_{-\infty}^{\infty}\psi^*\frac{\mathrm{d}^2\psi}{\mathrm{d}q^2}\,\mathrm{d}q + \tfrac{1}{2}\int_{-\infty}^{\infty}\psi^* q^2\psi\,\mathrm{d}q\,.\end{aligned}$$

Integrating the first term on the right by parts,

$$\langle\lambda\rangle = -\tfrac{1}{2}\left[\psi^*\frac{\mathrm{d}\psi}{\mathrm{d}q}\right]_{-\infty}^{\infty} + \tfrac{1}{2}\int_{-\infty}^{\infty}\frac{\mathrm{d}\psi^*}{\mathrm{d}q}\frac{\mathrm{d}\psi}{\mathrm{d}q}\,\mathrm{d}q + \tfrac{1}{2}\int_{-\infty}^{\infty}|\psi|^2 q^2\,\mathrm{d}q\,.$$

The first term vanishes since ψ must vanish as $|q| \to \infty$, and the remaining terms give

$$\langle\lambda\rangle = \tfrac{1}{2}\int_{-\infty}^{\infty}\left(\left|\frac{\mathrm{d}\psi}{\mathrm{d}q}\right|^2 + |\psi|^2 q^2\right)\mathrm{d}q \geq 0\,,$$

since the integrand is positive definite. There exists then a lowest possible value of $\lambda_n = \lambda_0$ and a corresponding wave function such that

$$\hat{a}\psi_0 = 0\,, \tag{2.74}$$

since any further lowering would lead to $\lambda < 0$. We can establish this lowest eigenvalue from Equation (2.65), where

$$\hat{a}^\dagger\hat{a}\psi_0 = 0 = \left(\lambda_0 - \tfrac{1}{2}\right)\psi_0\,,$$

so that

$$\lambda_0 = \tfrac{1}{2} = \frac{E_0}{\hbar\omega}\,,$$

or

$$E_0 = \tfrac{1}{2}\hbar\omega\,. \tag{2.75}$$

Since all of the other eigenvalues increase by 1 each time we use the raising operator, the general expression for the eigenvalues is

$$\lambda_n = n + \tfrac{1}{2}\,, \qquad n = 0, 1, 2, \ldots, \tag{2.76}$$

which is equivalent to

$$E_n = \hbar\omega(n + \tfrac{1}{2})\,, \qquad n = 0, 1, 2, \ldots. \tag{2.77}$$

The lowest energy state of a system with discrete energy states is called a **ground state** and we refer to both the ground state eigenvalue and the ground state eigenfunction. The ground state eigenfunction for the harmonic oscillator may be obtained from the form of the lowering operator where

$$\hat{a}\psi_0 = \frac{1}{\sqrt{2}}\left(q + \frac{\mathrm{d}}{\mathrm{d}q}\right)\psi_0 = 0\,,$$

or

$$\frac{\mathrm{d}\psi_0}{\mathrm{d}q} + q\psi_0 = 0\,. \tag{2.78}$$

This is a first order differential equation whose solution is

$$\psi_0 = C_0\mathrm{e}^{-q^2/2}\,, \tag{2.79}$$

where the normalization constant is given by

$$\langle \psi_0 | \psi_0 \rangle = 1 = |C_0|^2 \int_{-\infty}^{\infty} \mathrm{e}^{-q^2}\, dq = |C_0|^2 \sqrt{\pi}\,,$$

so $C_0 = \pi^{-1/4}$.

While the successive application of the raising operator will produce all of the higher-order wave functions, they are not necessarily normalized. For the normalization constants, we define the normalization constants in terms of the raising operator such that

$$\psi_{n+1} = \frac{C_{n+1}}{C_n} \hat{a}^\dagger \psi_n\,. \tag{2.80}$$

Starting at the bottom,

$$\begin{aligned}
\psi_1 &= \frac{C_1}{C_0} \hat{a}^\dagger \psi_0 \\
\psi_2 &= \frac{C_2}{C_1} \hat{a}^\dagger \psi_1 = \frac{C_2}{C_0} (\hat{a}^\dagger)^2 \psi_0 \\
&\vdots \\
\psi_n &= \frac{C_n}{C_0} (\hat{a}^\dagger)^n \psi_0\,.
\end{aligned}$$

The steps then follow as

$$\langle \psi_n | \psi_n \rangle = \left| \frac{C_n}{C_{n-1}} \right|^2 \langle \hat{a}^\dagger \psi_{n-1} | \hat{a}^\dagger \psi_{n-1} \rangle \tag{2.81}$$

$$= \left| \frac{C_n}{C_{n-1}} \right|^2 \langle \hat{a} \hat{a}^\dagger \psi_{n-1} | \psi_{n-1} \rangle \tag{2.82}$$

$$= \left| \frac{C_n}{C_{n-1}} \right|^2 \left(\lambda_{n-1} + \tfrac{1}{2} \right) \langle \psi_{n-1} | \psi_{n-1} \rangle \tag{2.83}$$

$$= \left| \frac{C_n}{C_{n-1}} \right|^2 n \langle \psi_{n-1} | \psi_{n-1} \rangle \tag{2.84}$$

$$= \left| \frac{C_n}{C_{n-1}} \right|^2 n \left| \frac{C_{n-1}}{C_{n-2}} \right|^2 (n-1) \cdots \left| \frac{C_1}{C_0} \right|^2 \langle \psi_0 | \psi_0 \rangle \tag{2.85}$$

$$= \left| \frac{C_n}{C_0} \right|^2 n! |C_0|^2 \sqrt{\pi} = 1\,. \tag{2.86}$$

In going from Equation (2.81) to Equation (2.82), we used the fact that the adjoint of $\hat{a}^\dagger$ is $\hat{a}$, which may be verified by integrating by parts. In going from Equation (2.82) to Equation (2.83), we used Equation (2.64). In going from Equation (2.83) to Equation (2.84), we used $\lambda_n = 2n+1$. The result is that

$$C_n = \left(\frac{1}{n!\sqrt{\pi}}\right)^{1/2} . \tag{2.87}$$

The relationship to the Hermite polynomials is illustrated by the relations

$$\psi_n = N_n \mathrm{e}^{-q^2/2} H_n(q) = C_n (\hat{a}^\dagger)^n \mathrm{e}^{-q^2/2} = \frac{C_n}{\sqrt{2^n}} \left(q - \frac{\mathrm{d}}{\mathrm{d}q}\right)^n \mathrm{e}^{-q^2/2} , \tag{2.88}$$

where the Hermite polynomials are given by

$$H_n(q) = \mathrm{e}^{q^2/2} \left(q - \frac{\mathrm{d}}{\mathrm{d}q}\right)^n \mathrm{e}^{-q^2/2} . \tag{2.89}$$

By comparing Equation (2.87) and Equation (2.88), we find

$$N_n = \frac{C_n}{\sqrt{2^n}} = \left(\frac{1}{2^n n!\sqrt{\pi}}\right)^{1/2} . \tag{2.90}$$

2.4.3 Expectation Values

Returning to Equation (2.80), we may now write that equation as

$$\hat{a}^\dagger \psi_n = \frac{C_n}{C_{n+1}} \psi_{n+1} = \sqrt{n+1}\,\psi_{n+1} . \tag{2.91}$$

Lowering the index by one and then operating by $\hat{a}$ on the left leads to

$$\hat{a}\hat{a}^\dagger \psi_{n-1} = \sqrt{n}\,\hat{a}\psi_n = n\psi_{n-1} ,$$

where the latter equality is from Equation (2.64). This may be written as

$$\hat{a}\psi_n = \sqrt{n}\,\psi_{n-1} . \tag{2.92}$$

With these expressions, it is convenient to write expressions for $\hat{p}$ and $\hat{q}$ in terms of the raising and lowering operators, such that

$$\hat{q} = \frac{1}{\sqrt{2}}(\hat{a} + \hat{a}^\dagger) \tag{2.93}$$

$$\hat{p} = \frac{\mathrm{i}}{\sqrt{2}}(\hat{a}^\dagger - \hat{a}) . \tag{2.94}$$

We also note that using Equations (2.91) and (2.92) leads to

$$\begin{aligned}
\langle a \rangle &= \langle \psi_n | \hat{a}\psi_n \rangle = \sqrt{n}\langle \psi_n | \psi_{n-1} \rangle = 0 \\
\langle \hat{a}^\dagger \rangle &= \langle \psi_n | \hat{a}^\dagger \psi_n \rangle = \sqrt{n+1}\langle \psi_n | \psi_{n+1} \rangle = 0 \\
\langle \hat{a}\hat{a}^\dagger \rangle &= \langle \psi_n | \hat{a}\hat{a}^\dagger \psi_n \rangle \\
&= \sqrt{n+1}\langle \psi_n | \hat{a}\psi_{n+1} \rangle = (n+1)\langle \psi_n | \psi_n \rangle = n+1 \\
\langle \hat{a}^\dagger \hat{a} \rangle &= \langle \psi_n | \hat{a}^\dagger \hat{a}\psi_n \rangle \\
&= \sqrt{n}\langle \psi_n | \hat{a}^\dagger \psi_{n-1} \rangle = n\langle \psi_n | \psi_n \rangle = n .
\end{aligned}$$

From these we may readily establish that:

$$\begin{aligned}
\langle q\rangle &= \frac{1}{\sqrt{2}}(\langle\hat{a}\rangle + \langle\hat{a}^\dagger\rangle) = 0\\
\langle p\rangle &= \frac{\mathrm{i}}{\sqrt{2}}(\langle\hat{a}^\dagger\rangle - \langle\hat{a}\rangle) = 0\\
\langle q^2\rangle &= \tfrac{1}{2}(\langle\hat{a}\hat{a}\rangle + \langle\hat{a}\hat{a}^\dagger\rangle + \langle\hat{a}^\dagger\hat{a}\rangle + \langle\hat{a}^\dagger\hat{a}^\dagger\rangle)\\
&= \tfrac{1}{2}(2n+1) = n + \tfrac{1}{2}\\
\langle p^2\rangle &= -\tfrac{1}{2}(\langle\hat{a}^\dagger\hat{a}^\dagger\rangle - \langle\hat{a}\hat{a}^\dagger\rangle - \langle\hat{a}^\dagger\hat{a}\rangle + \langle\hat{a}\hat{a}\rangle)\\
&= n + \tfrac{1}{2}\,.
\end{aligned}$$

Therefore,

$$\begin{aligned}
\langle x\rangle &= \frac{\langle q\rangle}{\alpha} = 0\\
\langle p_x\rangle &= \hbar\alpha\langle p\rangle = 0\\
\langle x^2\rangle &= \frac{\langle q^2\rangle}{\alpha^2} = \frac{1}{\alpha^2}(n+\tfrac{1}{2})\\
\langle p_x^2\rangle &= \hbar^2\alpha^2\langle p^2\rangle = \hbar^2\alpha^2(n+\tfrac{1}{2})\\
\langle T\rangle &= \frac{\langle p_x^2\rangle}{2m} = \tfrac{1}{2}\hbar\omega(n+\tfrac{1}{2})\\
\langle V\rangle &= \frac{k\langle x^2\rangle}{2} = \tfrac{1}{2}\hbar\omega(n+\tfrac{1}{2})\\
\langle E\rangle &= \hbar\omega\langle\lambda\rangle = \tfrac{1}{2}\hbar\omega(\langle p^2\rangle + \langle q^2\rangle) = \hbar\omega(n+\tfrac{1}{2})\,.
\end{aligned}$$

Problem 2.22 *Expectation values from the raising and lowering operators.*

(a) Use the raising and lowering operators to evaluate $\langle\psi_n|x|\psi_m\rangle$. (For a given n, find for which values of m there is a nonzero result, and find the result for those special cases.)

(b) Use the raising and lowering operators to evaluate $\langle\psi_n|x^2|\psi_m\rangle$.

(c) Use the raising and lowering operators to prove

$$\begin{aligned}
\langle\psi_n|x^3|\psi_m\rangle = \frac{1}{2\sqrt{2}\alpha^3}&[\sqrt{(n+3)(n+2)(n+1)}\delta_{n,m-3}\\
&+3(n+1)^{3/2}\delta_{n,m-1} + 3n^{3/2}\delta_{n,m+1}\\
&+\sqrt{n(n-1)(n-2)}\delta_{n,m+3}].
\end{aligned} \tag{2.95}$$

(d) Use the raising and lowering operators to evaluate $\langle\psi_n|x^4|\psi_m\rangle$. [The result should be similar in form to Equation (2.95), but with more terms.]

Problem 2.23 *Uncertainty product.* Find the minimum uncertainty product $(\Delta x)(\Delta p_x)$ for the nth state of the harmonic oscillator.

Problem 2.24 *Adjoint relationships.* Using the definition of the adjoint of an operator, and integrating by parts,

(a) Prove that $\hat{a}^\dagger$ is the adjoint of $\hat{a}$.

(b) Prove that $\hat{a}$ is the adjoint of $\hat{a}^\dagger$.

2.4.3.1 Concluding Remarks

Since we have seen that the ground state of the harmonic oscillator has a gaussian wave function, we can now develop a physical interpretation for the wave packet in Section 2.1.2 and Figure 2.1. If we imagine that we are moving at velocity $v_0 = p_0/m$ past a stationary harmonic oscillator system (or that the system were moving by *us*, it's all just relative motion) and that suddenly at time $t = 0$, the spring breaks so that the particle is suddenly a free particle, the wave function will evolve in time as in the figure.

If instead, we were standing still with respect to the system when the spring broke, the packet would simply spread in time ($p_0 = 0$), retaining its gaussian envelope that gets broader and lower over time so that the area is conserved. In such a case, there would neither be a phase velocity nor a group velocity.

3

The Schrödinger Equation in Three Dimensions

In three dimensions (3-D), the ordinary differential equations of the previous chapter become partial differential equations. There are two general methods for solving partial differential equations: (1) use the separation of variables, or (2) use numerical methods. Although many realistic problems require numerical techniques, we will restrict our attention to those simpler problems that allow the separation of variables so that we obtain a set of three ordinary differential equations. The complications of the inseparable problems may often be managed through perturbation methods if the problems are *nearly* separable. Our first example demonstrates the technique for the separation of variables and leads us to an already solved one-dimension problem.

3.1 The Free Particle in Three Dimensions

It is not difficult to extend the analysis of the free particle in one dimension to three dimensions. We begin with the three-dimensional time-independent Schrödinger equation,

$$-\frac{\hbar^2}{2m}\nabla^2\psi - E\psi = 0\,,$$

where the potential energy is taken to be zero and E is constant. For this case, we *assume* that we can separate variables, such that

$$\psi(x,y,z) = X(x)Y(y)Z(z)\,.$$

Further assuming that the wave function does not vanish, we may divide by ψ and obtain the equation

$$\frac{1}{X}\frac{\mathrm{d}^2X}{\mathrm{d}x^2} + \frac{1}{Y}\frac{\mathrm{d}^2Y}{\mathrm{d}y^2} + \frac{1}{Z}\frac{\mathrm{d}^2Z}{\mathrm{d}z^2} + k^2 = 0 \tag{3.1}$$

with $k^2 = 2mE/\hbar^2$. Since each of the first three terms in Equation (3.1) depends only on a single variable, it follows that each of these three terms must remain constant as we change the variables, since all of the remaining terms remain constant as we vary any one variable. We write this result as

$$\frac{1}{X}\frac{\mathrm{d}^2X}{\mathrm{d}x^2} = -k_x^2\,, \qquad \frac{1}{Y}\frac{\mathrm{d}^2Y}{\mathrm{d}y^2} = -k_y^2\,, \qquad \frac{1}{Z}\frac{\mathrm{d}^2Z}{\mathrm{d}z^2} = -k_z^2\,,$$

where

$$k_x^2 + k_y^2 + k_z^2 = k^2\,. \tag{3.2}$$

Each individual solution may be written in complex exponential form as

$$\begin{aligned} X(x) &= A\mathrm{e}^{\mathrm{i}k_x x} + A'\mathrm{e}^{-\mathrm{i}k_x x} \\ Y(y) &= B\mathrm{e}^{\mathrm{i}k_y y} + B'\mathrm{e}^{-\mathrm{i}k_y y} \\ Z(z) &= C\mathrm{e}^{\mathrm{i}k_z z} + C'\mathrm{e}^{-\mathrm{i}k_z z} \end{aligned}$$

or in the form

$$\psi(x, y, z) = A\exp\left(\mathrm{i}\boldsymbol{k}\cdot\boldsymbol{r}\right), \tag{3.3}$$

where $\boldsymbol{k} = k_x\boldsymbol{e}_x + k_y\boldsymbol{e}_y + k_z\boldsymbol{e}_z$. This form shows that the solutions are in the form of **plane waves** that may travel in arbitrary directions, and $\boldsymbol{k}$ is called the **propagation vector**. It is possible to construct three-dimensional wave packets just as we did for one dimension.

Problem 3.1 Show that the solution in Equation (3.3) is an eigenfunction of $\hat{p}_x$, $\hat{p}_y$, $\hat{p}_z$ and hence of $\hat{\boldsymbol{p}}$. Find the momentum $\boldsymbol{p}$ in terms of the propagation vector.

3.2 Particle in a Three-Dimensional Box

Just as we considered a one-dimensional free particle and then found the solution for a particle in a one-dimensional box, we may extend the three-dimensional solution to consider a particle in a three-dimensional box by letting the potential at the edges of the box approach infinity. By considering a potential in the Schrödinger equation and then letting it approach infinity, it is clear that the wave function must approach zero as a boundary condition at each wall. If we take the box to have sides extending from $x = 0$ to $x = a$, $y = 0$ to $y = b$, and $z = 0$ to $z = c$, the solution must be a standing wave form of Equation (3.3), or

$$\psi(\boldsymbol{r}) = A\sin\frac{\ell\pi x}{a}\sin\frac{m\pi y}{b}\sin\frac{n\pi z}{c}\,, \tag{3.4}$$

where ℓ, m, and n are integers 1, 2, 3, ... (solutions with negative integers are not independent, since they simply change the sign of the wave function).

Another kind of related problem can also be treated by solutions of this form, where, for instance, the box is so large that the walls cannot materially affect the problem. For this case, we simply let the solution be periodic over a distance L_x where $L_x = N_x a$ with N_x an integer, and $N_x \gg 1$. This means that the solution at $x = L_x$ is the same as the solution at $x = 0$, with similar adjustments for the other two dimensions. In this case, the solution of Equation (3.4) becomes

$$\psi(\boldsymbol{r}) = A \exp\left[2\pi\mathrm{i}\left(\frac{\ell x}{a} + \frac{my}{b} + \frac{nz}{c}\right)\right] = A \exp\left(\mathrm{i}\boldsymbol{K}\cdot\boldsymbol{r}\right), \qquad (3.5)$$

where for this case $\boldsymbol{K} = 2\pi[(\ell/a)\boldsymbol{e}_x + (m/b)\boldsymbol{e}_y + (n/c)\boldsymbol{e}_z]$ and ℓ, m, and n are positive *or negative* integers.

Problem 3.2 Find the energy eigenvalues corresponding to the eigenfunction of Equation (3.4). *Ans.*

$$E_{\ell mn} = \frac{h^2}{8m}\left(\frac{\ell^2}{a^2} + \frac{m^2}{b^2} + \frac{n^2}{c^2}\right). \qquad (3.6)$$

Problem 3.3 Find the absolute value of the normalization constant for a particle in a three-dimensional box.

Problem 3.4 The energy for a particle in a 3-D cube of dimension a is found to be $E = 14\pi^2\hbar^2/2ma^2$. How many states have this energy and which states are they (give ℓ, m, and n for each state)?

3.3 The One-Electron Atom

The Schrödinger equation for the one-electron atom in three dimensions can be broken into two parts, the motion of the center of mass and the motion about the center of mass. The motion of the center of mass is essentially that of a free particle of mass $m_e + M_i$, and will not be considered further. The motion about the center of mass is similar to the Bohr atom problem in that the equation of motion is the same as if the rotation were about a stationary nucleus except that the effective mass of the electron is the reduced mass, $\mu_e = m_e/(1 + m_e/M_i)$. The Schrödinger equation (time independent) then becomes

$$-\frac{\hbar^2}{2\mu_e}\nabla^2\psi - \frac{Ze^2}{4\pi\epsilon_0 r}\psi = E\psi\,, \qquad (3.7)$$

where we have assumed the charge on the nucleus is $+Ze$ with Z a positive integer.

3.3.1 Separating Variables

In spherical coordinates, the Laplacian operator is

$$\nabla^2 = \frac{1}{r^2}\frac{\partial}{\partial r}\left(r^2\frac{\partial}{\partial r}\right) + \frac{1}{r^2\sin\theta}\frac{\partial}{\partial\theta}\left(\sin\theta\frac{\partial}{\partial\theta}\right) + \frac{1}{r^2\sin^2\theta}\frac{\partial^2}{\partial\phi^2}\,. \tag{3.8}$$

The first step in separating variables is to divide Equation (3.7) by ψ, assuming a separable solution of the form

$$\psi(r,\theta,\phi) = R(r)\Theta(\theta)\Phi(\phi)\,. \tag{3.9}$$

The result is, after further multiplying by $-(2\mu_e/\hbar^2)r^2\sin^2\theta$,

$$\frac{\sin^2\theta}{R}\frac{\partial}{\partial r}\left(r^2\frac{\partial R}{\partial r}\right) + \frac{\sin\theta}{\Theta}\frac{\partial}{\partial\theta}\left(\sin\theta\frac{\partial\Theta}{\partial\theta}\right) + \frac{1}{\Phi}\frac{\partial^2\Phi}{\partial\phi^2} + r^2\sin^2\theta\left(\frac{2\mu_e E}{\hbar^2} + \frac{2\mu_e Ze^2}{4\pi\epsilon_0\hbar^2 r}\right) = 0\,. \tag{3.10}$$

It is apparent from this that the third term depends only on the variable ϕ, so is independent of r and θ. The rest are independent of ϕ, so the third term must be a constant that we will define by

$$\frac{1}{\Phi}\frac{\mathrm{d}^2\Phi}{\mathrm{d}\phi^2} \equiv -m^2\,. \tag{3.11}$$

This equation is an ordinary differential equation now since Φ depends only on ϕ. Using this value in Equation (3.10) and then dividing by $\sin^2\theta$, the resulting two-dimensional equation is

$$\frac{1}{R}\frac{\partial}{\partial r}\left(r^2\frac{\partial R}{\partial r}\right) + \frac{1}{\Theta\sin\theta}\frac{\partial}{\partial\theta}\left(\sin\theta\frac{\partial\Theta}{\partial\theta}\right) - \frac{m^2}{\sin^2\theta} + r^2\left(\frac{2\mu_e E}{\hbar^2} + \frac{2\mu_e Ze^2}{4\pi\epsilon_0\hbar^2 r}\right) = 0\,. \tag{3.12}$$

In this equation, the second and third terms depend only on θ and the remaining terms depend only on r, so each sum must be constant, which we define by

$$\frac{1}{\Theta\sin\theta}\frac{\mathrm{d}}{\mathrm{d}\theta}\left(\sin\theta\frac{\mathrm{d}\Theta}{\mathrm{d}\theta}\right) - \frac{m^2}{\sin^2\theta} \equiv -\Lambda\,, \tag{3.13}$$

which is again an ordinary differential equation. This leaves the radial equation in the form

$$\frac{1}{R}\frac{\mathrm{d}}{\mathrm{d}r}\left(r^2\frac{\mathrm{d}R}{\mathrm{d}r}\right) - \Lambda + r^2\left(\frac{2\mu_e E}{\hbar^2} + \frac{2\mu_e Ze^2}{4\pi\epsilon_0\hbar^2 r}\right) = 0\,. \tag{3.14}$$

The separation of variables is therefore complete, since we now have three ordinary differential equations instead of one partial differential equation.

3.3.2 Solution of the $\Phi(\phi)$ Equation

Writing Equation (3.11) as

$$\frac{\mathrm{d}^2\Phi}{\mathrm{d}\phi^2} = -m^2\Phi\,, \tag{3.15}$$

the solution is simply

$$\Phi(\phi) = A_{\pm}\mathrm{e}^{\pm im\phi}\,, \tag{3.16}$$

where the general solution is the sum of the two solutions with $\pm m$. If this solution is to be single-valued, we must demand that after going around the polar axis and returning to the same angle, the wave function must be unchanged. This requires

$$\Phi(\phi + 2\pi) = \Phi(\phi)\,,$$

or that

$$\mathrm{e}^{im(\phi+2\pi)} = \mathrm{e}^{im\phi}\,,$$

which in turn requires

$$\mathrm{e}^{\mathrm{i}2\pi m} = 1\,.$$

The only values of m admissible from this constraint are integers, so that we have for m the result

$$m = 0, \pm 1, \pm 2, \ldots \tag{3.17}$$

The application of the simple requirement that the wave function be single-valued has thus led us to our first of three eigenvalues for this problem, which is a simple integer. The normalization is trivial in this case, with

$$\int_0^{2\pi} |\Phi|^2 \mathrm{d}\phi = |A|^2 \int_0^{2\pi} \mathrm{d}\phi = |A|^2 2\pi = 1\,,$$

so the normalized eigenfunctions are given by

$$\Phi_m(\phi) = \frac{1}{\sqrt{2\pi}}\mathrm{e}^{\pm im\phi}\,. \tag{3.18}$$

3.3.3 Orbital Angular Momentum

The solution of the equation for $\Theta(\theta)$ is intimately connected with orbital angular momentum, so we will take a first look at the quantum mechanical implications for angular momentum. A more complete examination of the subject follows in Chapter 4.

3.3.3.1 Angular Momentum in Rectangular and Spherical Coordinates

The definition of orbital angular momentum comes from the classical formula

$$\boldsymbol{L} = \boldsymbol{r} \times \boldsymbol{p}\,, \tag{3.19}$$

and in quantum mechanics, the order is important since $\hat{\boldsymbol{r}}$ and $\hat{\boldsymbol{p}}$ do not commute. Changing this into the quantum mechanical operator for angular momentum, we have

$$\hat{\boldsymbol{L}} = \boldsymbol{r} \times \frac{\hbar}{\mathrm{i}} \nabla \,, \tag{3.20}$$

or in component form,

$$\hat{L}_x = \hat{y}\hat{p}_z - \hat{z}\hat{p}_y = \frac{\hbar}{\mathrm{i}} \left(y \frac{\partial}{\partial z} - z \frac{\partial}{\partial y} \right) \tag{3.21}$$

$$\hat{L}_y = \hat{z}\hat{p}_x - \hat{x}\hat{p}_z = \frac{\hbar}{\mathrm{i}} \left(z \frac{\partial}{\partial x} - x \frac{\partial}{\partial z} \right) \tag{3.22}$$

$$\hat{L}_z = \hat{x}\hat{p}_y - \hat{y}\hat{p}_x = \frac{\hbar}{\mathrm{i}} \left(x \frac{\partial}{\partial y} - y \frac{\partial}{\partial x} \right) . \tag{3.23}$$

We need angular momentum in spherical coordinates, however, so using the transformation

$$\begin{aligned} x &= r \sin\theta \cos\phi \\ y &= r \sin\theta \sin\phi \\ z &= r \cos\theta \,, \end{aligned}$$

we can convert from rectangular to spherical coordinates. It is generally easier to go the other way, however, as seen by the example,

$$\begin{aligned} \frac{\partial}{\partial \phi} &= \frac{\partial x}{\partial \phi}\frac{\partial}{\partial x} + \frac{\partial y}{\partial \phi}\frac{\partial}{\partial y} + \frac{\partial z}{\partial \phi}\frac{\partial}{\partial z} \\ &= -r \sin\theta \sin\phi \frac{\partial}{\partial x} + r \sin\theta \cos\phi \frac{\partial}{\partial y} \\ &= -y \frac{\partial}{\partial x} + x \frac{\partial}{\partial y} \,. \end{aligned}$$

From this result, we have

$$\hat{L}_z = \frac{\hbar}{\mathrm{i}} \left(x \frac{\partial}{\partial y} - y \frac{\partial}{\partial x} \right) = \frac{\hbar}{\mathrm{i}} \frac{\partial}{\partial \phi} \,. \tag{3.24}$$

Now for any **central potential**, of the form $V(r)$, then, the eigenfunctions of the Hamiltonian are of the form

$$\psi(r, \theta, \phi) = A R(r) \Theta(\theta) \mathrm{e}^{\mathrm{i} m \phi} \,, \tag{3.25}$$

so we have

$$\hat{L}_z \psi = \frac{\hbar}{\mathrm{i}} \frac{\partial \psi}{\partial \phi} = m\hbar\psi \,, \tag{3.26}$$

so ψ, which is an eigenfunction of the Hamiltonian, is also an eigenfunction of $\hat{L}_z$ with eigenvalue $m\hbar$. Since $\hat{\mathcal{H}}$ and $\hat{L}_z$ have simultaneous eigenfunctions, then $[\hat{\mathcal{H}}, \hat{L}_z] = 0$ [see Equation (1.55)] and thus it is possible to make simultaneous measurements of both the energy and the z-component of angular momentum.

For the total angular momentum, we have

$$\begin{aligned}\hat{L}^2 &= \hat{L}_x^2 + \hat{L}_y^2 + \hat{L}_z^2 \\ &= -\hbar^2 \left[\frac{1}{\sin\theta} \frac{\partial}{\partial\theta} \left(\sin\theta \frac{\partial}{\partial\theta} \right) + \frac{1}{\sin^2\theta} \frac{\partial^2}{\partial\phi^2} \right], \end{aligned} \tag{3.27}$$

but this operator appears in the Laplacian of Equation (3.8) that forms part of the one-electron atom Hamiltonian, such that we may write

$$\hat{\mathcal{H}} = -\frac{\hbar^2}{2m} \frac{1}{r^2} \frac{\partial}{\partial r} \left(r^2 \frac{\partial}{\partial r} \right) + \frac{\hat{L}^2}{2mr^2} + V(r) . \tag{3.28}$$

In separating the variables, we found from the angular part of that problem that

$$\frac{1}{\sin\theta} \frac{\mathrm{d}}{\mathrm{d}\theta} \left(\sin\theta \frac{\mathrm{d}\Theta}{\mathrm{d}\theta} \right) + \left(\Lambda - \frac{m^2}{\sin^2\theta} \right) \Theta = 0 , \tag{3.29}$$

with Λ being the eigenvalue. Thus we have

$$\hat{L}^2 \psi = \Lambda \hbar^2 \psi , \tag{3.30}$$

so again ψ is an eigenfunction of $\hat{\mathcal{H}}$, $\hat{L}_z$ *and* $\hat{L}^2$ simultaneously, so all three of these operators commute with one another and their expectation values can be simultaneously measured. We note, however, that $\hat{L}_x$ and $\hat{L}_y$ *do not commute* with $\hat{L}_z$, so not all of the information about the angular momentum can be measured simultaneously.

Problem 3.5 *Angular momentum in spherical coordinates.*

(a) Using Equation (3.21) and Equation (3.22), find $\hat{L}_x$ and $\hat{L}_y$ in spherical coordinates.

(b) Using the expressions for $\hat{L}_x$, $\hat{L}_y$, and $\hat{L}_z$ in spherical coordinates, show that $\hat{L}^2$ is given by Equation (3.27).

3.3.3.2 General $\hat{L}_x$, $\hat{L}_y$, $\hat{L}_z$ Commutation Relations

We already know that $[\hat{L}^2, \hat{L}_z] = 0$ so these commute. We also know the basic commutation relations,

$$[\hat{x}, \hat{p}_x] = [\hat{y}, \hat{p}_y] = [\hat{z}, \hat{p}_z] = \mathrm{i}\hbar ,$$

and that $[\hat{y}, \hat{p}_x] = [\hat{x}, \hat{p}_y] = 0$, etc., so that

$$[\hat{L}_x, \hat{L}_y] = (\hat{y}\hat{p}_z - \hat{z}\hat{p}_y)(\hat{z}\hat{p}_x - \hat{x}\hat{p}_z) - (\hat{z}\hat{p}_x - \hat{x}\hat{p}_z)(\hat{y}\hat{p}_z - \hat{z}\hat{p}_y)$$

$$\begin{aligned}
&= \hat{y}\hat{p}_x\hat{p}_z\hat{z} - \hat{y}\hat{x}\hat{p}_z^2 - \hat{z}^2\hat{p}_y\hat{p}_x + \hat{z}\hat{x}\hat{p}_y\hat{p}_z \\
&\quad -\hat{z}\hat{y}\hat{p}_x\hat{p}_z + \hat{z}^2\hat{p}_x\hat{p}_y + \hat{x}\hat{y}\hat{p}_z^2 - \hat{x}\hat{p}_y\hat{p}_z\hat{z} \\
&= \hat{y}\hat{p}_x(\hat{p}_z\hat{z} - \hat{z}\hat{p}_z) + \hat{x}\hat{p}_y(\hat{z}\hat{p}_z - \hat{p}_z\hat{z}) \\
&= \mathrm{i}\hbar(\hat{x}\hat{p}_y - \hat{y}\hat{p}_x) = \mathrm{i}\hbar\hat{L}_z\,. \qquad (3.31)
\end{aligned}$$

Similarly,

$$[\hat{L}_y, \hat{L}_z] = \mathrm{i}\hbar\hat{L}_x \qquad (3.32)$$

$$[\hat{L}_z, \hat{L}_x] = \mathrm{i}\hbar\hat{L}_y\,, \qquad (3.33)$$

or combining these, we may write

$$\hat{\boldsymbol{L}} \times \hat{\boldsymbol{L}} = \mathrm{i}\hbar\hat{\boldsymbol{L}}\,. \qquad (3.34)$$

This relation is obviously false for any ordinary vector, but it is valid for the angular momentum operator because of the noncommuting operators it contains.

In order to find an expression for $[\hat{L}^2, \hat{L}_x]$ or $[\hat{L}^2, \hat{L}_y]$, we examine

$$[\hat{L}_x, \hat{L}^2] = [\hat{L}_x, \hat{L}_x^2] + [\hat{L}_x, \hat{L}_y^2] + [\hat{L}_x, \hat{L}_z^2]\,.$$

Now for any two operators, $\hat{A}$, $\hat{B}$,

$$\begin{aligned}
[\hat{A}, \hat{B}^2] &= \hat{A}\hat{B}\hat{B} - \hat{B}\hat{A}\hat{B} + \hat{B}\hat{A}\hat{B} - \hat{B}\hat{B}\hat{A} \\
&= (\hat{A}\hat{B} - \hat{B}\hat{A})\hat{B} + \hat{B}(\hat{A}\hat{B} - \hat{B}\hat{A}) \\
&= [\hat{A}, \hat{B}]\hat{B} + \hat{B}[\hat{A}, \hat{B}]\,, \qquad (3.35)
\end{aligned}$$

where we added and subtracted $\hat{B}\hat{A}\hat{B}$ on the first line, so

$$\begin{aligned}
[\hat{L}_x, \hat{L}_y^2] &= [\hat{L}_x, \hat{L}_y]\hat{L}_y + \hat{L}_y[\hat{L}_x, \hat{L}_y] \\
&= \mathrm{i}\hbar[\hat{L}_z\hat{L}_y + \hat{L}_y\hat{L}_z]
\end{aligned}$$

and

$$\begin{aligned}
[\hat{L}_x, \hat{L}_z^2] &= [\hat{L}_x, \hat{L}_z]\hat{L}_z + \hat{L}_z[\hat{L}_x, \hat{L}_z] \\
&= -\mathrm{i}\hbar[\hat{L}_y\hat{L}_z + \hat{L}_z\hat{L}_y]\,,
\end{aligned}$$

but these are equal and opposite, and $[\hat{L}_x, \hat{L}_x^2] = 0$, so adding, our result is

$$[\hat{L}_x, \hat{L}^2] = 0\,. \qquad (3.36)$$

Similarly, $[\hat{L}_y, \hat{L}^2] = 0$, so we may write $[\hat{\boldsymbol{L}}, \hat{L}^2] = 0$. This really means that any individual component of the angular momentum vector may be measured simultaneously with the magnitude, but since the individual components do not commute, the vector may not be completely specified without any uncertainty.

Problem 3.6 *Angular momentum commutators.* Using the cross product representation of a determinant, show that Equation (3.34) properly represents the commutation relations of Equations (3.31) through (3.33).

3.3.3.3 Raising and Lowering Operators

Suppose for a moment that we had not looked at the spherical coordinates representations of the angular momentum operators and identified the eigenvalues with those from the solution of the one-electron atom. Suppose instead that we had only Equation (3.31) through Equation (3.33). Since $\hat{L}_z$ commutes with $\hat{L}^2$, they have simultaneous eigenvalues, which means that there exists a $\psi_{\Lambda,m}$ such that

$$\hat{L}^2\psi_{\Lambda,m} = \Lambda\hbar^2\psi_{\Lambda,m} \tag{3.37}$$

$$\hat{L}_z\psi_{\Lambda,m} = m\hbar\psi_{\Lambda,m}\,, \tag{3.38}$$

where Λ and m are unknown dimensionless constants. We then examine the operator

$$\hat{\mathcal{L}} = \hat{L}_x + b\hat{L}_y\,,$$

where b is a constant, so that

$$\begin{aligned}[\hat{L}_z, \hat{L}_x + b\hat{L}_y] &= [\hat{L}_z, \hat{L}_x] + b[\hat{L}_z, \hat{L}_y] \\ &= \mathrm{i}\hbar\hat{L}_y - \mathrm{i}b\hbar\hat{L}_x\,. \end{aligned} \tag{3.39}$$

We want to choose b so that $\hat{\mathcal{L}}$ is a **raising** or **lowering** operator, or so that

$$\hat{\mathcal{L}}\psi_{\Lambda,m} = N_{\Lambda,m}\psi_{\Lambda,m+a}\,,$$

where $\psi_{\Lambda,m+a}$ is a normalized eigenfunction of $\hat{L}^2$ and $\hat{L}_z$ and $N_{\Lambda,m}$ is a normalizing constant, since the operator may not automatically produce a normalized eigenfunction. Thus, in addition to Equations (3.37) and (3.38), we require

$$\begin{aligned}\hat{L}^2\psi_{\Lambda,m+a} &= \Lambda\hbar^2\psi_{\Lambda,m+a} \\ \hat{L}_z\psi_{\Lambda,m+a} &= (m+a)\hbar\psi_{\Lambda,m+a}\,. \end{aligned}$$

Then it follows that

$$\begin{aligned}[\hat{L}_z, \hat{\mathcal{L}}]\psi_{\Lambda,m} &= \hat{L}_z\hat{\mathcal{L}}\psi_{\Lambda,m} - \hat{\mathcal{L}}\hat{L}_z\psi_{\Lambda,m} \\ &= \hat{L}_z N_{\Lambda,m}\psi_{\Lambda,m+a} - \hat{\mathcal{L}}m\hbar\psi_{\Lambda,m} \\ &= (m+a)\hbar N_{\Lambda,m}\psi_{\Lambda,m+a} - m\hbar N_{\Lambda,m}\psi_{\Lambda,m+a} \\ &= a\hbar N_{\Lambda,m}\psi_{\Lambda,m+a} \\ &= a\hbar\hat{\mathcal{L}}\psi_{\Lambda,m}\,. \end{aligned}$$

The commutator can thus be written

$$[\hat{L}_z, \hat{\mathcal{L}}] = a\hbar\hat{\mathcal{L}}\,. \tag{3.40}$$

Comparing this result with Equation (3.39), we find the equality

$$\mathrm{i}\hat{L}_y - \mathrm{i}b\hat{L}_x = a(\hat{L}_x + b\hat{L}_y)\,,$$

so we have $ab = \mathrm{i}$ and $a = -\mathrm{i}b$, so $b = \mathrm{i}a$ and $ab = \mathrm{i}a^2 = \mathrm{i}$ so $a^2 = 1$. This has two roots:

I. $a = 1$, $b = \mathrm{i}$, $\hat{\mathcal{L}} \equiv \hat{L}_+ = \hat{L}_x + \mathrm{i}\hat{L}_y$. This is a raising operator since $m \to m + 1$.

II. $a = -1$, $b = -\mathrm{i}$, $\hat{\mathcal{L}} \equiv \hat{L}_- = \hat{L}_x - \mathrm{i}\hat{L}_y$. This is a lowering operator since $m \to m - 1$.

One may ask if the eigenvalues may be raised or lowered indefinitely, since it appears the operators may be applied an arbitrary number of times. To prove that this is not possible, we examine

$$\begin{aligned}\langle L^2\rangle &= \langle L_x^2\rangle + \langle L_y^2\rangle + \langle L_z^2\rangle \\ \Lambda\hbar^2 &= \langle L_x^2\rangle + \langle L_y^2\rangle + m^2\hbar^2 ,\end{aligned}$$

but since $\hat{L}_x$ and $\hat{L}_y$ are Hermitian operators, $\langle L_x^2\rangle + \langle L_y^2\rangle \geq 0$, so $\Lambda \geq m^2$. This means that the raising and lowering operators cannot move up and down the ladder too far or $|m|$ would get too large. This implies the existence of an upper limit to m, called m_T, and a lower limit to m, called m_B, such that

$$\hat{L}_+\psi_{\Lambda,m_T} = 0, \qquad \hat{L}_-\psi_{\Lambda,m_B} = 0 .$$

Since the raising and lowering operators always change m by one unit ($|a| = 1$), it follows that m_T must differ from m_B by an integer, or that

$$m_T - m_B = n, \qquad n = 0, 1, 2, \ldots \tag{3.41}$$

Then we examine the products of the raising and lowering operators to find

$$\begin{aligned}\hat{L}_+\hat{L}_- &= (\hat{L}_x + \mathrm{i}\hat{L}_y)(\hat{L}_x - \mathrm{i}\hat{L}_y) \\ &= \hat{L}_x^2 + \hat{L}_y^2 - \mathrm{i}[\hat{L}_x, \hat{L}_y] \\ &= \hat{L}^2 - \hat{L}_z^2 + \hbar\hat{L}_z \end{aligned} \tag{3.42}$$

$$\begin{aligned}\hat{L}_-\hat{L}_+ &= (\hat{L}_x - \mathrm{i}\hat{L}_y)(\hat{L}_x + \mathrm{i}\hat{L}_y) \\ &= \hat{L}_x^2 + \hat{L}_y^2 + \mathrm{i}[\hat{L}_x, \hat{L}_y] \\ &= \hat{L}^2 - \hat{L}_z^2 - \hbar\hat{L}_z .\end{aligned} \tag{3.43}$$

By solving these two expressions for $\hat{L}^2$, we obtain the two relations

$$\hat{L}^2 = \hat{L}_+\hat{L}_- + \hat{L}_z^2 - \hbar\hat{L}_z \tag{3.44}$$

$$\quad = \hat{L}_-\hat{L}_+ + \hat{L}_z^2 + \hbar\hat{L}_z . \tag{3.45}$$

Applying Equation (3.45) to ψ_{Λ,m_T}, we find

$$\begin{aligned}\hat{L}^2\psi_{\Lambda,m_T} &= (\hat{L}_-\hat{L}_+ + \hat{L}_z^2 + \hbar\hat{L}_z)\psi_{\Lambda,m_T} \\ &= (0 + m_T^2 + m_T)\hbar^2\psi_{\Lambda,m_T} \\ &= \Lambda\hbar^2\psi_{\Lambda,m_T} ,\end{aligned}$$

so

$$\Lambda = m_T(m_T + 1)\,. \tag{3.46}$$

Then applying Equation (3.44) to ψ_{Λ,m_B}, we find

$$\begin{aligned}\hat{L}^2\psi_{\Lambda,m_B} &= (\hat{L}_+\hat{L}_- + \hat{L}_z^2 - \hbar\hat{L}_z)\psi_{\Lambda,m_B}\\ &= (0 + m_B^2 - m_B)\hbar^2\psi_{\Lambda,m_B}\\ &= \Lambda\hbar^2\psi_{\Lambda,m_B}\,,\end{aligned}$$

so

$$\Lambda = m_B(m_B - 1)\,. \tag{3.47}$$

Equating these, we obtain

$$m_T^2 + m_T = m_B^2 - m_B\,,$$

whose quadratic solutions are

$$m_T = m_B - 1,\ \ -m_B\,,$$

The first root is inconsistent with Equation (3.41) since it implies $m_T - m_B = -1$, so we must have

$$m_T = -m_B \equiv \ell\,, \tag{3.48}$$

from which Equation (3.46) gives

$$\Lambda = \ell(\ell + 1)\,. \tag{3.49}$$

This establishes that ℓ is a nonnegative number and that $\ell \geq |m|$ since m lies between $-\ell$ and ℓ. The values of m range from $-\ell$ to ℓ in integer steps so there will be $2\ell + 1$ values of m for each value of ℓ.

The consequences of these results may seem surprising. The trivial case, where $\ell = m = 0$ indicates there is no angular momentum at all. Since the ground state of the one-electron atom has $\ell = m = 0$, there is no angular momentum in the ground state, *in direct contradiction to the Bohr postulate.* In fact, we see that angular momentum is related to the quantum numbers ℓ and m, and not the principal quantum number n, which is the only number in the Bohr theory, so the angular momentum postulate of Bohr is *never* correct, since we will discover that $n > \ell$ always.

For $\ell = 1$, the magnitude of the angular momentum vector is $\sqrt{\ell(\ell+1)}\hbar = \sqrt{2}\hbar$, and the z-component is either 0 or $\pm\hbar$. The vectors are shown in Figure 3.1a, where the figure should be understood to be rotated about the z-axis so that the vectors lie somewhere along a cone ($|m| > 0$) or in a plane ($m = 0$), but L_x and L_y cannot be known simultaneously.

For $\ell = 2$, the magnitude of the angular momentum vector is $L = \sqrt{6}\hbar$, and the z-component is either 0, $\pm\hbar$, or $\pm 2\hbar$. The vectors are shown in Figure 3.1b.

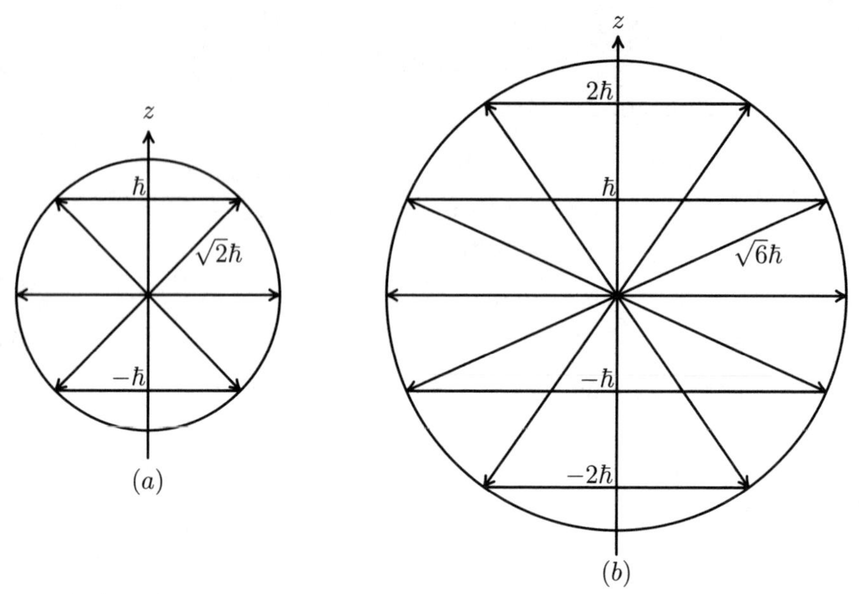

FIGURE 3.1
Angular momentum vectors for (a) $\ell = 1$, and (b) $\ell = 2$.

We now wish to evaluate the normalization constant, since we have no assurance that raising a normalized eigenfunction with $\hat{L}_+$ will produce a normalized eigenfunction with the new eigenvalue. The normalizing parameter is defined such that

$$\hat{L}_+\psi_{\ell,m} \equiv N^+_{\ell,m}\psi_{\ell,m+1}\,. \tag{3.50}$$

We require $\psi_{\ell,m+1}$ to be normalized if $\psi_{\ell,m}$ is. Hence,

$$|N^+_{\ell,m}|^2\int|\psi_{\ell,m+1}|^2\,\mathrm{d}\tau = \int[(\hat{L}_x + \mathrm{i}\hat{L}_y)\psi_{\ell,m}]^*(\hat{L}_x + \mathrm{i}\hat{L}_y)\psi_{\ell,m}\,\mathrm{d}\tau\,,$$

where $\mathrm{d}\tau = r^2\sin\theta\,\mathrm{d}r\,\mathrm{d}\theta\,\mathrm{d}\phi$ represents the volume element. Since $\hat{L}_x$ and $\hat{L}_y$ are Hermitian, we have

$$\int(\hat{L}_x\psi_{\ell,m})^*\chi\,\mathrm{d}\tau = \int\psi^*_{\ell,m}\hat{L}_x\chi\,\mathrm{d}\tau$$
$$\int(\mathrm{i}\hat{L}_y\psi_{\ell,m})^*\chi\,\mathrm{d}\tau = -\mathrm{i}\int\psi^*_{\ell,m}\hat{L}_y\chi\,\mathrm{d}\tau$$

so

$$\begin{aligned}|N^+_{\ell,m}|^2\int|\psi_{\ell,m+1}|^2\,\mathrm{d}\tau &= \int\psi^*_{\ell,m}(\hat{L}_x - \mathrm{i}\hat{L}_y)(\hat{L}_x + \mathrm{i}\hat{L}_y)\psi_{\ell,m}\,\mathrm{d}\tau \\ &= \int\psi^*_{\ell,m}(\hat{L}^2 - \hat{L}_z^2 - \hbar\hat{L}_z)\psi_{\ell,m}\,\mathrm{d}\tau\end{aligned}$$

$$= \hbar^2[\ell(\ell+1) - m^2 - m] \int |\psi_{\ell,m}|^2 \, d\tau$$
$$= \hbar^2[\ell(\ell+1) - m(m+1)] \,,$$

where we used Equation (3.43) to obtain the second line. The resulting normalization constant is

$$N^+_{\ell,m} = \hbar\sqrt{\ell(\ell+1) - m(m+1)} \,. \tag{3.51}$$

Similarly, we define

$$\hat{L}_- \psi_{\ell,m} \equiv N^-_{\ell,m} \psi_{\ell,m-1} \,. \tag{3.52}$$

with the corresponding result for the other normalization constant,

$$N^-_{\ell,m} = \hbar\sqrt{\ell(\ell+1) - m(m-1)} \,. \tag{3.53}$$

Problem 3.7 *Angular momentum angles.* Sketch the figure corresponding to Figure 3.1 for $\ell = 3$ and calculate the angle between each allowable vector and the z-axis.

Problem 3.8 *Uncertainty in angular momentum.*

(a) Express the operators for $\hat{L}_x$ and $\hat{L}_y$ in terms of $\hat{L}_+$ and $\hat{L}_-$.

(b) Find the expectation values $\langle L_x \rangle$, $\langle L_y \rangle$, $\langle L_z \rangle$, $\langle L_x^2 \rangle$, $\langle L_y^2 \rangle$, $\langle L_z^2 \rangle$, for a general wave function $\psi_{\ell m}$ and then find the uncertainty in each of the components.

(c) If $\langle L_z^2 \rangle = \langle L^2 \rangle$ for some m, what would it imply about the uncertainty in L_x and L_y?

Problem 3.9 Find expressions for $\hat{L}_\pm$ in spherical coordinates.

Problem 3.10 Fill in the steps leading to Equation (3.53).

3.3.3.4 The Angular Wave Functions

Although we have already discovered the eigenvalues for the functions of θ and ϕ, we still need the form of the $\Theta(\theta)$ in order to completely define the wave function. We begin with Equation (3.13) after a change of variables such that $\mu = \cos\theta$, in which case we have

$$\frac{d}{d\mu}\left[(1-\mu^2)\frac{d\Theta}{d\mu}\right] + \left[\ell(\ell+1) - \frac{m^2}{1-\mu^2}\right]\Theta = 0 \,, \tag{3.54}$$

With $m = 0$, this is the **Legendre equation** whose bounded solutions are polynomials in μ that form an orthogonal set over the range $-1 \le \mu \le 1$. With $m \ne 0$, this is the **associated Legendre equation** whose solutions are denoted by $P_\ell^m(\mu)$ or $P_\ell^m(\cos\theta)$. The first few polynomials with $m = 0$

are $P_0 = 1$, $P_1 = \mu$, and the entire set can be obtained from the **Rodrigues formula**:

$$P_\ell(\mu) = \frac{1}{2^\ell \ell!}\frac{\mathrm{d}^\ell}{\mathrm{d}\mu^\ell}(\mu^2 - 1)^\ell . \tag{3.55}$$

The polynomials are of order ℓ and even or odd in μ as ℓ is even or odd. When $m \neq 0$, we have the **associated Legendre functions**, and these are not polynomials in μ if m is odd, since these functions may be obtained from

$$P_\ell^m(\mu) = (1 - \mu^2)^{|m|/2}\frac{\mathrm{d}^{|m|}P_\ell}{\mathrm{d}\mu^{|m|}} . \tag{3.56}$$

The first several associated Legendre functions are listed in Table 3.1.

TABLE 3.1
The first few Legendre polynomials ($m = 0$) and associated Legendre functions, $P_\ell^m(\mu)$ where $\mu = \cos\theta$ and $s = (1 - \mu^2)^{1/2} = \sin\theta$.

$m\backslash\ell$	0	1	2	3	4
0	1	μ	$\frac{1}{2}(3\mu^2 - 1)$	$\frac{1}{2}(5\mu^3 - 3\mu)$	$\frac{1}{8}(35\mu^4 - 30\mu^2 + 3)$
1		s	$3\mu s$	$\frac{3}{2}(5\mu^2 - 1)s$	$\frac{5}{2}(7\mu^3 - 3\mu)s$
2			$3s^2$	$15\mu s^2$	$\frac{15}{2}(7\mu^2 - 1)s^2$
3				$15s^3$	$105\mu s^3$
4					$105s^4$

The associated Legendre functions as defined above are not normalized, but the normalization is established in Appendix B. The two angular wave functions are commonly combined to form the **Spherical Harmonics**, which are given by

$$Y_\ell^m(\theta, \phi) = \left[\frac{(2\ell + 1)(\ell - m)!}{4\pi(\ell + m)!}\right]^{1/2} P_\ell^m(\cos\theta)\mathrm{e}^{\mathrm{i}m\phi} . \tag{3.57}$$

The first few of the spherical harmonics are given in Table 3.2.

Problem 3.11 *Orthogonality.*

(a) Show that each of the Legendre polynomials ($m = 0$) listed in Table 3.1 are mutually orthogonal.

(b) Show that *not* all of the Associated Legendre polynomials listed in Table 3.1 with $\ell = 2$ are mutually orthogonal.

(c) Show that each of the Y_ℓ^m from Equation (3.57) with different values of m are mutually orthogonal.

TABLE 3.2
The first few spherical harmonics, $Y_\ell^m(\mu,\phi)$, where $\mu = \cos\theta$ and $s = \mp(1-\mu^2)^{1/2} = \mp\sin\theta$.

$\ell\backslash m$	0	±1	±2	±3
0	$\frac{1}{\sqrt{4\pi}}$			
1	$\sqrt{\frac{3}{4\pi}}\mu$	$\sqrt{\frac{3}{8\pi}}s\mathrm{e}^{\pm\mathrm{i}\phi}$		
2	$\sqrt{\frac{5}{16\pi}}(3\mu^2-1)$	$\sqrt{\frac{15}{8\pi}}s\mu\mathrm{e}^{\pm\mathrm{i}\phi}$	$\sqrt{\frac{15}{32\pi}}s^2\mathrm{e}^{\pm2\mathrm{i}\phi}$	
3	$\sqrt{\frac{7}{16\pi}}(5\mu^3-3\mu)$	$\sqrt{\frac{21}{128\pi}}s(5\mu^2-1)\mathrm{e}^{\pm\mathrm{i}\phi}$	$\sqrt{\frac{105}{32\pi}}s^2\mu\mathrm{e}^{\pm2\mathrm{i}\phi}$	$\sqrt{\frac{35}{64\pi}}s^3\mathrm{e}^{\pm3\mathrm{i}\phi}$

Problem 3.12 *Normalization.*

(a) Show that each of the spherical harmonics in Table 3.2 with $m = 0$ are normalized.

(b) Show that each of the spherical harmonics in Table 3.2 with $\ell = 2$ are normalized.

3.3.4 Solving the Radial Equation

In the solution of the radial equation, we will use the same procedures we used in solving the harmonic oscillator problem as listed on page 57.

3.3.4.1 Changing to Dimensionless Variables

Using the appropriate value of Λ from the previous section, Equation (3.14) may be written as

$$\frac{1}{r^2}\frac{\mathrm{d}}{\mathrm{d}r}\left(r^2\frac{\mathrm{d}R}{\mathrm{d}r}\right) + \left[\frac{2\mu_e E}{\hbar^2} + \frac{2\mu_e Ze^2}{4\pi\epsilon_0\hbar^2 r} - \frac{\ell(\ell+1)}{r^2}\right]R = 0\,. \tag{3.58}$$

Changing to dimensionless variables, we let $\rho = \alpha r$, $\mathrm{d}/\mathrm{d}r \to \alpha\mathrm{d}/\mathrm{d}\rho$, so that Equation (3.58) becomes (after dividing by α^2)

$$\frac{1}{\rho^2}\frac{\mathrm{d}}{\mathrm{d}\rho}\left(\rho^2\frac{\mathrm{d}R}{\mathrm{d}\rho}\right) + \left[\frac{2\mu_e E}{\alpha^2\hbar^2} + \frac{2\mu_e Ze^2}{4\pi\epsilon_0\hbar^2\alpha\rho} - \frac{\ell(\ell+1)}{\rho^2}\right]R = 0\,. \tag{3.59}$$

If we now choose

$$\frac{2\mu_e E}{\alpha^2\hbar^2} = -\frac{1}{4}\,, \qquad \lambda = \frac{2\mu_e Ze^2}{4\pi\epsilon_0\hbar^2\alpha}\,,$$

then $\alpha^2 = -8\mu_e E/\hbar^2$ and Equation (3.59) becomes

$$R'' + \frac{2}{\rho}R' + \left[-\frac{1}{4} + \frac{\lambda}{\rho} - \frac{\ell(\ell+1)}{\rho^2}\right]R = 0\,. \tag{3.60}$$

3.3.4.2 Factoring out the Asymptotic Behavior

As $\rho \to \infty$, Equation (3.60) is dominated by

$$R'' - \tfrac{1}{4}R \approx 0\,,$$

with solutions of the form

$$R \approx \mathrm{e}^{\pm\rho/2}\,.$$

We cannot accept the growing exponential, so we will choose

$$R(\rho) = \mathrm{e}^{-\rho/2}F(\rho)\,. \tag{3.61}$$

Differentiating,

$$\begin{aligned} R' &= -\tfrac{1}{2}\mathrm{e}^{-\rho/2}F + \mathrm{e}^{-\rho/2}F' \\ R'' &= \frac{\mathrm{e}^{-\rho/2}F}{4} - \mathrm{e}^{-\rho/2}F' + \mathrm{e}^{-\rho/2}F''\,, \end{aligned}$$

so, factoring out the common exponential, Equation (3.60) becomes

$$F'' + \left(\frac{2}{\rho} - 1\right)F' + \left[\frac{\lambda-1}{\rho} - \frac{\ell(\ell+1)}{\rho^2}\right]F = 0\,. \tag{3.62}$$

3.3.4.3 Power Series Solution

Since there is a possible singularity at the origin, we multiply the usual power series expression by ρ^s to avoid the singularity, so that the solution is assumed to be of the form

$$F(\rho) = \rho^s \sum_{k=0}^{\infty} a_k \rho^k\,. \tag{3.63}$$

The differential equation then becomes

$$\sum_{k=0}^{\infty}[(s+k)(s+k-1)a_k\rho^{s+k-2} + 2(s+k)a_k\rho^{s+k-2} - (s+k)a_k\rho^{s+k-1} \\ +(\lambda-1)a_k\rho^{s+k-1} - \ell(\ell+1)a_k\rho^{s+k-2}] = 0\,.$$

Breaking out the terms proportional to ρ^{s-2}, we have

$$s(s-1)a_0 + 2sa_0 - \ell(\ell+1)a_0 = [s(s+1) - \ell(\ell+1)]a_0 = 0\,,$$

which requires $s = \ell$ or $s = -(\ell+1)$, and we must reject the negative root for the function to be well-behaved at the origin. Adjusting the indices so that exponents $s+k-2 \to s+t-1$ or $k \to t+1$ and $s+k-1 \to s+t-1$ or $k \to t$, we may write

$$\sum_{t=0}^{\infty}[(s+t+1)(s+t)a_{t+1} + 2(s+t+1)a_{t+1} \\ -(s+t)a_t + (\lambda-1)a_t - \ell(\ell+1)a_{t+1}]\rho^{s+t-1} = 0\,.$$

The coefficients must all vanish, so we have (using $s = \ell$)

$$\frac{a_{t+1}}{a_t} = \frac{\ell + t + 1 - \lambda}{(\ell + t + 2)(\ell + t + 1) - \ell(\ell + 1)} \to \frac{1}{t} \quad \text{as } t \to \infty\,. \tag{3.64}$$

From the recursion formula, it is plain that the series converges, but the ratio of successive terms for e^{ρ} has the same behavior, so the asymptotic form of the series is e^{ρ}, and $R = e^{-\rho/2}F \sim e^{-\rho/2}e^{\rho} \sim e^{\rho/2}$ so the function is unbounded as $\rho \to \infty$. Therefore we must truncate the series by the proper choice of λ. The series will truncate if we choose

$$\lambda \equiv n \geq \ell + 1\,, \tag{3.65}$$

where the condition is necessary to ensure that the truncation will occur for some positive value of t. We thus end up with another integer and a corresponding set of polynomials.

3.3.4.4 The Energy Eigenvalues

From Equation (3.65) and the transformations that gave us the dimensionless variables, we have

$$\lambda = n = \frac{2\mu_e Z e^2}{4\pi\epsilon_0 \hbar^2 \alpha}\,,$$

which we can solve for $\alpha(\to \alpha_n)$ so that

$$\alpha_n = \frac{2\mu_e Z e^2}{4\pi\epsilon_0 \hbar^2 n} = \frac{2Z}{n a_0'} \tag{3.66}$$

with $a_0' = 4\pi\epsilon_0\hbar^2/\mu_e e^2$, and from $2\mu_e E/\alpha^2\hbar^2 = -1/4$, we find

$$E = -\frac{\hbar^2\alpha^2}{8\mu_e} = -\frac{\mu_e Z^2 e^4}{32\pi^2\epsilon_0^2\hbar^2 n^2}\,, \tag{3.67}$$

or

$$E_n = -\frac{E_0}{n^2}\,, \qquad E_0 = \frac{\mu_e Z^2 e^4}{2(4\pi\epsilon_0)^2\hbar^2}\,. \tag{3.68}$$

In terms of the dimensionless fine-structure constant, $\alpha \equiv e^2/4\pi\epsilon_0\hbar c \sim 1/137$ (do not get confused between this universal dimensionless constant and the constant we used to convert the radial equation to dimensionless variables), we may write these as

$$a_0' = \frac{\hbar}{\mu_e c\alpha}\,, \qquad E_0 = \frac{\mu_e c^2 Z^2 \alpha^2}{2}\,. \tag{3.69}$$

Note that the Bohr radius, $a_0 = \hbar/m_e c\alpha$ does *not* use the reduced mass while the energy does and is equivalent to the Bohr result.

3.3.4.5 Laguerre and Associated Laguerre Polynomials

We note from Equation (3.63) and the fact that $s = \ell$ that the solution is now a polynomial times ρ^ℓ, so we define $F(\rho) = \rho^\ell L(\rho)$ and find that $L(\rho)$ satisfies

$$\rho L'' + (2\ell + 2 - \rho)L' + (n - 1 - \ell)L = 0\,, \tag{3.70}$$

which is the equation for the associated Laguerre polynomials,

$$\rho L_j^{k''} + (k + 1 - \rho)L_j^{k'} + (j - k)L_j^k = 0\,, \tag{3.71}$$

with $k = 2\ell + 1$ and $j = n + \ell$. The Laguerre polynomials satisfy Equation (3.71) with $k = 0$, so that

$$\rho L_j{}'' + (1 - \rho)L_j{}' + jL_j = 0\,. \tag{3.72}$$

It may be shown by direct substitution that the solution of Equation (3.72) is

$$L_j(\rho) = \mathrm{e}^\rho \frac{\mathrm{d}^j}{\mathrm{d}\rho^j}(\rho^j \mathrm{e}^{-\rho}) = \sum_{n=0}^{j} \left(\frac{j!}{n!}\right)^2 \frac{(-1)^n}{(j-n)!}\rho^n\,, \tag{3.73}$$

which are the Laguerre polynomials, and that

$$L_j^k(\rho) = (-1)^k \frac{\mathrm{d}^k}{\mathrm{d}\rho^k} L_j(\rho) = \sum_{n=k}^{j} \frac{(j!)^2(-1)^{n+k)}}{n!(j-n)!(n-k)!}\rho^{n-k} \tag{3.74}$$

is a solution of Equation (3.71) and forms the associated Laguerre polynomials.

The radial solution is then given by

$$R_{n\ell}(\rho) = A_{n\ell}\mathrm{e}^{-\rho/2}\rho^\ell L_{n+\ell}^{2\ell+1}(\rho)\,. \tag{3.75}$$

A brief table of Laguerre (without the superscript) and associated Laguerre (with the superscripts) polynomials is given in Table 3.3.

TABLE 3.3
The first several Laguerre and associated Laguerre polynomials.

$L_0 = 1$			
$L_1 = 1 - \rho$	$L_1^1 = 1$		
$L_2 = 2 - 4\rho + \rho^2$	$L_2^1 = 4 - 2\rho$	$L_2^2 = 2$	
$L_3 = 6 - 18\rho + 9\rho^2 - \rho^3$	$L_3^1 = 18 - 18\rho + 3\rho^2$	$L_3^2 = 18 - 6\rho$	$L_3^3 = 6$

Problem 3.13 Show that $F(\rho) = \rho^\ell L_{n+\ell}^{2\ell+1}(\rho)$ is a solution of Equation (3.62).

Problem 3.14 (a) Prove that the polynomials defined by Equation (3.73) satisfy Equation (3.72). (*Hint:* It may be useful to first prove $D^j(\rho U) = \rho D^j U + jD^{j-1}U$ where $D \equiv \mathrm{d}/\mathrm{d}\rho$.)

(b) Prove that the polynomials defined by Equation (3.74) satisfy Equation (3.71).

3.3.4.6 Normalization

It is established in Section B.3.2 of the Appendix through the use of the generating function that the normalization constants for the associated Laguerre polynomials are given by

$$A_{n\ell} = \left[\left(\frac{2Z}{na_0'} \right)^3 \frac{(n-\ell-1)!}{2n[(n+\ell)!]^3} \right]^{1/2} . \tag{3.76}$$

3.3.5 Normalized Wave Functions

The first several normalized eigenfunctions $\psi_{n\ell m}(r,\theta,\phi)$ are

$$\psi_{100} = \frac{1}{\sqrt{\pi}} \left(\frac{Z}{a_0'} \right)^{3/2} \mathrm{e}^{-Zr/a_0'} \tag{3.77}$$

$$\psi_{200} = \frac{1}{\sqrt{32\pi}} \left(\frac{Z}{a_0'} \right)^{3/2} \left(2 - \frac{Zr}{a_0'} \right) \mathrm{e}^{-Zr/2a_0'} \tag{3.78}$$

$$\psi_{210} = \frac{1}{\sqrt{32\pi}} \left(\frac{Z}{a_0'} \right)^{3/2} \frac{Zr}{a_0'} \mathrm{e}^{-Zr/2a_0'} \cos\theta \tag{3.79}$$

$$\psi_{21\pm1} = \frac{1}{\sqrt{64\pi}} \left(\frac{Z}{a_0'} \right)^{3/2} \frac{Zr}{a_0'} \mathrm{e}^{-Zr/2a_0'} \sin\theta \mathrm{e}^{\pm \mathrm{i}\phi} \tag{3.80}$$

$$\psi_{300} = \frac{1}{81\sqrt{3\pi}} \left(\frac{Z}{a_0'} \right)^{3/2} \left(27 - 18\frac{Zr}{a_0'} + 2\frac{Z^2r^2}{a_0'^2} \right) \mathrm{e}^{-Zr/3a_0'} \tag{3.81}$$

$$\psi_{310} = \frac{\sqrt{2}}{81\sqrt{\pi}} \left(\frac{Z}{a_0'} \right)^{3/2} \left(6 - \frac{Zr}{a_0'} \right) \frac{Zr}{a_0'} \mathrm{e}^{-Zr/3a_0'} \cos\theta \tag{3.82}$$

$$\psi_{31\pm1} = \frac{1}{81\sqrt{\pi}} \left(\frac{Z}{a_0'} \right)^{3/2} \left(6 - \frac{Zr}{a_0'} \right) \frac{Zr}{a_0'} \mathrm{e}^{-Zr/3a_0'} \sin\theta \mathrm{e}^{\pm \mathrm{i}\phi} \tag{3.83}$$

$$\psi_{320} = \frac{1}{81\sqrt{6\pi}} \left(\frac{Z}{a_0'} \right)^{3/2} \frac{Z^2r^2}{a_0'^2} \mathrm{e}^{-Zr/3a_0'} (3\cos^2\theta - 1) \tag{3.84}$$

$$\psi_{32\pm1} = \frac{1}{81\sqrt{\pi}} \left(\frac{Z}{a_0'} \right)^{3/2} \frac{Z^2r^2}{a_0'^2} \mathrm{e}^{-Zr/3a_0'} \sin\theta \cos\theta \mathrm{e}^{\pm \mathrm{i}\phi} \tag{3.85}$$

$$\psi_{32\pm2} = \frac{1}{162\sqrt{\pi}} \left(\frac{Z}{a_0'} \right)^{3/2} \frac{Z^2r^2}{a_0'^2} \mathrm{e}^{-Zr/3a_0'} \sin^2\theta \mathrm{e}^{\pm 2\mathrm{i}\phi} . \tag{3.86}$$

The behavior of the radial portion of the probability densities for the wave functions listed above. $\rho^2|R_{nl}(\rho)|^2$, are shown in Figure 3.2 where for this plot, $\rho = r/a_0'$.

Example 3.1

Spread in radius. The quantity, $\langle r\rangle$, and the most probable value of r, $r_{\rm max}$, are generally different. For the ψ_{320} state, is the difference within the spread, (Δr)?

1. We first calculate $\langle r\rangle$ from the integral

$$
\begin{aligned}
\langle r\rangle_{320} &= \int_0^\infty\int_0^\pi\int_0^{2\pi} r|\psi_{320}|^2 r^2 \sin\theta\, \mathrm{d}r\mathrm{d}\theta\mathrm{d}\phi \\
&= \frac{1}{(81)^2 6\pi}\left(\frac{Z}{a_0'}\right)^3 \int_0^\infty\int_0^\pi\int_0^{2\pi} r\left(\frac{Zr}{a_0'}\right)^4 \mathrm{e}^{-2Zr/3a_0'} \\
&\quad \times(3\cos^2\theta - 1)^2 r^2 \sin\theta\, \mathrm{d}r\mathrm{d}\theta\mathrm{d}\phi \\
&= \frac{1}{3(81)^2}\left(\frac{a_0'}{Z}\right)\int_0^\infty \rho^7 \mathrm{e}^{-2\rho/3}\mathrm{d}\rho \int_{-1}^1 (3\mu^2-1)^2 \mathrm{d}\mu \\
&= \frac{7!a_0'}{3(81)^2 Z}\left(\frac{3}{2}\right)^8 \frac{8}{5} \\
&= \frac{21a_0'}{2Z}
\end{aligned}
$$

where we changed variables such that $\rho = Zr/a_0'$ and $\mu = \cos\theta$. We also used the general integral

$$
\int_0^\infty x^n \mathrm{e}^{-ax}\mathrm{d}x = n!/a^{n+1}\,.
$$

2. For the most probable value of r, we use $P(r) = r^2|\psi_{320}|^2$ (the leading r^2 comes from the volume element) and for this part of the problem, we can neglect the angular part of ψ, so we have

$$
P_{320}(r) = Ar^6\mathrm{e}^{-2Zr/3a_0'}
$$

and the maximum occurs where

$$
\frac{\mathrm{d}P}{\mathrm{d}r} = A\left[6r^5 - \frac{2Z}{3a_0'}r^6\right]\mathrm{e}^{-2Zr/3a_0'}\bigg|_{r=r_{\rm max}} = 0
$$

so $r_{\rm max} = 9a_0'/Z$.

3. For the spread, we need $\langle r^2\rangle$ or

$$
\langle r^2\rangle_{320} = \int_0^\infty\int_0^\pi\int_0^{2\pi} r^2|\psi_{320}|^2 r^2\sin\theta\,\mathrm{d}r\mathrm{d}\theta\mathrm{d}\phi
$$

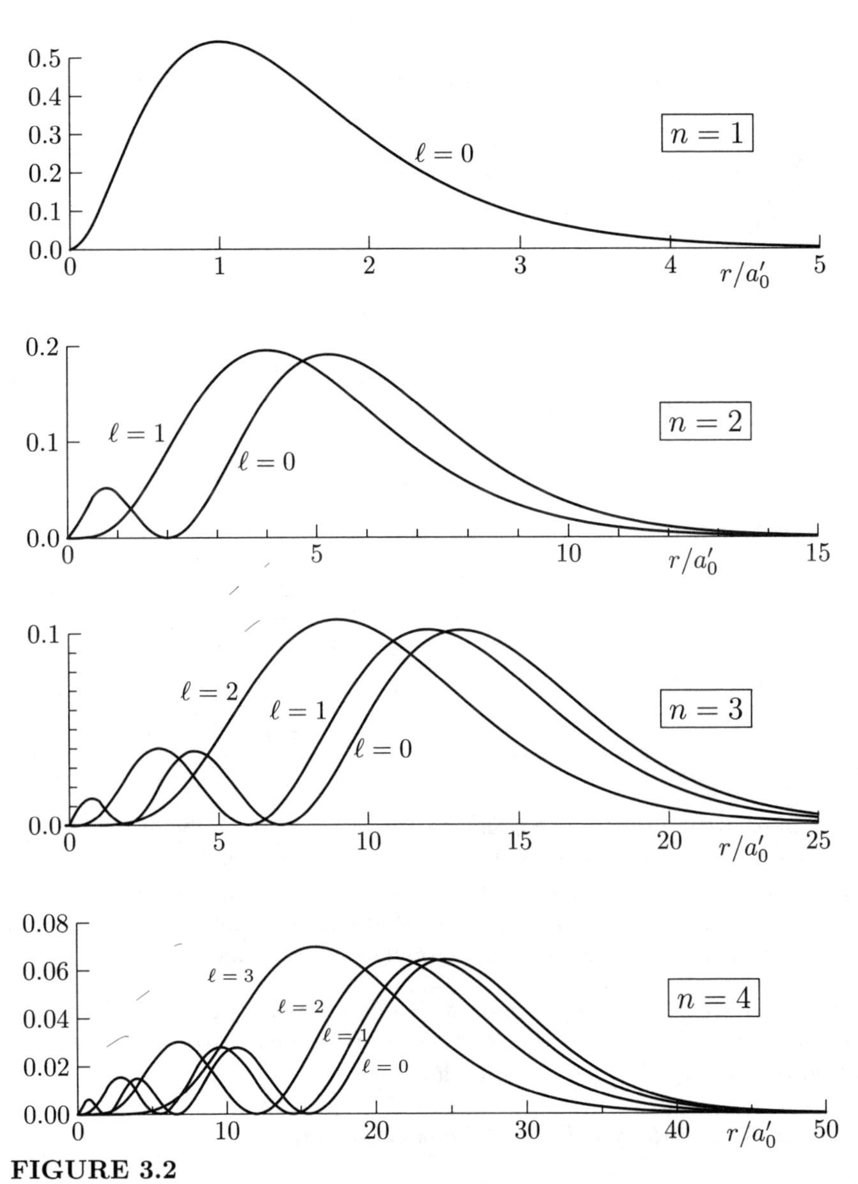

FIGURE 3.2
Probability densities, $\rho^2[R_{nl}(\rho)]^2$ with $\rho = r/a'_0$ for $n = 1, 2, 3, 4$.

$$= \frac{1}{(81)^2 6\pi}\left(\frac{Z}{a'_0}\right)^3 \int_0^\infty \int_0^\pi \int_0^{2\pi} r^2 \left(\frac{Zr}{a'_0}\right)^4 \mathrm{e}^{-2Zr/3a'_0}$$
$$\times (3\cos^2\theta - 1)^2 r^2 \sin\theta \, \mathrm{d}r \mathrm{d}\theta \mathrm{d}\phi$$
$$= \frac{1}{3(81)^2}\left(\frac{a'_0}{Z}\right)^2 \int_0^\infty \rho^8 \mathrm{e}^{-2\rho/3} \mathrm{d}\rho \int_{-1}^1 (3\mu^2 - 1)^2 \mathrm{d}\mu$$
$$= \frac{8! a_0'^2}{3(81)^2 Z^2}\left(\frac{3}{2}\right)^9 \frac{8}{5}$$
$$= 126 \frac{a_0'^2}{Z^2}$$

so

$$(\Delta r) = \sqrt{\langle r^2\rangle - \langle r\rangle^2} = \sqrt{15.75} a'_0/Z = 3.97 a'_0/Z$$

so since $\langle r\rangle - r_{\max} = 1.5 a'_0/Z$, the two are within the width of the probability distribution.

▯

3.3.5.1 Orthogonality

The orthogonality of the Laguerre polynomials *cannot* be established by simple use of the generating function. On the other hand, however, it may be established from the summation in Equation (B.47) and the expression for I in Equation (B.48) that

$$\int_0^\infty \mathrm{e}^{-\rho} \rho^{s+1} L_r^s(\rho) L_{r+1}^s(\rho) \, \mathrm{d}\rho = -\frac{[(r+1)!]^2 r!}{(r-s)!} \tag{3.87}$$

so they are certainly not orthogonal in this sense. The difficulty is that in changing to dimensionless variables through the constant α_n, the change is different for each order, since α_n depends on n. The general proof of orthogonality comes from the earlier proof that eigenfunctions of an Hermitian operator with different eigenvalues are orthogonal [see Equation (1.57)].

Problem 3.15 Find the *most probable* value of r that will be found for an electron with $\ell = m = n - 1$. (*Hint*: The probability distribution function will be $W_r(r) = 2\pi r^2 \int_0^\pi |\psi_{n\ell m}|^2 \sin\theta \, \mathrm{d}\theta$.)

Problem 3.16 *Orthogonality and nonorthogonality.*

(a) Show explicitly that Equation (3.87) is satisfied for $L_2^1(\rho)$ and $L_3^1(\rho)$.

(b) Evaluate the corresponding orthogonality integral in r [$\rho = (2Z/3a'_0)r$ for R_{30} and $\rho = (Z/a'_0)r$ for R_{20}] and show that this does lead to orthogonality.

Problem 3.17 *Electrons inside the nucleus.* (*Hint:* You may approximate that the exponential factor is unity since $R \ll a_0'$.)

(a) If the radius of the nucleus is given approximately by $R = R_0 n^{1/3}$ with $R_0 = 1.3 \times 10^{-15}$ m, where n is the number of nucleons (protons + neutrons), find the probability that an electron in the innermost state (assume a one-electron atom in the ground state) is inside the nucleus of $^{238}U_{92}$.

(b) Repeat the calculation for tritium, $^{3}H_{1}$.

3.3.5.2 Quantum Numbers

The relationships between the quantum numbers, $|m| \leq \ell$, and $\ell \leq n - 1$, enable us to determine how many states there are for each value of ℓ or n. Since for each value of ℓ, $-\ell \leq m \leq \ell$, there are always $2\ell + 1$ possible m states for each value of ℓ. Then for each value of n, we sum over the possible values of ℓ and find

$$\sum_{\ell=0}^{n-1}(2\ell + 1) = n^2 \,. \tag{3.88}$$

If we then assume for the moment that we may put *two* electrons in each of these states, then the numbers of electrons in the various n-states are $2, 8, 18, 32, \ldots$, so the patterns of the periodic table are already apparent just from the quantum numbers.

Problem 3.18 Prove Equation (3.88).

Problem 3.19 *Zeros of* $R_{31}(r)$.

(a) From Equation (3.75), we deduce that $R_{31}(r)$ comes from $L_4^3(\rho)$. Calculate $L_4^3(\rho)$.

(b) $L_4^3(\rho)$ has a zero at $\rho = 4$ while $R_{31}(r)$ has a zero at $r = 6a_0'/Z$. Reconcile these results.

Problem 3.20 *Mean values of* r.

(a) Calculate $\langle r \rangle$ for the ψ_{100}, ψ_{210}, and ψ_{320} states and compare with the Bohr theory.

(b) Without calculating it, guess the value of $\langle r \rangle_{430}/r_4$.

3.4 Central Potentials

In addition to the Coulomb potential, there are a variety of other central potentials that are of interest, although some of them may only be solved numerically. Examples of such potentials are molecular potentials, which exhibit relatively weak attraction far away but strong repulsion when very close, and nuclear potentials, where neutrons are essentially particles in a spherical box and quarks appear to have a constant attractive force which leads to a linear potential, at least in the limit where the separation becomes large.

In treating these more general spherical potential problems it is useful to first note the identity

$$\frac{1}{r^2}\frac{\mathrm{d}}{\mathrm{d}r}\left(r^2\frac{\mathrm{d}R}{\mathrm{d}r}\right) \equiv \frac{1}{r}\frac{\mathrm{d}^2}{\mathrm{d}r^2}(rR)\,.$$

With this identity, we may write the general radial equation as

$$-\frac{\hbar^2}{2\mu}\frac{\mathrm{d}^2}{\mathrm{d}r^2}(rR) + \left[V(r) + \frac{\hbar^2}{2\mu}\frac{\ell(\ell+1)}{r^2}\right](rR) = E(rR)\,, \qquad (3.89)$$

where we have multiplied by r. Letting $u = rR$, this result may be written more conveniently as

$$u'' + \left[\frac{2\mu}{\hbar^2}(E - V) - \frac{\ell(\ell+1)}{r^2}\right]u = 0\,. \qquad (3.90)$$

3.4.1 Nuclear Potentials

The simplest model of neutrons is that they are particles in a box with a finite radius and finite but constant potential in the box. Another model is to consider the nucleus to be a water droplet, where there are no forces between particles except for the surface tension when they try to escape. These two models are similar, in that the potential inside is constant, but the former considers a rigid spherical box, while the water droplet model constrains the volume to remain fixed, but the surface can be deformed from a spherical shape. The latter case is outside the scope of this book, but the former, with $\ell = 0$, leads simply to

$$u'' + \frac{2\mu}{\hbar^2}[E - V(r)]u = 0\,, \quad \text{with} \quad V(r) = \begin{cases} \infty, & r < 0, \\ -V_0, & 0 \le r \le a, \\ 0, & r \ge a. \end{cases}$$

The nuclear potential sketched in Figure 3.3 for neutrons is nearly the same as the particle in a finite well as sketched in Figure 2.3, except that in this case the region $r < 0$ is excluded by an infinite potential barrier at the origin (to get to the other side, one goes around in *angle*, not by passing through the origin to $-r$), and outside the well the potential is taken to be zero so that outside the neutron is a free particle. Once a proton or alpha particle leaves the nucleus, however, it experiences a Coulomb repulsion, so its well is effectively deeper, but there is a finite chance of tunneling through the barrier if $E > 0$.

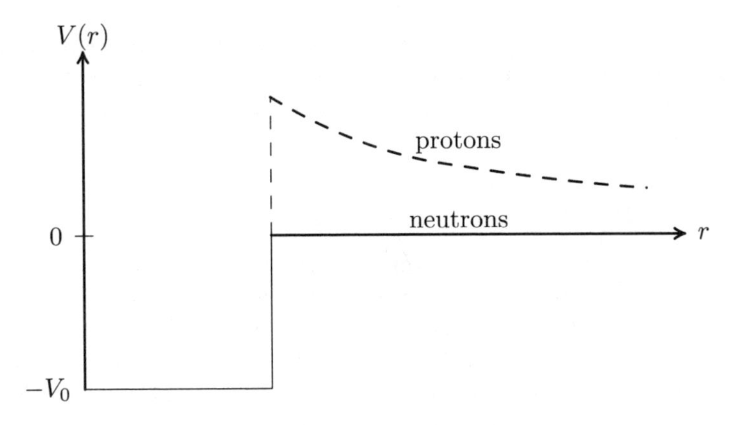

FIGURE 3.3
A nuclear potential well, showing the additional Coulomb potential for protons or alphas.

Problem 3.21 The neutron potential problem is one we have solved already, with a few observations. If you were to put an impenetrable barrier in the center of the rectangular well problem of Figure 2.3, relate the energies of this neutron well problem of Figure 3.3 to those of that problem. (*Hint:* The symmetry of the wave function, i.e., even or odd parity, and the boundary condition at the origin are important in making the connection between the problems.)

Problem 3.22 *Neutron potential well depth.* A neutron is in the ground state of a well as depicted in Figure 3.3. The lowest possible energy of a gamma ray that can remove the neutron from the well is 8.0 MeV. The radius of the well is $3.0 \cdot 10^{-15}$ m. Obtain the depth V_0 in MeV.

Problem 3.23 *Deuteron potential well depth.* The deuteron is the ground state of a bound proton and neutron in a potential well as depicted in Figure 3.3 (solid line since the neutron is neutral). Assuming that the neutron is just barely bound ($E = V_0 - \epsilon$, $\epsilon \sim 0$), estimate the depth of the potential well if the width is $r_0 \sim 10^{-15}$ m and the proton and neutron masses are each $m = 1.67 \times 10^{-27}$ kg. Express the depth in MeV (1 MeV $= 10^6$ eV or 1.602×10^{-13} J).

(a) Solve for the radial function $u(r) = rR(r)$ for $0 \leq r \leq r_0$ and for $r > r_0$. Match boundary conditions at r_0 so that both $u(r)$ and $u'(r)$ are continuous and assume $V = \infty$ for $r \leq 0$.

(b) From these boundary conditions, show that we find a relationship like Equation (2.21). Since there is only one bound state and since it is just barely bound, estimate the value of k' with $E \sim V_0$ and then estimate V_0 (in MeV).

Problem 3.24 *Spherical harmonic oscillator.* A particle is bound in a three dimensional symmetric harmonic oscillator well with $V = \frac{1}{2}Kr^2$. If it is in the the ground state with $\ell = 0$, find the normalized eigenfunction $R(r)$ and the energy eigenvalue.

3.4.2 Quarks and Linear Potentials

There are a number of physical problems where the force is constant, giving rise to linear potentials. Examples include uniform electric fields or gravitation near the earth's surface, but one of particular interest is the force between quarks. At least at large separations (but still less than the size of a nucleus), the force between two quarks appears to be constant and attractive. This force is due to "color charge" and entirely independent of the usual electrostatic forces and electric charge. We can then model the force between two color charges by $\boldsymbol{F} = -\nabla V = -g\boldsymbol{e}_r$ and the corresponding potential is $V(r) = gr - V_0$ where V_0 gives the potential at the bottom of the potential well. With this potential, the radial equation of Equation (3.90) becomes

$$u'' + \frac{2\mu}{\hbar^2}(E + V_0 - gr)u = 0\,.$$

If we change to dimensionless variables so that $\xi = \alpha[r - (E + V_0)/g]$, and set $\alpha^3 = 2\mu g/\hbar^2$, then we obtain

$$u'' - \xi u = 0\,. \qquad \text{Airy equation} \tag{3.91}$$

Equation (3.91) is called the Airy equation, and although deceptively simple in appearance, its solutions may not be represented in terms of elementary functions. The solutions can be related to Bessel functions of one-third integral order, but this is of little direct use. It has general properties such that the solutions are either exponentially growing or decaying for $\xi > 0$, while the solutions oscillate for $\xi < 0$. The two linearly independent solutions are usually designated $\mathrm{Ai}(\xi)$ and $\mathrm{Bi}(\xi)$, where $\mathrm{Ai}(\xi)$ is exponentially decaying for positive ξ and $\mathrm{Bi}(\xi)$ is exponentially growing for positive ξ, and is therefore inadmissible for our purposes, since we require bounded solutions. Many properties of the $\mathrm{Ai}(\xi)$ solution are evident in Figure 3.4.

While the study of Airy functions is generally very difficult, the solution can be written as an integral, such that

$$\mathrm{Ai}(\xi) \equiv \frac{1}{\pi} \lim_{\epsilon \to 0} \int_0^\infty \mathrm{e}^{-\epsilon u} \cos\left(\frac{u^3}{3} + \xi u\right) \, du. \tag{3.92}$$

Direct substitution into the Airy equation, interchanging the order of integration and differentiation, leads to

$$\frac{\mathrm{d}^2}{\mathrm{d}\xi^2}\pi \mathrm{Ai}(\xi) - \xi\pi\mathrm{Ai}(\xi)$$

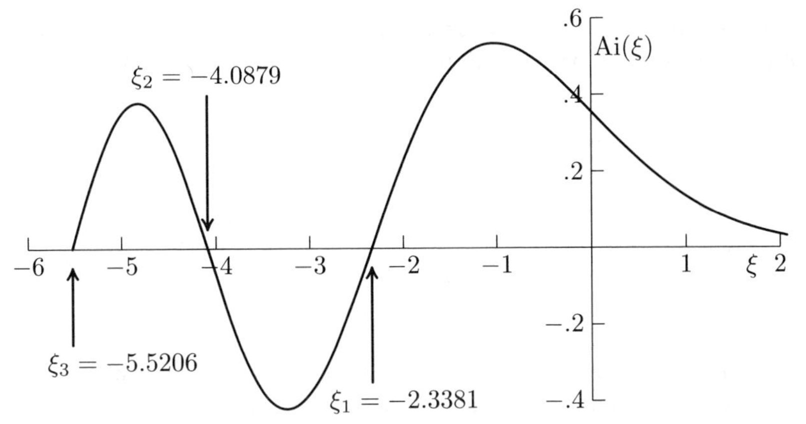

FIGURE 3.4
The Airy function, Ai(ξ), showing the first three zeros.

$$= \lim_{\epsilon\to 0}\int_0^\infty \mathrm{e}^{-\epsilon u}\left[\frac{\partial^2}{\partial\xi^2}\cos\left(\frac{u^3}{3}+\xi u\right) - \xi\cos\left(\frac{u^3}{3}+\xi u\right)\right] du$$
$$= -\lim_{\epsilon\to 0}\int_0^\infty \mathrm{e}^{-\epsilon u}(u^2+\xi)\cos\left(\frac{u^3}{3}+\xi u\right) du$$

$$= -\lim_{\epsilon\to 0}\int_0^\infty \mathrm{e}^{-\epsilon u}\frac{\partial}{\partial u}\sin\left(\frac{u^3}{3}+\xi u\right) du$$
$$= -\lim_{\epsilon\to 0}\left[\mathrm{e}^{-\epsilon u}\sin\left(\frac{u^3}{3}+\xi u\right)\right]_0^\infty - \lim_{\epsilon\to 0}\epsilon\int_0^\infty \mathrm{e}^{-\epsilon u}\sin\left(\frac{u^3}{3}+\xi u\right) du$$
$$= 0\,.$$

The only purpose of the $\mathrm{e}^{-\epsilon u}$ factor was to guarantee convergence as $u \to \infty$, and the factor is frequently omitted with the understanding that the contribution from the upper limit vanishes. From the form of the argument of the oscillatory term, it is clear that the oscillation frequency increases in the large argument limit so that the integral is summing rapidly oscillating contributions of substantially equal amplitudes, so it is evident that the integral will converge even without the exponential convergence factor.

For large arguments, i.e., as $|\xi| \to \infty$, the asymptotic forms are given by

$$\mathrm{Ai}(\xi) \sim \begin{cases} \dfrac{1}{2\sqrt{\pi}\xi^{1/4}}\mathrm{e}^{-2\xi^{3/2}/3}\,, & \xi \to +\infty\,, \\ \dfrac{1}{\sqrt{\pi}(-\xi)^{1/4}}\sin\left[2(-\xi)^{3/2}/3+\pi/4\right], & \xi \to -\infty\,. \end{cases} \tag{3.93}$$

These expressions show the rapid convergence for positive ξ, as well as the oscillatory character with a slowly decreasing amplitude for negative ξ. While this is not an elementary function, it is a tabulated function [1].

Returning to our quark potential problem, we note that at $r = 0$ we require $u = 0$, but in terms of the variable ξ, we must have $u(\xi_n) = 0$ where

$$\xi_n = \left(0 - \frac{E_n + V_0}{g}\right)\left(\frac{2\mu g}{\hbar^2}\right)^{1/3},$$

and ξ_n is one of the zeros of the Airy function. Hence for the energy eigenvalues, we have

$$E_n = |\xi_n|\left(\frac{g^2\hbar^2}{2\mu}\right)^{1/3} - V_0\,. \tag{3.94}$$

If our model of the quark force and potential are reasonable approximations to the real forces and potentials, then we should be able to find systems of two quarks (a quark and an antiquark form a meson) and our candidate is the J/ψ meson made of a pair of charmed quarks. This meson has more massive quarks than those found in protons and neutrons and hence the relativistic corrections are expected to be smaller and provide a better test of the model than a pair of up or down quarks that form π mesons. When the mass of the two quarks is equal, the reduced mass is just half the mass of either quark, and the system is analogous to positronium, which is a bound system of a positron and an electron. The only thing we can measure about the meson is its mass in its lowest and some of its first few excited states. If we call the total rest energy of the meson M_nc^2, then this energy is comprised of

$$M_nc^2 = 2Mc^2 + |\xi_n|\left(\frac{g^2\hbar^2}{M}\right)^{1/3} - V_0\,, \tag{3.95}$$

where the total energy is made up of the rest energy of the two charmed quarks with mass M plus the potential energy. We do not know M, g or V_0, but there are really only two unknown constants, $2Mc^2 - V_0$ and $g^2\hbar^2/M$. By doing a least squares fit to the first four energy levels of charmonium (the ground state of charmonium is the J/ψ), which are 3.097 GeV, 3.686 GeV, 4.1 GeV, and 4.414 GeV, with

$$M_nc^2 = A + B|\xi_n|\,,$$

we find $A = 2.42$ GeV and $B = 0.300$ GeV. The fit with just these two parameters leads to values of M_nc^2 that are within about 1% or better of the measured values. That one can fit all four levels within a percent with only two constants indicates that the model has some validity, especially when one considers that relativity, spin and other effects are neglected.

Problem 3.25 *Charmonium energy levels.*

(a) Find the percent error for the M_nc^2 for each of the first four states between the calculated and measured values.

(b) With $Mc^2 = 1.84$ GeV, evaluate V_0 in GeV and g in GeV/f, where f ($\equiv 10^{-15}$m) is a Fermi.

(c) If one considers that the classical turning point occurs when the kinetic energy vanishes, estimate the classical radius of the turning point in Fermis.

(d) Using half this classical radius value (the average is approximately half the maximum), estimate the energy of charmonium when $\ell = 1$ by adding the additional potential term in Equation (3.89) to the ground state energy. Compare this value with the average of the measured values of the first three charmonium values with $\ell = 1$, which are 3.413 GeV, 3.508 GeV, and 3.554 GeV.

Problem 3.26 *Charmonium levels — Bohr model.* Solve the Charmonium problem using the Bohr method, where $L = n\hbar$ for the angular momentum. The answer may be written in the form

$$E_n^{(B)} + V_0 = aE_0 n^{2/3}\,, \tag{3.96}$$

where $E_0 \equiv (g^2\hbar^2/2\mu)^{1/3}$ so that the solution from Equation (3.94) is written

$$E_n + V_0 = E_0|\xi_n|\,.$$

(a) Find a in Equation (3.96).

(b) Find the percent error for each of the two lowest states (percentage $E_n^{(B)}$ is *below* E_n).

3.4.3 Potentials for Diatomic Molecules*

When two neutral atoms approach one another, the weak interactions between the electrons cause slight rearrangements of the charge distribution that leads to an attractive force which varies as r^{-6}. On the other hand, when atoms get very close so that there is a lot of overlapping, there is a strong repulsion from the nuclei of the atomic cores, and there is therefore some equilibrium distance where there is a potential minimum. There is a variety of models for this molecular potential, but the first one we note is of the form

$$V(r) = \frac{A}{r^{12}} - \frac{B}{r^6}\,. \tag{3.97}$$

The r^{-12} term represents the repulsive core while the r^{-6} term dominates at large distance and gives the attraction that draws the atoms into a molecular configuration. This potential is difficult to solve directly, but if one only needs to know the energy levels of the first few states, we can expand about the minimum, and represent the potential as

$$V(r) = \tfrac{1}{2}k(r - r_0)^2 - V_0 + \mathcal{O}(r^3)\,. \tag{3.98}$$

If we restrict our attention to the lowest lying states where $\ell = 0$, then this is just the harmonic oscillator potential, where

$$u'' + \frac{2\mu}{\hbar^2}[E + V_0 - \tfrac{1}{2}k(r - r_0)^2]u = 0\,.$$

The energy eigenvalues of this problem are then given simply by

$$E_n = (n + \tfrac{1}{2})\hbar\omega - V_0\,, \tag{3.99}$$

where $\omega^2 = k/\mu$.

Another model potential for a diatomic molecule is called the Morse potential, and is given by

$$V(r) = V_0[1 - \mathrm{e}^{-a(r-r_0)}]^2 - V_0\,. \tag{3.100}$$

This potential has the property of having a minimum of $-V_0$ at $r = r_0$, and goes to zero exponentially as $r \to \infty$. The potential at the origin is finite, whereas it should be infinite, but this is generally insignificant since the potential is quite large at the origin. With this potential, we will keep all the terms in Equation (3.90), obtaining

$$u'' + \left\{\frac{2\mu}{\hbar^2}[E + 2V_0\mathrm{e}^{-a(r-r_0)} - V_0\mathrm{e}^{-2a(r-r_0)}] - \frac{\ell(\ell+1)}{r^2}\right\}u = 0\,. \tag{3.101}$$

At this point, we change variables to

$$y = \mathrm{e}^{-a(r-r_0)}\,,$$

and define $A \equiv \hbar^2\ell(\ell+1)/2\mu r_0^2$. The radial equation then becomes

$$\frac{\mathrm{d}^2u}{\mathrm{d}y^2} + \frac{1}{y}\frac{\mathrm{d}u}{\mathrm{d}y} + \frac{2\mu}{\hbar^2a^2}\left(\frac{E}{y^2} + \frac{2V_0}{y} - V_0 - \frac{Ar_0^2}{y^2r^2}\right)u = 0\,. \tag{3.102}$$

We then approximate the r_0^2/r^2 term by expanding about $y = 1$ (which is equivalent to expanding r about r_0) and find that

$$\frac{r_0^2}{r^2} = \frac{1}{\left(1 - \frac{\ln y}{r_0a}\right)^2} = 1 + \frac{2}{r_0a}(y-1) + \left(-\frac{1}{r_0a} + \frac{3}{r_0^2a^2}\right)(y-1)^2 + \mathcal{O}[(y-1)^3]$$

so that Equation (3.102) becomes

$$\frac{\mathrm{d}^2u}{\mathrm{d}y^2} + \frac{1}{y}\frac{\mathrm{d}u}{\mathrm{d}y} + \frac{2\mu}{\hbar^2a^2}\left(\frac{E - c_0}{y^2} + \frac{2V_0 - c_1}{y} - V_0 - c_2\right)u = 0\,, \tag{3.103}$$

where

$$c_0 = A\left(1 - \frac{3}{r_0a} + \frac{3}{r_0^2a^2}\right),$$

$$c_1 = A\left(\frac{4}{r_0a} - \frac{6}{r_0^2a^2}\right),$$

$$c_2 = A\left(-\frac{1}{r_0a} + \frac{3}{r_0^2a^2}\right).$$

We could factor out the asymptotic form and examine the resulting power series, but if we simply assume a solution of the form (which is the same form as the radial solution for the one-electron atom)

$$u = \mathrm{e}^{-\alpha z} z^m F(z)\,,$$

with $z = \beta y$, then Equation (3.103) can be written as

$$F'' + \left(\frac{2m+1}{z} - 2\alpha\right) F' + \left[\frac{m^2 + \frac{2\mu}{\hbar^2 a^2}(E - c_0)}{z^2} + \frac{\frac{2\mu(2V_0 - c_1)}{\hbar^2 a^2 \beta} - \alpha(2m+1)}{z} + \alpha^2 - \frac{2\mu(V_0 + c_2)}{\beta^2 \hbar^2 a^2}\right] F = 0\,. \qquad (3.104)$$

Then if we choose

$$\begin{aligned} \alpha &= \tfrac{1}{2}\,, \\ \beta^2 &= \frac{8\mu}{\hbar^2 a^2}(V_0 + c_2)\,, \\ m^2 &= \frac{2\mu}{\hbar^2 a^2}(c_0 - E)\,, \end{aligned}$$

Equation (3.104) can be written simply as

$$zF'' + (2m + 1 - z)F' + nF = 0\,,$$

which is the associated Laguerre equation [which is not surprising since the form of $u(z)$ was the same as $R(\rho)$] if we let

$$n = \frac{2\mu}{\hbar^2 a^2 \beta}(2V_0 - c_1) - m - \tfrac{1}{2}\,, \qquad (3.105)$$

so that with $2m = k$ and n an integer, $F = L^k_{n+k}(z)$. Rearranging Equation (3.105) and eliminating m, we find

$$\left[n + \tfrac{1}{2} - \frac{2\mu}{\hbar^2 a^2 \beta}(2V_0 - c_1)\right]^2 = m^2 = \frac{2\mu}{\hbar^2 a^2}(c_0 - E)\,,$$

and solving for the energy, we find

$$E_n = c_0 - \frac{\hbar^2 a^2}{2\mu}(n + \tfrac{1}{2})^2 + \frac{\hbar a(2V_0 - c_1)}{\sqrt{2\mu(V_0 + c_2)}}(n + \tfrac{1}{2}) - \frac{(2V_0 - c_1)^2}{4(V_0 + c_2)}\,. \qquad (3.106)$$

Expanding in the smallness of c_1/V_0 and c_2/V_0, with $\omega \equiv a\sqrt{2V_0/\mu}$, the result may be expressed as

$$E_n = \hbar\omega(n + \tfrac{1}{2})\left[1 - \frac{3\hbar^2 C}{2IV_0}\ell(\ell+1) - \frac{\hbar\omega}{4V_0}(n + \tfrac{1}{2})\right] - V_0 + \frac{\hbar^2}{2I}\ell(\ell+1)\,, \qquad (3.107)$$

where $I = \mu r_0^2$ is the rotational moment of inertia and $C = 1/ar_0 - 1/a^2r_0^2$.

From a comparison of Equation (3.107) to Equation (3.99), it is apparent that with $\ell = 0$, the solutions are nearly identical except for a term of the order of $\hbar\omega/V_0$ and the inclusion of angular momentum terms. We will later be able to show that including cubic and quartic terms in the expansion of Equation (3.98) will bring the two results even closer.

Problem 3.27 For the potential of Equation (3.97),

(a) Find r_0, k, and V_0 in Equation (3.98) in terms of A and B.

(b) Graph the potentials of Equations (3.97) and (3.98) on the same graph, matching the values of V_0, r_0, and k for the two cases.

(c) If, for a particular molecule, $V_0 = 2$ eV and $r_0 = 1.5 \cdot 10^{-10}$ m, find A, B, and $\frac{1}{2}\hbar\omega$ (in eV) for a hydrogen molecule.

Problem 3.28 Fill in the missing steps starting with Equation (3.90) with the potential from Equation (3.100) and ending with Equation (3.107).

4

Total Angular Momentum

The concept and much of the mathematical treatment of orbital angular momentum was treated in Chapter 3. An unexpected result from that treatment is that using the result of Equation (3.41) along with the result of Equation (3.48), we find that

$$m_T - m_B = 2\ell = n\,, \tag{4.1}$$

so ℓ is either an integer or *a half integer*! The integers correspond to orbital angular momentum, while the half-integer values correspond to **intrinsic angular momentum**, or **spin angular momentum**. There is no classical model for spin angular momentum, but the most fundamental particles (electrons, quarks, protons, and neutrons) all exhibit this type of intrinsic angular momentum.

4.1 Orbital and Spin Angular Momentum

The coordinates for orbital angular momentum are just the coordinates of 3-space, but the variables for intrinsic angular momentum are spin variables, which we shall examine more closely later in this chapter, but it is important to note that these spaces are independent so that spin angular momentum operators will invariably commute with orbital angular momentum operators, since they have independent variables. It is common to use the notation $\hat{\boldsymbol{L}}$ for orbital angular momentum operators only, and $\hat{\boldsymbol{S}}$ for spin angular momentum, and introduce $\hat{\boldsymbol{J}}$ for the total angular momentum operator, where $\hat{\boldsymbol{J}} = \hat{\boldsymbol{L}} + \hat{\boldsymbol{S}}$. The eigenvalues for the total angular momentum are given by

$$\begin{aligned}
\hat{J}^2\psi_{j,m} &= j(j+1)\hbar^2\psi_{j,m} \\
\hat{J}_z\psi_{j,m} &= m\hbar\psi_{j,m} \\
\hat{J}_\pm\psi_{j,m} &= N^{\pm}_{j,m}\psi_{j,m\pm 1}\,.
\end{aligned}$$

Since $\hat{\boldsymbol{J}}$ is still an angular momentum operator, it follows the same rules as orbital angular momentum, so that $-j \leq m \leq j$, and the normalization constants are the same with j substituting for ℓ. For this case, however, j and m can be either integers ($j = 0, 1, 2, \ldots$) or half integers ($j = \frac{1}{2}, \frac{3}{2}, \frac{5}{2}, \ldots$).

4.1.1 Eigenfunctions of $\hat{J}_x$ and $\hat{J}_y$

In general, a set of all eigenfunctions with a fixed value for the total angular momentum is a complete set of functions for that value, so that every possible eigenfunction with that value for the total angular momentum can be represented as a linear combination of the members of the set. In particular, we know there are $2j+1$ different eigenfunctions for a fixed value of j, and we may represent the members of this set by the different values of m, $-j \le m \le j$, so the complete set is (changing to Dirac notation, where the wave functions are represented by either a ket or a bra and their eigenvalues)

$$|j,-j\rangle,\ |j,-j+1\rangle, \cdots, |j,j-1\rangle, |j,j\rangle\,.$$

Since every possible eigenfunction can be represented by a linear combination of the members this set, it must be possible to construct the eigenfunctions of J_x and J_y from this set. We may write this as

$$\hat{J}_x\chi = \alpha\hbar\chi \tag{4.2}$$

$$\chi = \sum_{m=-j}^{j} a_m|j,m\rangle\,,$$

where the a_m are constants to be determined. The raising and lowering operators for total angular momentum are

$$\begin{aligned}\hat{J}_+ &= \hat{J}_x + \mathrm{i}\hat{J}_y\\ \hat{J}_- &= \hat{J}_x - \mathrm{i}\hat{J}_y\end{aligned}$$

in analogy to the raising and lowering operators for orbital angular momentum on page 82. Adding these and solving for $\hat{J}_x$, we may write $\hat{J}_x$ as

$$\hat{J}_x = \tfrac{1}{2}(\hat{J}_+ + \hat{J}_-)\,,$$

so that Equation (4.2) becomes

$$\begin{aligned}\tfrac{1}{2}(\hat{J}_+ + \hat{J}_-)\sum_{m=-j}^{j} a_m|j,m\rangle &= \tfrac{1}{2}\sum_{m=-j}^{j-1} a_m N^+_{j,m}|j,m+1\rangle\\ &\quad+\tfrac{1}{2}\sum_{m=-j+1}^{j} a_m N^-_{j,m}|j,m-1\rangle\\ &= \alpha\hbar\sum_{m=-j}^{j} a_m|j,m\rangle\,,\end{aligned}$$

since $\hat{J}_+|j,j\rangle = 0$ and $\hat{J}_-|j,-j\rangle = 0$, and α is a dimensionless constant to be determined. If we multiply on the left by $\langle j,k|$ and use the orthogonality relations,

$$\langle j,k|j,m\rangle = \delta_{k,m} \equiv \begin{cases}1 & k=m\,,\\ 0 & k\neq m\,,\end{cases}$$

we find

$$\tfrac{1}{2}(a_{k-1}N^{+}_{j,k-1} + a_{k+1}N^{-}_{j,k+1}) = \alpha\hbar a_k \,,$$

or

$$\alpha a_k = \tfrac{1}{2}[a_{k-1}\sqrt{j(j+1)-k(k-1)} + a_{k+1}\sqrt{j(j+1)-k(k+1)}] \,. \qquad (4.3)$$

Example 4.1

$j = 1$ *case.* We first consider the case for $j = 1$, so $k = -1, 0, 1$. Thus

$$\begin{aligned} \alpha a_{-1} &= \frac{a_0}{\sqrt{2}} \\ \alpha a_0 &= \frac{1}{\sqrt{2}}(a_1 + a_{-1}) \\ \alpha a_1 &= \frac{a_0}{\sqrt{2}} \,, \end{aligned}$$

which may be written in matrix form as

$$\begin{pmatrix} 0 & \frac{1}{\sqrt{2}} & 0 \\ \frac{1}{\sqrt{2}} & 0 & \frac{1}{\sqrt{2}} \\ 0 & \frac{1}{\sqrt{2}} & 0 \end{pmatrix} \begin{pmatrix} a_1 \\ a_0 \\ a_{-1} \end{pmatrix} = \alpha \begin{pmatrix} a_1 \\ a_0 \\ a_{-1} \end{pmatrix} .$$

Subtracting the term on the right from both sides, this becomes

$$\begin{pmatrix} -\alpha & \frac{1}{\sqrt{2}} & 0 \\ \frac{1}{\sqrt{2}} & -\alpha & \frac{1}{\sqrt{2}} \\ 0 & \frac{1}{\sqrt{2}} & -\alpha \end{pmatrix} \begin{pmatrix} a_1 \\ a_0 \\ a_{-1} \end{pmatrix} = 0 \,.$$

In order to have a nontrivial solution, we require the determinant of coefficients to vanish, or

$$\begin{vmatrix} -\alpha & \frac{1}{\sqrt{2}} & 0 \\ \frac{1}{\sqrt{2}} & -\alpha & \frac{1}{\sqrt{2}} \\ 0 & \frac{1}{\sqrt{2}} & -\alpha \end{vmatrix} = \alpha(1-\alpha^2) = 0 \,.$$

The roots of this are easily found to be $0, \pm 1$, so the eigenvalues of $\hat{J}_x$ are $0, \pm\hbar$. For $\alpha = 0$, it is clear that $a_0 = 0$, so that $a_1 = -a_{-1}$. Normalizing, $a_1^2 + a_{-1}^2 = 2a_1^2 = 1$, so $a_1 = 1/\sqrt{2}$. For $\alpha = 1$,

$$\begin{aligned} a_1 &= \frac{a_0}{\sqrt{2}} \\ a_0 &= \frac{1}{\sqrt{2}}(a_1 + a_{-1}) \\ a_{-1} &= \frac{a_0}{\sqrt{2}} \,. \end{aligned}$$

Normalizing leads to $a_0 = 1/\sqrt{2}$, $a_{\pm 1} = \frac{1}{2}$. For $\alpha = -1$,

$$\begin{aligned} -a_1 &= \frac{a_0}{\sqrt{2}} \\ -a_0 &= \frac{1}{\sqrt{2}}(a_1 + a_{-1}) \\ -a_{-1} &= \frac{a_0}{\sqrt{2}}, \end{aligned}$$

so $a_0 = 1/\sqrt{2}$, $a_{\pm 1} = -\frac{1}{2}$. We summarize these results by writing

$$\begin{aligned} \chi_1 &= \tfrac{1}{2}|1,1\rangle + \tfrac{1}{\sqrt{2}}|1,0\rangle + \tfrac{1}{2}|1,-1\rangle \\ \chi_0 &= \tfrac{1}{\sqrt{2}}(|1,1\rangle - |1,-1\rangle) \\ \chi_{-1} &= -\tfrac{1}{2}|1,1\rangle + \tfrac{1}{\sqrt{2}}|1,0\rangle - \tfrac{1}{2}|1,-1\rangle . \end{aligned}$$

□

Problem 4.1 *Eigenfunctions for $\hat{J}_y$.*

(a) Find the eigenfunctions for $\hat{J}_y$ for $j = 1$.

(b) Find the eigenvalues and eigenfunctions for $\hat{J}_y$ for $j = \frac{3}{2}$.

4.2 Half-Integral Spin Angular Momentum

Half-integral spin is a nonclassical phenomenon that cannot be described by simple operators in configuration space. From Section 3.3.3.1, orbital angular momentum can be represented by operators in spherical or rectangular coordinates, but the eigenvalues are all integers, and there is no room in that space for half-integral quantum numbers. For the present, we shall proceed blithely ahead as if this were no problem, examining the implications of a space where $j = s = \frac{1}{2}$, and the angular momentum operator is designated $\hat{\boldsymbol{S}}$ with $\hat{S}^2$ and $\hat{S}_z$ representing the total spin angular momentum and its z-component.

If $s = \frac{1}{2}$, there are only two states where $m_s = \pm\frac{1}{2}$. We shall label these two spin functions (or **spinors**) $\chi_\pm$ so that they satisfy

$$\hat{S}^2 \chi_\pm = \tfrac{3}{4}\hbar^2 \chi_\pm \tag{4.4}$$

$$\hat{S}_z \chi_\pm = \pm\tfrac{1}{2}\hbar \chi_\pm . \tag{4.5}$$

We may also denote the spin functions by $\chi_+ = |\uparrow\rangle$ and $\chi_- = |\downarrow\rangle$.

Because these are angular momentum operators, they satisfy the commutation relations

$$[\hat{S}_x, \hat{S}_y] = \mathrm{i}\hbar\hat{S}_z\,, \qquad [\hat{S}_y, \hat{S}_z] = \mathrm{i}\hbar\hat{S}_x\,, \qquad [\hat{S}_z, \hat{S}_x] = \mathrm{i}\hbar\hat{S}_y\,.$$

We similarly have the raising and lowering operators, so that

$$\hat{S}_+ = \hat{S}_x + \mathrm{i}\hat{S}_y\,, \qquad \hat{S}_- = \hat{S}_x - \mathrm{i}\hat{S}_y\,,$$

and in terms of these, we may write

$$\hat{S}_x = \tfrac{1}{2}(\hat{S}_+ + \hat{S}_-)\,, \qquad \hat{S}_y = \tfrac{1}{2\mathrm{i}}(\hat{S}_+ - \hat{S}_-)\,.$$

The normalization constants associated with $\hat{S}_\pm$ are almost trivial in this case, since

$$\hat{S}_+\chi_+ = 0 \tag{4.6}$$

$$\hat{S}_+\chi_- = N^+_{\frac{1}{2},-\frac{1}{2}}\chi_+ = \hbar\chi_+ \tag{4.7}$$

$$\hat{S}_-\chi_+ = N^-_{\frac{1}{2},\frac{1}{2}}\chi_- = \hbar\chi_- \tag{4.8}$$

$$\hat{S}_-\chi_- = 0\,, \tag{4.9}$$

since

$$N^+_{\frac{1}{2},-\frac{1}{2}} = \hbar\sqrt{\tfrac{1}{2}(\tfrac{1}{2}+1) + \tfrac{1}{2}(-\tfrac{1}{2}+1)} = \hbar\,,$$

$$N^-_{\frac{1}{2},\frac{1}{2}} = \hbar\sqrt{\tfrac{1}{2}(\tfrac{1}{2}+1) - \tfrac{1}{2}(\tfrac{1}{2}-1)} = \hbar$$

and $N^+_{\frac{1}{2},\frac{1}{2}} = N^-_{\frac{1}{2},-\frac{1}{2}} = 0$. From these relations, we can immediately see that

$$\begin{aligned} \hat{S}_x\chi_+ &= \tfrac{1}{2}\hbar\chi_-\,, & \hat{S}_x\chi_- &= \tfrac{1}{2}\hbar\chi_+\,, \\ \hat{S}_y\chi_+ &= \tfrac{\mathrm{i}}{2}\hbar\chi_-\,, & \hat{S}_y\chi_- &= -\tfrac{\mathrm{i}}{2}\hbar\chi_+\,. \end{aligned} \tag{4.10}$$

Furthermore, since χ_+ and χ_- span the space, any function in the space must be a linear combination of these, so we may write

$$\chi = C_+\chi_+ + C_-\chi_-$$

for any state. This state must be normalized, so that

$$\langle\chi|\chi\rangle = 1\,.$$

The normalization implies

$$|C_+|^2 + |C_-|^2 = 1\,, \tag{4.11}$$

since the basis functions are themselves normalized and orthogonal, $\langle\chi_+|\chi_+\rangle = \langle\chi_-|\chi_-\rangle = 1$, and $\langle\chi_+|\chi_-\rangle = 0$. We also have the prescription for expectation values

$$\langle\Omega\rangle = \langle\chi|\hat{\Omega}|\chi\rangle\,.$$

For example, if we choose $\chi = \chi_-$, then

$$\langle S_z\rangle = \langle\chi_-|\hat{S}_z|\chi_-\rangle = \langle\chi_-| - \tfrac{1}{2}\hbar|\chi_-\rangle = -\tfrac{1}{2}\hbar\,,$$
$$\langle S_x\rangle = \langle\chi_-|\hat{S}_x|\chi_-\rangle = \langle\chi_-|\hbar|\chi_+\rangle = 0\,.$$

Example 4.2

Eigenfunctions and eigenvalues for $\hat{S}_x$. If we want the eigenfunctions and eigenvalues for $\hat{S}_x$, then

$$\begin{aligned}\hat{S}_x(C_+\chi_+ + C_-\chi_-) &= \tfrac{1}{2}\hbar(C_+\chi_- + C_-\chi_+)\\ &= \alpha\hbar(C_+\chi_+ + C_-\chi_-)\,,\end{aligned}$$

where the first line uses Equation (4.10) and the second line is the eigenvalue expression. Since χ_+ and χ_- are independent, we must have

$$\tfrac{1}{2}C_+ - \alpha C_- = 0 \quad \text{and} \quad \tfrac{1}{2}C_- - \alpha C_+ = 0\,,$$

so $4\alpha^2 = 1$ and $\alpha = \pm\frac{1}{2}$. For $\alpha = \frac{1}{2}$, $C_+ = C_-$, while for $\alpha = -\frac{1}{2}$, $C_- = -C_+$. From Equation (4.11), this leads to

$$\begin{aligned}\chi &= \tfrac{1}{\sqrt{2}}(\chi_+ + \chi_-)\,, \qquad & \alpha &= \tfrac{1}{2}\,,\\ \chi &= \tfrac{1}{\sqrt{2}}(\chi_+ - \chi_-)\,, \qquad & \alpha &= -\tfrac{1}{2}\,.\end{aligned}$$

□

It is clear that any quantity involving these spin operators and eigenfunctions can be obtained *without any kind of representation*, as we have not needed to say anything about the nature of χ_+ or χ_- in making the above calculations. It is *convenient*, however, to introduce a representation, and a particularly simple one is

$$\chi_+ = \begin{pmatrix}1\\0\end{pmatrix}, \; \chi_- = \begin{pmatrix}0\\1\end{pmatrix}, \; \chi = C_+\begin{pmatrix}1\\0\end{pmatrix} + C_-\begin{pmatrix}0\\1\end{pmatrix} = \begin{pmatrix}C_+\\C_-\end{pmatrix}. \tag{4.12}$$

We then use the rules of matrix manipulation to establish the other results. The orthogonality condition then becomes

$$\langle\chi_+|\chi_-\rangle = (1,0)\begin{pmatrix}0\\1\end{pmatrix} = 0\,, \qquad \langle\chi_-|\chi_+\rangle = (0,1)\begin{pmatrix}1\\0\end{pmatrix} = 0\,.$$

In order to obtain an explicit form for the spin operators, we introduce the **Pauli spin matrices** through the relation

$$\hat{\mathbf{S}} = \tfrac{1}{2}\hbar\hat{\boldsymbol{\sigma}}\,, \tag{4.13}$$

with components

$$\hat{S}_x = \tfrac{1}{2}\hbar\hat{\sigma}_x\,, \qquad \hat{S}_y = \tfrac{1}{2}\hbar\hat{\sigma}_y\,, \qquad \hat{S}_z = \tfrac{1}{2}\hbar\hat{\sigma}_z\,.$$

Then, for Equation (4.10), we have

$$\hat{S}_x\chi_+ = \tfrac{1}{2}\hbar\chi_-\,, \implies \hat{\sigma}_x\chi_+ = \chi_-\,,$$

which, when the $\hat{\sigma}$ operators are written in terms of 2×2 matrices with as yet undetermined elements, we may write this last result as

$$\begin{pmatrix} \sigma_{x11} & \sigma_{x12} \\ \sigma_{x21} & \sigma_{x22} \end{pmatrix} \begin{pmatrix} 1 \\ 0 \end{pmatrix} = \begin{pmatrix} 0 \\ 1 \end{pmatrix}, \implies \begin{matrix} \sigma_{x11} = 0\,, \\ \sigma_{x21} = 1\,. \end{matrix}$$

and the other case leads to

$$\hat{S}_x\chi_- = \tfrac{1}{2}\hbar\chi_+\,, \implies \hat{\sigma}_x\chi_- = \chi_+\,,$$

or

$$\begin{pmatrix} \sigma_{x11} & \sigma_{x12} \\ \sigma_{x21} & \sigma_{x22} \end{pmatrix} \begin{pmatrix} 0 \\ 1 \end{pmatrix} = \begin{pmatrix} 1 \\ 0 \end{pmatrix}, \implies \begin{matrix} \sigma_{x12} = 1\,, \\ \sigma_{x22} = 0\,. \end{matrix}$$

Therefore, this spin matrix may be written as

$$\hat{\sigma}_x = \begin{pmatrix} 0 & 1 \\ 1 & 0 \end{pmatrix}. \tag{4.14}$$

Problem 4.2 *Spin matrices.* Use a similar approach to show that

$$\hat{\sigma}_y = \begin{pmatrix} 0 & -\mathrm{i} \\ \mathrm{i} & 0 \end{pmatrix}. \tag{4.15}$$

and

$$\hat{\sigma}_z = \begin{pmatrix} 1 & 0 \\ 0 & -1 \end{pmatrix}. \tag{4.16}$$

Problem 4.3 *Operators for $\hat{S}_+$ and $\hat{S}_-$.*

(a) Use Equation (4.13) to find the 2×2 matrices for $\hat{S}_+$ and $\hat{S}_-$.

(b) Using these matrix forms, prove Equations (4.6) through (4.9).

Problem 4.4 *Eigenfunctions of $\hat{S}_x$ and $\hat{S}_y$.* Find the column vector representations for the eigenfunctions of $\hat{S}_x$ (denoted as $|\rightarrow\rangle$ and $|\leftarrow\rangle$) and $\hat{S}_y$ (denoted as $|\nearrow\rangle$ and $|\swarrow\rangle$).

Partial ans. $|\rightarrow\rangle = \dfrac{1}{\sqrt{2}}\begin{pmatrix} 1 \\ 1 \end{pmatrix}$.

Example 4.3
Spin state probabilities. A particle is initially in the $|\uparrow\rangle$ state when a measurement is made of the x-component of the spin. What is the probability that the particle will be found to be in the $|\rightarrow\rangle$ state?

The probability is given by the magnitude squared of the matrix element

$$\begin{aligned} P &= |\langle \rightarrow | \uparrow\rangle|^2 \\ &= \left| \frac{1}{\sqrt{2}}(1 \;\; 1) \begin{pmatrix} 1 \\ 0 \end{pmatrix} \right|^2 \\ &= \left| \frac{1}{\sqrt{2}}(1) \right|^2 \\ &= \frac{1}{2} \end{aligned}$$

□

Problem 4.5 *Basic properties of the Pauli spin matrices.*

(a) Show that

$$\hat{\sigma}_x^2 = \hat{\sigma}_y^2 = \hat{\sigma}_z^2 = \mathsf{I} = \begin{pmatrix} 1 & 0 \\ 0 & 1 \end{pmatrix} .$$

(b) Write $\hat{\mathbf{S}}$ as a matrix, and show that

$$\hat{S}^2 \chi = \tfrac{3}{4}\hbar^2 \chi ,$$

for any χ.

(c) Show explicitly that

$$[\hat{S}_x, \hat{S}_y] = \mathrm{i}\hbar \hat{S}_z ,$$

and find the corresponding commutator relations for the Pauli matrices. (i.e., $\hat{\boldsymbol{\sigma}} \times \hat{\boldsymbol{\sigma}} = ?$).

Problem 4.6 *Spin $\frac{3}{2}$ system.* When the total spin is $S = \frac{3}{2}$ instead of $\frac{1}{2}$, there are four states instead of two, $\chi_{3/2}$, $\chi_{1/2}$, $\chi_{-1/2}$, and $\chi_{-3/2}$ where the label refers to M_S.

(a) Work out the expressions corresponding to Equation (4.10) for each of the four spin wave functions. (Remember that $\hat{S}_\pm$ are entirely equivalent to $\hat{L}_\pm$ and $\hat{J}_\pm$ except for the symbol since all are angular momentum operators and follow the same rules.)

(b) Representing the four spin wave functions by the column vectors,

$$\chi_{3/2} = \begin{pmatrix} 1 \\ 0 \\ 0 \\ 0 \end{pmatrix}, \; \chi_{1/2} = \begin{pmatrix} 0 \\ 1 \\ 0 \\ 0 \end{pmatrix}, \; \chi_{-/12} = \begin{pmatrix} 0 \\ 0 \\ 1 \\ 0 \end{pmatrix}, \; \chi_{-3/2} = \begin{pmatrix} 0 \\ 0 \\ 0 \\ 1 \end{pmatrix},$$

and using Equation (4.13), find the 4×4 matrices representing $\hat{S}_x$, $\hat{S}_y$, and $\hat{S}_z$.

Problem 4.7 *Spin state probability.* A particle is initially in the mixed spin state $|\psi\rangle = A(|\uparrow\rangle + 2\mathrm{i}|\downarrow\rangle)$.

(a) Find $|A|$.

(b) After a measurement of the x-component of the spin, find the probability that the particle will be in the $|\leftarrow\rangle$ state.

4.3 Addition of Angular Momenta

Whenever we consider angular momentum, it is always more complicated than we have indicated thus far because even in a one-electron atom, the electron has both orbital angular momentum *and* spin angular momentum, and so far we have treated them separately whereas the total angular momentum is a combination of both. In more complicated atoms, the total angular momentum is the sum over all of the electrons (plus that of the nucleus which generally has spin angular momentum), and the wave functions are products of all the individual electron wave functions, which are themselves products of space and spin wave functions. This can get very complicated in many-electron atoms. The first step in understanding what effects these considerations have on atomic structure is to consider adding only two angular momenta. We will first consider orbital angular momentum only for two different electrons and then consider a single electron with both orbital and spin angular momentum.

4.3.1 Adding Orbital Angular Momenta for Two Electrons

The wave function for two electrons is made from products of the wave functions for each electron, such that a general example would be written as (neglecting any radial dependencies)

$$Y_{\ell_1,\ell_2}^{m_1,m_2}(1,2) = Y_{\ell_1}^{m_1}(\theta_1,\phi_1)Y_{\ell_2}^{m_2}(\theta_2,\phi_2)\,, \tag{4.17}$$

where we have used the notation of Equation (3.57) and the coordinates of the two electrons are independent of one another. We already know that $-\ell_1 \leq m_1 \leq \ell_1$ and $-\ell_2 \leq m_2 \leq \ell_2$ so that there are $(2\ell_1+1)(2\ell_2+1)$ states of this kind for each choice of ℓ_1 and ℓ_2. For the variety of such states, we will abbreviate the notation to (m_1, m_2) to designate which state we are referring to. We already know that the z-component of angular momentum may be found from

$$\hat{J}_z = \hat{L}_{z1} + \hat{L}_{z2}\,,$$

so that $m = m_1 + m_2$ for the total z-component. The maximum value of m is $\ell_1 + \ell_2$ as we can see from the calculation of $\hat{J}^2$ for the state $Y_{\ell_1,\ell_2}^{\ell_1,\ell_2}$, which has the maximum total angular momentum since m_1 and m_2 have their maximum values. Before we calculate this specific value, we note that with $\hat{\boldsymbol{J}} = \hat{\boldsymbol{L}}_1 + \hat{\boldsymbol{L}}_2$, we have

$$\begin{aligned} \hat{J}^2 &= (\hat{\boldsymbol{L}}_1 + \hat{\boldsymbol{L}}_2) \cdot (\hat{\boldsymbol{L}}_1 + \hat{\boldsymbol{L}}_2) \\ &= \hat{L}_1^2 + \hat{L}_2^2 + 2\hat{\boldsymbol{L}}_1 \cdot \hat{\boldsymbol{L}}_2 \\ &= \hat{L}_1^2 + \hat{L}_2^2 + 2\hat{L}_{1z}\hat{L}_{2z} + \hat{L}_{1+}\hat{L}_{2-} + \hat{L}_{1-}\hat{L}_{2+}\,, \end{aligned} \tag{4.18}$$

remembering that $[\hat{\boldsymbol{L}}_1, \hat{\boldsymbol{L}}_2] = 0$ since the coordinates are independent for the two electrons. Using this general expression, we find

$$\begin{aligned} \hat{J}^2 Y_{\ell_1,\ell_2}^{\ell_1,\ell_2} &= (\hat{L}_1^2 + \hat{L}_2^2 + 2\hat{L}_{1z}\hat{L}_{2z} + \hat{L}_{1+}\hat{L}_{2-} + \hat{L}_{1-}\hat{L}_{2+})Y_{\ell_1,\ell_2}^{\ell_1,\ell_2} \\ &= \hbar^2[\ell_1(\ell_1+1) + \ell_2(\ell_2+1) + 2\ell_1\ell_2)Y_{\ell_1,\ell_2}^{\ell_1,\ell_2} \\ &= \hbar^2[(\ell_1+\ell_2)(\ell_1+\ell_2+1)Y_{\ell_1,\ell_2}^{\ell_1,\ell_2}\,, \end{aligned} \tag{4.19}$$

since

$$\begin{aligned} \hat{L}_{1+}\hat{L}_{2-}Y_{\ell_1,\ell_2}^{\ell_1,\ell_2} &= 0\,, \\ \hat{L}_{1-}\hat{L}_{2+}Y_{\ell_1,\ell_2}^{\ell_1,\ell_2} &= 0\,, \end{aligned}$$

as one cannot raise m_1 above ℓ_1 in the first case and one cannot raise m_2 above ℓ_2 in the second case. From Equation (4.19), it is apparent that $j = \ell_1 + \ell_2$ since we always have $\hat{J}^2\psi = \hbar^2 j(j+1)\psi$ for the total angular momentum. It follows that $-(\ell_1 + \ell_2) \le m \le \ell_1 + \ell_2$ so that there are $2j + 1 = 2\ell_1 + 2\ell_2 + 1$ values of m for this particular case.

It does not follow that every combination of two electrons with wave functions having ℓ_1 and ℓ_2 will have $j = \ell_1 + \ell_2$. After all, it is quite possible that the orbital angular momentum of one electron could be in the opposite direction from the orbital angular momentum of the other. Generally speaking, j may vary over the range $|\ell_1 - \ell_2| \le j \le \ell_1 + \ell_2$ and for any particular j, we obviously have $-j \le m \le j$. For each value of j and m, the wave function is actually a linear combination of states with the appropriate ℓ_1 and ℓ_2 but with every possible combination of the allowable values of m_1 and m_2 such that $m = m_1 + m_2$. The kinds of combinations are illustrated by Table 4.1 where the various combinations for two electrons having $\ell_1 = 3$ and $\ell_2 = 2$ are listed along with the number of distinct states.

Adding up the number of various combinations in Table 4.1 (the column on the right), we find that there are $35 = (2 \cdot 3 + 1)(2 \cdot 2 + 1)$ total. When there are multiple combinations for a particular value of m, then the linear combination that makes up the eigenfunction for the total angular momentum takes the form

$$Y_j^m(1)(2) = \sum_{m_1, m_2 = m - m_1} C[jm, \ell_1 m_1 \ell_2 m_2] Y_{\ell_1}^{m_1} Y_{\ell_2}^{m_2}\,, \tag{4.20}$$

TABLE 4.1
Various combinations possible with $\ell_1 = 3$ and $\ell_2 = 2$.

m value	(m_1, m_2) combinations	states
5	(3,2)	1
4	(3,1) (2,2)	2
3	(3,0) (2,1) (1,2)	3
2	$(3,-1)(2,0)(1,1)(0,2)$	4
1	$(3,-2)(2,-1)(1,0)(0,1)(-1,2)$	5
0	$(2,-2)(1,-1)(0,0)(-1,1)(-2,2)$	5
-1	$(1,-2)(0,-1)(-1,0)(-2,1)(-3,2)$	5
-2	$(0,-2)(-1,-1)(-2,0)(-3,1)$	4
-3	$(-1,-2)(-2,-1)(-3,0)$	3
-4	$(-2,-2)(-3,-1)$	2
-5	$(-3,-2)$	1

where the $C[jm, \ell_1 m_1 \ell_2 m_2]$ are called the Clebsch–Gordan coefficients and are tabulated for many values of the parameters, but too complicated to consider more than a few here.

Example 4.4

Clebsch–Gordan coefficients. We can evaluate the Clebsch–Gordan coefficients by use of the raising or lowering operators. If we take an example with $\ell_1 = 2$ and $\ell_2 = 1$, then we may start with the highest state in the table that corresponds to Table 4.1 such that we begin with $Y_3^3(1,2) = Y_2^2(1)Y_1^1(2)$. The lowering operator is $\hat{J}_- = \hat{J}_{1-} + \hat{J}_{2-}$, so that

$$\begin{aligned}\hat{J}_- Y_3^3(1,2) &= N_{2,2}^- Y_2^1(1)Y_1^1(2) + Y_2^2(1)N_{1,1}^- Y_1^0(2)\\ &= \hbar[2Y_2^1(1)Y_1^1(2) + \sqrt{2}Y_2^2(1)Y_1^0(2)]\\ &= N_{3,3}^- Y_3^2(1,2) = \sqrt{6}\hbar Y_3^2(1,2)\end{aligned}$$

so

$$Y_3^2(1,2) = \sqrt{\frac{2}{3}}Y_2^1(1)Y_1^1(2) + \frac{1}{\sqrt{3}}Y_2^2(1)Y_1^0(2)$$

and the Clebsch–Gordan coefficients for this case are $C[32,2111] = \sqrt{2/3}$ and $C[32,2210] = 1/\sqrt{3}$. Lowering once more, we find

$$\begin{aligned}\hat{J}_- Y_3^2(1,2) &= \sqrt{\frac{2}{3}}N_{2,1}^- Y_2^0(1)Y_1^1(2) + \sqrt{\frac{2}{3}}Y_2^1(1)N_{1,1}^- Y_1^0(2)\\ &\quad + \frac{1}{\sqrt{3}}N_{2,2}^- Y_2^1(1)Y_1^0(2) + \frac{1}{\sqrt{3}}Y_2^2(1)N_{1,0}^- Y_1^{-1}\\ &= \hbar\left[2Y_2^0(1)Y_1^1(2) + \frac{4}{\sqrt{3}}Y_2^1(1)Y_1^0(2) + \sqrt{\frac{2}{3}}Y_2^2(1)Y_1^{-1}(2)\right]\\ &= N_{3,2}^- Y_3^1(1,2) = \sqrt{10}\hbar Y_3^1(1,2)\end{aligned}$$

so

$$Y_3^1(1,2) = \sqrt{\frac{2}{5}} Y_2^0(1)Y_1^1(2) + \frac{2\sqrt{2}}{\sqrt{15}} Y_2^1(1)Y_1^0(2) + \frac{1}{\sqrt{15}} Y_2^2(1)Y_1^{-1}(2)$$

and the Clebsch–Gordan coefficients for this case are $C[31,2011] = \sqrt{2/5}$, $C[31,2110] = 2\sqrt{2/15}$, and $C[31,221(-1)] = \sqrt{1/15}$. There are altogether $(2\ell_1+1)(2\ell_2+1) = 5 \cdot 3 = 15$ states, so there will be 15 Clebsch–Gordan coefficients for this example, but the first and last are unity. If $\ell_1 = \ell_2$, There is symmetry when $m_1 \leftrightarrow m_2$ and this reduces the number of independent coefficients, and $C[\ell(-m), \ell_1(-m_1)\ell_2(-m_2)] = C[\ell m, \ell_1 m_1 \ell_2 m_2]$, but in general the calculation of the coefficients is tedious. □

Problem 4.8 *Clebsch–Gordan coefficients.*

(a) Use the lowering operator on $Y_3^1(1,2)$ of Example 4.4 to find $Y_3^0(1,2)$ and its Clebsch–Gordan coefficients.

(b) Using the symmetries of the Clebsch–Gordan coefficients, write down the wave functions, $Y_3^{-1}(1,2)$, $Y_3^{-2}(1,2)$, and $Y_3^{-3}(1,2)$ for the example above.

4.3.2 Adding Orbital and Spin Angular Momenta

For a single electron, the wave function is of the form $Y_\ell^m \chi_\pm$ and $\hat{\boldsymbol{J}} = \hat{\boldsymbol{L}} + \hat{\boldsymbol{S}}$. For the eigenfunction of $\hat{J}^2$ with projection $m + \frac{1}{2}$, we have the linear combination

$$\psi_{j,m+\frac{1}{2}} = \alpha Y_\ell^m \chi_+ + \beta Y_\ell^{m+1} \chi_- \,, \tag{4.21}$$

since both terms give $\hat{J}_z \psi = (m + \frac{1}{2})\hbar\psi$.

The constants α and β must be chosen so that the wavefunction is an eigenfunction of $\hat{J}^2$. In this case, the operator may be obtained from Equation (4.18) by letting $\hat{\boldsymbol{L}}_1 \to \hat{\boldsymbol{L}}$ and $\hat{\boldsymbol{L}}_2 \to \hat{\boldsymbol{S}}$ so that

$$\hat{J}^2 = \hat{L}^2 + \hat{S}^2 + 2\hat{\boldsymbol{L}} \cdot \hat{\boldsymbol{S}} = \hat{L}^2 + \hat{S}^2 + 2\hat{L}_z\hat{S}_z + \hat{L}_+\hat{S}_- + \hat{L}_-\hat{S}_+ \,.$$

For the last two terms, we find

$$\begin{aligned} \hat{L}_+ Y_\ell^m &= \hbar\sqrt{\ell(\ell+1) - m(m+1)}\, Y_\ell^{m+1} = \hbar\sqrt{(\ell+m+1)(\ell-m)}\, Y_\ell^{m+1} \,, \\ \hat{L}_- Y_\ell^{m+1} &= \hbar\sqrt{\ell(\ell+1) - (m+1)m}\, Y_\ell^m = \hbar\sqrt{(\ell+m+1)(\ell-m)}\, Y_\ell^m \,. \end{aligned}$$

Then, since $\hat{S}_+\chi_+ = \hat{S}_-\chi_- = 0$, we have

$$\begin{aligned} \hat{J}^2 \psi_{j,m+\frac{1}{2}} &= j(j+1)\hbar^2[\alpha Y_\ell^m \chi_+ + \beta Y_\ell^{m+1} \chi_-] \,, \\ &= \alpha\hbar^2\{[\ell(\ell+1) + \tfrac{3}{4} + 2m(\tfrac{1}{2})] Y_\ell^m \chi_+ \\ &\quad + \sqrt{(\ell+m+1)(\ell-m)}\, Y_\ell^{m+1} \chi_-\} \\ &\quad + \beta\hbar^2\{[\ell(\ell+1) + \tfrac{3}{4} + 2(m+1)(-\tfrac{1}{2})] Y_\ell^{m+1} \chi_- \\ &\quad + \sqrt{(\ell+m+1)(\ell-m)}\, Y_\ell^m \chi_+\} \,, \end{aligned} \tag{4.22}$$

where the first line is simply the eigenvalue statement,

$$\hat{j}^2\psi_{j,m+\frac{1}{2}} = j(j+1)\hbar^2\psi_{j,m+\frac{1}{2}},$$

and the other terms on the right make up Problem 4.9. Since the two separate wave functions, $Y_\ell^m\chi_+$ and $Y_\ell^{m+1}\chi_-$, are orthogonal to one another, the coefficients of each wave function must individually satisfy this equation, leading to the pair of equations

$$\alpha j(j+1) = \alpha[\ell(\ell+1) + \tfrac{3}{4} + m] + \beta\sqrt{(\ell+m+1)(\ell-m)}\,, \tag{4.23}$$

$$\beta j(j+1) = \beta[\ell(\ell+1) + \tfrac{3}{4} - m - 1] + \alpha\sqrt{(\ell+m+1)(\ell-m)}\,. \tag{4.24}$$

We may write this pair of equations as

$$\begin{pmatrix} a_{11} & a_{12} \\ a_{21} & a_{22} \end{pmatrix} \begin{pmatrix} \alpha \\ \beta \end{pmatrix} = j(j+1)\begin{pmatrix} \alpha \\ \beta \end{pmatrix}, \tag{4.25}$$

where

$$\begin{aligned} a_{11} &= \ell(\ell+1) + \tfrac{3}{4} + m\,, \\ a_{22} &= \ell(\ell+1) + \tfrac{3}{4} - m - 1\,, \\ a_{12} = a_{21} &= \sqrt{(\ell+m+1)(\ell-m)}\,. \end{aligned}$$

Letting $j(j+1) = \lambda$, Equation (4.25) may be written as

$$\begin{pmatrix} a_{11}-\lambda & a_{12} \\ a_{21} & a_{22}-\lambda \end{pmatrix} \begin{pmatrix} \alpha \\ \beta \end{pmatrix} = 0\,, \tag{4.26}$$

so that for a nontrivial solution we must have

$$\begin{vmatrix} a_{11}-\lambda & a_{12} \\ a_{21} & a_{22}-\lambda \end{vmatrix} = 0\,, \tag{4.27}$$

so that

$$\lambda^2 - (a_{11}+a_{22})\lambda + a_{11}a_{22} - a_{12}a_{21} = 0\,, \tag{4.28}$$

and

$$\begin{aligned} a_{11} + a_{22} &= 2\ell(\ell+1) + \tfrac{1}{2}\,, \\ a_{11}a_{22} - a_{12}a_{21} &= \ell^2(\ell+1)^2 - \tfrac{1}{2}\ell(\ell+1) - \tfrac{3}{16}\,. \end{aligned}$$

Using these, we find

$$\lambda = j(j+1) = \ell(\ell+1) + \tfrac{1}{4} \pm (\ell + \tfrac{1}{2}) = \begin{cases} \ell^2 + 2\ell + \tfrac{3}{4}, \\ \ell^2 - \tfrac{1}{4}. \end{cases} \tag{4.29}$$

Solving for j, there are two values of j for each of the two roots above for λ. Using the first root for λ, we find

$$j = -\tfrac{1}{2} \pm (\ell+1) = \begin{cases} \ell + \tfrac{1}{2}, \\ -\ell - \tfrac{3}{2}, \end{cases} \tag{4.30}$$

but j is nonnegative, so the second root must be excluded. Similarly for the second root for λ, we find

$$j = -\tfrac{1}{2} \pm \ell = \begin{cases} \ell - \tfrac{1}{2}, \\ -\ell - \tfrac{1}{2}, \end{cases} \tag{4.31}$$

and again we must exclude the negative result, leaving us with the values $j = \ell \pm \frac{1}{2}$. These values are consistent with the previous case with orbital angular momentum only, where $\ell_1 \to \ell$ for the orbital angular momentum and $\ell_2 \to \frac{1}{2}$ for the spin angular momentum.

In order to find the values of α and β for each value of j, we return to Equation (4.23) with $j = \ell + \frac{1}{2}$, so that

$$\beta = \alpha \frac{(\ell + \frac{1}{2})(\ell + \frac{3}{2}) - \ell(\ell + 1) - m - \frac{3}{4}}{\sqrt{(\ell + m + 1)(\ell - m)}} = \alpha \sqrt{\frac{\ell - m}{\ell + m + 1}} . \tag{4.32}$$

Then using the normalization condition that $\alpha^2 + \beta^2 = 1$, we find

$$\alpha^2 + \beta^2 = \alpha^2 \left[1 + \frac{\ell - m}{\ell + m + 1} \right] = 1 ,$$

so

$$\alpha = \sqrt{\frac{\ell + m + 1}{2\ell + 1}} , \tag{4.33}$$

and using Equation (4.32) we also have

$$\beta = \sqrt{\frac{\ell + m + 1}{2\ell + 1}} \sqrt{\frac{\ell - m}{\ell + m + 1}} = \sqrt{\frac{\ell - m}{2\ell + 1}} . \tag{4.34}$$

The eigenfunction for $j = \ell + \frac{1}{2}$ is therefore

$$\psi_{\ell + \frac{1}{2}, m + \frac{1}{2}} = \sqrt{\frac{\ell + m + 1}{2\ell + 1}} Y_\ell^m \chi_+ + \sqrt{\frac{\ell - m}{2\ell + 1}} Y_\ell^{m+1} \chi_- . \tag{4.35}$$

We do not need to do all of this work over again for the coefficients for the $j = \ell - \frac{1}{2}$ eigenfunction, since we know it must be orthogonal to the $j = \ell + \frac{1}{2}$ eigenfunction. Thus we can guess it to be

$$\psi_{\ell - \frac{1}{2}, m + \frac{1}{2}} = \sqrt{\frac{\ell - m}{2\ell + 1}} Y_\ell^m \chi_+ - \sqrt{\frac{\ell + m + 1}{2\ell + 1}} Y_\ell^{m+1} \chi_- . \tag{4.36}$$

Problem 4.9 *Adding orbital and spin angular momenta.* Verify each term on the right-hand side of Equation (4.22) (10 total).

Problem 4.10 *Adding angular momentum.* Find the corresponding one-electron eigenvalues and eigenfunctions for a case with $j_z = m - \frac{1}{2}$.

4.4 Interacting Spins for Two Particles

Let us suppose that two particles, each with 1/2-integral spins, interact only through their spins, such that the interaction energy is given by

$$\hat{\mathcal{H}} = \lambda \hat{\boldsymbol{S}}_1 \cdot \hat{\boldsymbol{S}}_2$$

where λ is a real parameter. The system of the two particles has spin of either $s = 1$ with $m_s = -1, 0, 1$ (a triplet) or $s = 0$ with $m_s = 0$ (a singlet). The eigenfunctions must be of the form $|s\ m_s\rangle$ since neither s_{1z} nor s_{2z} will be good quantum numbers. It is convenient to note that

$$\hat{S}^2 = (\hat{\boldsymbol{S}}_1 + \hat{\boldsymbol{S}}_2)^2 = \hat{S}_1^2 + \hat{S}_2^2 + 2\hat{\boldsymbol{S}}_1 \cdot \hat{\boldsymbol{S}}_2$$

so that

$$\hat{\boldsymbol{S}}_1 \cdot \hat{\boldsymbol{S}}_2 = \tfrac{1}{2}(\hat{S}^2 - \hat{S}_1^2 - \hat{S}_2^2)$$

The eigenvalue equation may therefore be written as

$$\begin{aligned} \hat{\mathcal{H}}|s, m_s\rangle &= \tfrac{\lambda}{2}(\hat{S}^2 - \hat{S}_1^2 - \hat{S}_2^2)|s, m_s\rangle \\ &= \frac{\lambda\hbar^2}{2}\left[s(s+1) - \frac{1}{2}\left(\frac{1}{2}+1\right) - \frac{1}{2}\left(\frac{1}{2}+1\right)\right]|s, m_s\rangle \\ &= E|s, m_s\rangle \end{aligned} \tag{4.37}$$

For the two cases with $s = 1$ and $s = 0$, we find

$$E = \frac{\lambda\hbar^2}{2}\left(\frac{1}{2}\right) = \frac{\lambda\hbar^2}{4}, \quad \text{triplet,} \tag{4.38}$$

$$E = \frac{\lambda\hbar^2}{2}\left(-\frac{3}{2}\right) = -\frac{3\lambda\hbar^2}{4}, \quad \text{singlet.} \tag{4.39}$$

4.4.1 Magnetic Moments

There is a simple relationship from electromagnetic theory that indicates a circulating current produces a magnetic moment that is equal to the current in the loop times the area of the loop, and points in the direction perpendicular to the area as determined from the right-hand rule with the fingers following the current. If we imagine a simple picture of an electron in a Bohr orbit, we have $\mu = IA$ where the current is given by $I = q/t = qv/2\pi r$ and $A = \pi r^2$ so

$$\mu = \tfrac{1}{2}qvr = \frac{qpr}{2m} = \frac{qL}{2m}$$

where $\boldsymbol{L} = \boldsymbol{r} \times \boldsymbol{p}$ is the angular momentum. The vector relationship for the electron is

$$\boldsymbol{\mu}_\ell = -\frac{e}{2m}\boldsymbol{L}\,.$$

For orbital angular momentum, we know $\boldsymbol{L} = m_\ell \hbar \hat{\boldsymbol{\ell}}$ so that, for example, for an $\ell = 1$ state, $\mu = -\mu_B \hat{\boldsymbol{\ell}}$ where $\hat{\boldsymbol{\ell}}$ is a unit vector and

$$\mu_B \equiv \frac{e\hbar}{2m_e}, \tag{4.40}$$

which is called the **Bohr magneton**. The ratio of the magnetic moment to the angular momentum is called the **gyromagnetic ratio** and is given by

$$\gamma_\ell = \left|\frac{\boldsymbol{\mu}_\ell}{\boldsymbol{L}}\right| = \frac{\mu_B}{\hbar}.$$

In addition to orbital angular momentum, the electron, proton, and neutron have spin angular momentum, so these also have a magnetic moment. The electron magnetic moment is *anomalous*, however, since the magnetic moment is approximately *twice* as large as expected. For this case, we have

$$|\mu_S| = \gamma_S |S| = \frac{g_e e}{2m_e} \cdot \frac{\hbar}{2} = \tfrac{1}{2} g_e \mu_B$$

where g_e is the electron *g-factor*. Classically, $g_e = 1$. The relativistic theory of Dirac finds $g_e = 2$, and the measured value is $g_e = 2.0024$. This means $|\mu_e| \simeq \mu_B$. This phenomenon is yet another example that shows that electron spin is *not* a classical effect.

If we turn our attention to the magnetic moment of the proton or neutron, the electron mass must be replaced by the appropriate nucleon mass, and we define a corresponding **nuclear magneton** such that

$$\mu_N \equiv \frac{e\hbar}{2M_p}. \tag{4.41}$$

The values of the magnetic moments for the proton, neutron, and deuteron are

$$\begin{aligned}
\mu_p &= 2.792776\mu_N \\
\mu_n &= -1.91304\mu_N \\
\mu_d &= 0.8576\mu_N .
\end{aligned}$$

It is interesting to note that the magnetic moment of the deuteron is *not* the sum of the two magnetic moments of the proton and neutron. Although the discrepancy is small, it was taken as an early indication that the proton and neutron are not spherically symmetric with a central force law, and with the advent of the quark theory, this discrepancy is essentially resolved.

Example 4.5

Magnetic moment interactions. Using some of the techniques above, we can examine the case where the interactions between two particles is due to their magnetic moments only. The interaction energy for two magnetic dipoles is

$$\hat{\mathcal{H}} = \frac{(\hat{\boldsymbol{\mu}}_1 \cdot \hat{\boldsymbol{\mu}}_2)}{r^3} - 3\frac{(\hat{\boldsymbol{\mu}}_1 \cdot \boldsymbol{r})(\hat{\boldsymbol{\mu}}_2 \cdot \boldsymbol{r})}{r^5}.$$

so considering two neutrons, where $\boldsymbol{\mu}_n = (g_n e/2m_p)\boldsymbol{S}$, where $g_n = -1.91304$, we have

$$\hat{\mathcal{H}} = \left(\frac{g_n e}{2m_p}\right)^2 \left[\frac{(\hat{\boldsymbol{S}}_1 \cdot \hat{\boldsymbol{S}}_2)}{r^3} - 3\frac{(\hat{\boldsymbol{S}}_1 \cdot \boldsymbol{r})(\hat{\boldsymbol{S}}_2 \cdot \boldsymbol{r})}{r^5}\right].$$

We may choose $\boldsymbol{r} = a\hat{\boldsymbol{e}}_z$ so that $r = a$, with the result being expressed as

$$\hat{\mathcal{H}} = \left(\frac{g_n e}{2m_p}\right)^2 \frac{1}{a^3}(\hat{\boldsymbol{S}}_1 \cdot \hat{\boldsymbol{S}}_2 - 3\hat{S}_{1z}\hat{S}_{2z}).$$

Using the same procedure for the product $\hat{S}_{1z}\hat{S}_{2z}$ as we did for $\hat{\boldsymbol{S}}_1 \cdot \hat{\boldsymbol{S}}_2$, such that

$$\hat{S}_z^2 = (\hat{S}_{1z} + \hat{S}_{2z})^2 = \hat{S}_{1z}^2 + \hat{S}_{2z}^2 + 2\hat{S}_{1z}\hat{S}_{2z}$$

we find

$$\hat{S}_{1z}\hat{S}_{2z} = \tfrac{1}{2}(\hat{S}_z^2 - \hat{S}_{1z}^2 - \hat{S}_{2z}^2)$$

so

$$\hat{\mathcal{H}} = \frac{g_n^2 e^2}{4m_p^2 a^3}\left[\frac{1}{2}(\hat{S}^2 - \hat{S}_1^2 - \hat{S}_2^2) - \frac{3}{2}(\hat{S}_z^2 - \hat{S}_{1z}^2 - \hat{S}_{2z}^2)\right].$$

If we apply this Hamiltonian to the state function $|s,\ m_s\rangle$, we finally obtain

$$\begin{aligned}\hat{\mathcal{H}}|s, m_s\rangle &= \frac{g_n^2 e^2}{4m_p^2 a^3}\left\{\frac{\hbar^2}{2}\left[s(s+1) - \frac{1}{2}\left(\frac{1}{2}+1\right) - \frac{1}{2}\left(\frac{1}{2}+1\right)\right]\right. \\ &\qquad \left. - \frac{3\hbar^2}{2}\left[m_s^2 - \left(\frac{1}{2}\right)^2 - \left(\frac{1}{2}\right)^2\right]\right\}|s, m_s\rangle \\ &= \frac{g_n^2 e^2 \hbar^2}{8m_p^2 a^3}[s(s+1) - 3m_s^2]|s, m_s\rangle \qquad (4.42)\end{aligned}$$

so

$$E = \begin{cases} -E_0\,, & m_s = \pm 1 \\ 2E_0\,, & m_s = 0 \end{cases} \qquad (4.43)$$

where $E_0 = g_n^2 e^2 \hbar^2/8m_p^2 a^3 = g_n^2 \mu_N^2/2a^3$ for the triplet $s = 1$ state and $E = 0$ for the singlet $s = 0$ state. □

5

Approximation Methods

5.1 Introduction — The Many-Electron Atom

When one includes not just one electron, but many electrons, there is a large variety of effects that affect the energy levels of the various states. Quite generally, these additional effects cannot be treated exactly as was done for the one-electron atom. Even with the one-electron atom, we did not include the effects of the spin magnetic moments of either the electron or the nucleus nor their interactions with the orbital magnetic moment of the electron, nor did we include any relativistic corrections, and any one of these additional effects complicates the problem to such an extent that exact results can no longer be obtained with nonrelativistic quantum mechanics. In addition to these effects, the many-electron atom has all of the electron–electron effects. We may represent these effects, however, through a very complicated potential, so that the Hamiltonian becomes

$$\mathcal{H} = T + V_{\text{en}} + V_{\text{ee}} + V_{\text{so}} + V_{\text{ss}} + V_{\text{oo}} + V_{\text{esns}} + V_{\text{nseo}} + V_{\text{rel}} + V_{\text{exch}} + V_{\text{etc}}\,, \quad (5.1)$$

where each term may be represented by the following expressions:

1. Kinetic energy:

$$T = \frac{p_n^2}{2m_n} + \sum_{i=1}^{N} \frac{p_i^2}{2m_e}\,. \quad (5.2)$$

2. Electron–nucleus electrostatic interaction:

$$V_{\text{en}} = -\sum_{i=1}^{N} \frac{Ze^2}{4\pi\epsilon_0 r_i}\,. \quad (5.3)$$

3. Electron–electron mutual electrostatic interaction (repulsion):

$$V_{\text{ee}} = \sum_{i=1}^{N}\sum_{j=1}^{i-1} \frac{e^2}{4\pi\epsilon_0 r_{ij}}\,. \quad (5.4)$$

4. Spin–orbit interaction (spin angular momentum — orbital angular momentum):

$$V_{\mathrm{so}} = -\sum_{i=1}^{N} \frac{\boldsymbol{\sigma}_i \cdot \boldsymbol{l}_i}{m^2 r_i c^2} \frac{\mathrm{d}V}{\mathrm{d}r_i} \,. \tag{5.5}$$

5. Spin–spin interaction (electron spin–electron spin):

$$V_{\mathrm{ss}} = \frac{\mu_0}{4\pi} \sum_{i=1}^{N} \sum_{j=1}^{i-1} \frac{e^2}{m^2} \left[\frac{\boldsymbol{\sigma}_i \cdot \boldsymbol{\sigma}_j}{r_{ij}^3} - 3 \frac{(\boldsymbol{\sigma}_i \cdot \boldsymbol{r}_{ij})(\boldsymbol{\sigma}_j \cdot \boldsymbol{r}_{ij})}{r_{ij}^5} \right] . \tag{5.6}$$

6. Orbit–orbit interaction (electron–electron orbital angular momentum):

$$V_{\mathrm{oo}} = \sum_{i=1}^{N} \sum_{j=1}^{i-1} C_{ij} \boldsymbol{l}_i \cdot \boldsymbol{l}_j \,. \tag{5.7}$$

7. Electron spin — nuclear magnetic moment:

$$V_{\mathrm{esns}} = \frac{\mu_0}{4\pi} \sum_{i=1}^{N} \frac{e}{m} \left[\frac{\boldsymbol{\mu}_n \cdot \boldsymbol{\sigma}_i}{r_i^3} - 3 \frac{(\boldsymbol{\mu}_n \cdot \boldsymbol{r}_i)(\boldsymbol{\sigma}_i \cdot \boldsymbol{r}_i)}{r_i^5} \right] . \tag{5.8}$$

8. Nuclear spin — electron orbital angular momentum:

$$V_{\mathrm{nseo}} = \frac{\mu_0}{4\pi} \sum_{i=1}^{N} \frac{e}{m} \left(\frac{\boldsymbol{\mu}_n \cdot \boldsymbol{l}_i}{2\pi r_i^3} \right) . \tag{5.9}$$

9. Relativistic correction (to the *kinetic* energy):

$$V_{\mathrm{rel}} = -\sum_{i=1}^{N} \frac{p_i^4}{8m^3c^2} \,. \tag{5.10}$$

10. "Exchange interaction." (Due to the Pauli exclusion principle, which is spin-dependent, there is a tendency to align the spins and effectively cause the electrons to "repel" one another.)

11. Miscellaneous other effects, such as quadrupole interactions, finite nuclear size, etc.

The dominant terms after the first two are generally term 10 and term 3, after which term 4 is next, and the remainder are usually negligible. For heavy atoms, sometimes term 4 predominates over terms 3 or 10.

In order to estimate the effects of these terms we cannot treat exactly, we need some methods to approximate the energy of the more complex system. If the corrections are in some sense *small*, then straightforward methods have been developed and will be treated in the next section. Other problems that are either difficult or impossible to solve analytically include transitions between two or more stationary states using the time-dependent Schrödinger equation, and the treatment of cases where the potential changes in time, either slowly or rapidly. Methods for dealing with each of these cases will be discussed in the following sections.

5.2 Nondegenerate Perturbation Theory

Few realistic problems in quantum mechanics can be solved exactly, and the many-electron atom described above is only one example, so we need some method of estimating an answer that is not exact. Frequently, we have an exact solution that is close enough to provide a first guess, so we only need a method to improve upon the guess. Two examples are (1) the energy levels of the Hydrogen atom, where effects of spin, electric and magnetic fields shift the energies a small amount from the one-electron atom model, and (2) a potential well that is nearly a harmonic oscillator potential but has either a weak asymmetry, or deviates from a quadratic well for large excursions. In both cases, we have a good first guess, and the purpose of perturbation theory is to provide a good first correction to that guess.

Formally, we will be dealing with a Hamiltonian that we cannot solve exactly, but the dominant portion presumably can be treated analytically. Therefore, we separate the Hamiltonian into a zero-order and a first-order part

$$\hat{\mathcal{H}} = \hat{\mathcal{H}}^{(0)} + \hat{\mathcal{H}}^{(1)}, \tag{5.11}$$

where $\hat{\mathcal{H}}^{(1)}$ is in some sense small, and the eigenfunctions and eigenvalues of $\hat{\mathcal{H}}^{(0)}$ are presumed known. In the formal development of the theory, we shall assume that we can "turn on" the perturbation slowly, so that Equation (5.11) may be written

$$\hat{\mathcal{H}} = \hat{\mathcal{H}}^{(0)} + \alpha\hat{\mathcal{H}}^{(1)}, \tag{5.12}$$

where $0 \le \alpha \le 1$ is a parameter. We shall expand both the eigenvalues and the eigenfunctions in this parameter, such that

$$E_n = E_n^{(0)} + \alpha E_n^{(1)} + \alpha^2 E_n^{(2)} + \cdots \tag{5.13}$$

$$\psi_n = \psi_n^{(0)} + \alpha\psi_n^{(1)} + \alpha^2\psi_n^{(2)} + \cdots, \tag{5.14}$$

where $E_n^{(0)}$ and $\psi_n^{(0)}$ correspond to the unperturbed system ($\alpha = 0$). We may then write the time-independent Schrödinger equation as

$$\begin{aligned}(\hat{\mathcal{H}} - E_n)\psi_n &= (\hat{\mathcal{H}}^{(0)} + \alpha\hat{\mathcal{H}}^{(1)} - E_n^{(0)} - \alpha E_n^{(1)} - \alpha^2 E_n^{(2)} - \cdots) \\ &\quad \times(\psi_n^{(0)} + \alpha\psi_n^{(1)} + \alpha^2\psi_n^{(2)} + \cdots) \\ &= 0,\end{aligned}$$

which becomes, after collecting terms of the same order in α,

$$
\begin{aligned}
&(\hat{\mathcal{H}}^{(0)} - E_n^{(0)})\psi_n^{(0)} + \alpha[(\hat{\mathcal{H}}^{(1)} - E_n^{(1)})\psi_n^{(0)} + (\hat{\mathcal{H}}^{(0)} - E_n^{(0)})\psi_n^{(1)}] \\
&\quad + \alpha^2[\hat{\mathcal{H}}^{(0)}\psi_n^{(2)} + \hat{\mathcal{H}}^{(1)}\psi_n^{(1)} - E_n^{(0)}\psi_n^{(2)} - E_n^{(1)}\psi_n^{(1)} - E_n^{(2)}\psi_n^{(0)}] + \cdots = 0\,.
\end{aligned}
$$

Since this must be valid for arbitrary values of α, the coefficients must individually vanish, so that

$$(\hat{\mathcal{H}}^{(0)} - E_n^{(0)})\psi_n^{(0)} = 0\,, \tag{5.15}$$

$$(\hat{\mathcal{H}}^{(1)} - E_n^{(1)})\psi_n^{(0)} + (\hat{\mathcal{H}}^{(0)} - E_n^{(0)})\psi_n^{(1)} = 0\,, \tag{5.16}$$

$$\hat{\mathcal{H}}^{(0)}\psi_n^{(2)} + \hat{\mathcal{H}}^{(1)}\psi_n^{(1)} - E_n^{(0)}\psi_n^{(2)} - E_n^{(1)}\psi_n^{(1)} - E_n^{(2)}\psi_n^{(0)} = 0\,, \text{ etc.} \tag{5.17}$$

Equation (5.15) is just the zero-order equation of the unperturbed system, and we assume that all of these zero-order quantities are known exactly.

5.2.1 Nondegenerate First-Order Perturbation Theory

In Equation (5.16), the first order quantities are related to the zero order quantities. To find the first order quantities, we take the scalar product of Equation (5.16) with $\psi_n^{(0)}$ so that

$$\langle\psi_n^{(0)}|\hat{\mathcal{H}}^{(1)}|\psi_n^{(0)}\rangle - E_n^{(1)}\langle\psi_n^{(0)}|\psi_n^{(0)}\rangle + \langle\psi_n^{(0)}|\hat{\mathcal{H}}^{(0)}|\psi_n^{(1)}\rangle - E_n^{(0)}\langle\psi_n^{(0)}|\psi_n^{(1)}\rangle = 0\,.$$

Using the Hermitian property of $\hat{\mathcal{H}}^{(0)}$ in the third term,

$$
\begin{aligned}
\langle\psi_n^{(0)}|\hat{\mathcal{H}}^{(0)}|\psi_n^{(1)}\rangle &= \langle\hat{\mathcal{H}}^{(0)}\psi_n^{(0)}|\psi_n^{(1)}\rangle \\
&= E_n^{(0)}\langle\psi_n^{(0)}|\psi_n^{(1)}\rangle\,,
\end{aligned}
$$

so that this term cancels the last term. Then, since we assume $\psi_n^{(0)}$ is normalized, we find

$$E_n^{(1)} = \langle\psi_n^{(0)}|\hat{\mathcal{H}}^{(1)}|\psi_n^{(0)}\rangle = \langle\hat{\mathcal{H}}^{(1)}\rangle\,, \tag{5.18}$$

so *the first-order correction to the energy is the average value of the perturbed energy in the unperturbed state.* We emphasize here that the first-order correction to the energy uses only the zero-order wave function.

For the first-order wave function, we may represent it in terms of the zero-order wave functions since they form a complete set, so that

$$\psi_n^{(1)} = \sum_i a_{ni}\psi_i^{(0)}\,, \tag{5.19}$$

and with this expression inserted into Equation (5.16), the first-order equation may be written

$$\hat{\mathcal{H}}^{(1)}\psi_n^{(0)} + \hat{\mathcal{H}}^{(0)}\sum_i a_{ni}\psi_i^{(0)} = E_n^{(1)}\psi_n^{(0)} + E_n^{(0)}\sum_i a_{ni}\psi_i^{(0)}\,.$$

The sum over i could just as well be written as a sum with $i \neq n$, since the first-order equation is satisfied with *any* value of a_{nn}, so one usually sets $a_{nn} = 0$. Then we take the scalar product with $\psi_k^{(0)} \equiv \langle k|$, so that this equation becomes

$$\langle k|\hat{\mathcal{H}}^{(1)}|n\rangle + \sum_i a_{ni}\langle k|\hat{\mathcal{H}}^{(0)}|i\rangle = E_n^{(1)}\langle k|n\rangle + E_n^{(0)}\sum_i a_{ni}\langle k|i\rangle$$

where three of the integrals immediately reduce to kronecker delta functions since the kets and bras are eigenfunctions, with the result

$$\langle k|\hat{\mathcal{H}}^{(1)}|n\rangle + \sum_i a_{ni}E_i^{(0)}\delta_{ki} = E_n^{(1)}\delta_{kn} + E_n^{(0)}\sum_i a_{ni}\delta_{ki}$$

and the δ_{ki} collapses the sums so that the first order coefficients are determined by

$$\mathcal{H}_{kn}^{(1)} + a_{nk}E_k^{(0)} = E_n^{(1)}\delta_{kn} + a_{nk}E_n^{(0)}\,, \tag{5.20}$$

where we have defined $\langle k|\hat{\mathcal{H}}^{(1)}|n\rangle \equiv \mathcal{H}_{kn}^{(1)}$. For $k = n$, this gives us the result of Equation (5.18), namely that $E_n^{(1)} = \mathcal{H}_{nn}^{(1)}$. Then, for $k \neq n$, Equation (5.20) gives

$$\mathcal{H}_{kn}^{(1)} = a_{nk}(E_n^{(0)} - E_k^{(0)})\,,$$

or solving for a_{nk}, we find

$$a_{nk} = \frac{\mathcal{H}_{kn}^{(1)}}{E_n^{(0)} - E_k^{(0)}}\,. \tag{5.21}$$

In the nondegenerate case, $E_k^{(0)} \neq E_n^{(0)}$, so this is well-defined. If $E_k^{(0)} = E_n^{(0)}$ for some k, then the system is said to be degenerate, and we must proceed differently.

5.2.2 Nondegenerate Second-Order Perturbation Theory

Sometimes, $\mathcal{H}_{nn}^{(1)} = 0$, so in order to find the lowest order nonzero correction, we must go to second order in perturbation theory. Here again, we will expand $\psi_n^{(2)}$ in the same basis set, so that

$$\psi_n^{(2)} = \sum_j b_{nj}\psi_j^{(0)}\,, \tag{5.22}$$

so that Equation (5.17) becomes

$$\hat{\mathcal{H}}^{(0)}\sum_j b_{nj}\psi_j^{(0)} + \hat{\mathcal{H}}^{(1)}\sum_i a_{ni}\psi_i^{(0)} - E_n^{(0)}\sum_j b_{nj}\psi_j^{(0)} - E_n^{(1)}\sum_i a_{ni}\psi_i^{(0)} - E_n^{(2)}\psi_n^{(0)} = 0\,.$$

Again we take the scalar product with $\langle k|$, so that the second-order equation becomes

$$\sum_j b_{nj}\langle k|\hat{\mathcal{H}}^{(0)}|j\rangle + \sum_i a_{ni}\langle k|\hat{\mathcal{H}}^{(1)}|i\rangle = E_n^{(0)}\sum_j b_{nj}\langle k|j\rangle + E_n^{(1)}\sum_i a_{ni}\langle k|i\rangle + E_n^{(2)}\langle k|n\rangle\,,$$

where the integrals again lead to kronecker delta functions such that

$$\sum_j b_{nj}E_j^{(0)}\delta_{kj} + \sum_i a_{ni}\mathcal{H}_{ki}^{(1)} = E_n^{(0)}\sum_j b_{nj}\delta_{kj} + E_n^{(1)}\sum_i a_{ni}\delta_{ki} + E_n^{(2)}\delta_{kn}\,,$$

where this time all the sums except one collapse, leading to

$$b_{nk}E_k^{(0)} + \sum_i a_{ni}\mathcal{H}_{ki}^{(1)} = E_n^{(0)}b_{nk} + E_n^{(1)}a_{nk} + E_n^{(2)}\delta_{kn}\,. \tag{5.23}$$

For $k = n$, Equation (5.23) gives the second-order energy correction as

$$\begin{aligned} E_n^{(2)} &= \sum_i a_{ni}\mathcal{H}_{ni}^{(1)} - a_{nn}E_n^{(1)} \\ &= \sum_i a_{ni}\mathcal{H}_{ni}^{(1)} - a_{nn}\mathcal{H}_{nn}^{(1)} \\ &= \sum_{i\neq n} a_{ni}\mathcal{H}_{ni}^{(1)}\,. \end{aligned}$$

Using Equation (5.21) for a_{ni}, this can also be written,

$$E_n^{(2)} = \sum_{i\neq n}\frac{\mathcal{H}_{in}^{(1)}\mathcal{H}_{ni}^{(1)}}{E_n^{(0)} - E_i^{(0)}} = \sum_{i\neq n}\frac{|\mathcal{H}_{ni}^{(1)}|^2}{E_n^{(0)} - E_i^{(0)}}\,. \tag{5.24}$$

We note here that the second-order energy correction has been obtained with only the first-order wave functions, just as the first-order energy correction was obtained with only the zero-order wave functions.

To obtain the second order wave function, we examine Equation (5.23) with $k \neq n$. This gives

$$b_{nk}(E_k^{(0)} - E_n^{(0)}) = a_{nk}\mathcal{H}_{nn}^{(1)} - \sum_i a_{ni}\mathcal{H}_{ki}^{(1)}\,,$$

or

$$b_{nk} = \sum_{i\neq n}\frac{\mathcal{H}_{in}^{(1)}\mathcal{H}_{ki}^{(1)}}{(E_n^{(0)} - E_k^{(0)})(E_n^{(0)} - E_i^{(0)})} - \frac{\mathcal{H}_{kn}^{(1)}\mathcal{H}_{nn}^{(1)}}{(E_n^{(0)} - E_k^{(0)})^2}\,. \tag{5.25}$$

It is evident that the higher order wave functions get complicated very quickly, such that

$$\psi_n = \psi_n^{(0)} + \sum_{k\neq n} a_{nk}\psi_k^{(0)} + \sum_{k\neq n} b_{nk}\psi_k^{(0)}\,,$$

so generally perturbation theory goes only to first order in wave function and second order in energy. There is no reason in principle, however, that one cannot proceed to even higher order in a fashion similar to that above.

Example 5.1

Anharmonic Oscillator. To illustrate nondegenerate perturbation theory, we take as an example the one-dimensional harmonic oscillator, since we know at the outset that all of the zero order energy eigenvalues are distinct, separated by $\hbar\omega$. For our perturbed Hamiltonian, we first pick a cubic correction, so that

$$\hat{\mathcal{H}} = \frac{\hat{p}^2}{2m} + \tfrac{1}{2}kx^2 + bx^3 = \hat{\mathcal{H}}^{(0)} + \hat{\mathcal{H}}^{(1)} ,$$

where it is apparent that $\hat{\mathcal{H}}^{(1)} = bx^3$. If b is small, then this is a perturbation, and we can proceed with the analysis. One might have expected our first example to be a linear or quadratic perturbation, but it is shown in Problem 5.1 that these corrections may be included exactly, so perturbation theory starts at the next higher order.

The first-order energy perturbation for a cubic correction is simply

$$E_n^{(1)} = \langle n|bx^3|n\rangle = 0 ,$$

since the raising and lowering operators for the harmonic oscillator indicate that x^3 will produce $n+3$, $n+1$, $n-1$, and $n-3$ states only, leading to the null result. This means we need to go to second order to get a nonzero result. From Equation (2.95) of Section 2.2 where we had $\alpha^2 = m\omega/\hbar^2$,

$$\begin{aligned}\langle m|x^3|n\rangle = \frac{1}{2\sqrt{2}\alpha^3}[&\sqrt{n(n-1)(n-2)}\delta_{m,n-3} + 3n^{3/2}\delta_{m,n-1} \\ &+3(n+1)^{3/2}\delta_{m,n+1} + \sqrt{(n+1)(n+2)(n+3)}\delta_{m,n+3}] ,\end{aligned}$$

so the second-order energy is given by the sum of the nonzero terms,

$$\begin{aligned} E_n^{(2)} &= \frac{|\mathcal{H}^{(1)}_{n+3,n}|^2}{E_n^{(0)} - E_{n+3}^{(0)}} + \frac{|\mathcal{H}^{(1)}_{n+1,n}|^2}{E_n^{(0)} - E_{n+1}^{(0)}} + \frac{|\mathcal{H}^{(1)}_{n-1,n}|^2}{E_n^{(0)} - E_{n-1}^{(0)}} + \frac{|\mathcal{H}^{(1)}_{n-3,n}|^2}{E_n^{(0)} - E_{n-3}^{(0)}} \\ &= \frac{b^2}{8\alpha^6}\left[\frac{(n+1)(n+2)(n+3)}{\hbar\omega(n+\frac{1}{2}) - \hbar\omega(n+3+\frac{1}{2})} + \frac{9(n+1)^3}{\hbar\omega(n+\frac{1}{2}) - \hbar\omega(n+1+\frac{1}{2})}\right. \\ &\quad \left. + \frac{9n^3}{\hbar\omega(n+\frac{1}{2}) - \hbar\omega(n-1+\frac{1}{2})} + \frac{n(n-1)(n-2)}{\hbar\omega(n+\frac{1}{2}) - \hbar\omega(n-3+\frac{1}{2})}\right] \\ &= \frac{b^2}{8\alpha^6\hbar\omega}\left[\frac{(n+1)(n+2)(n+3)}{-3} + \frac{9(n+1)^3}{-1} + \frac{9n^3}{1} + \frac{n(n-1)(n-2)}{3}\right] \\ &= \frac{-b^2}{8\alpha^6\hbar\omega}(30n^2 + 30n + 11) . \end{aligned} \tag{5.26}$$

The total energy is therefore given by

$$
\begin{aligned}
E_n &= \hbar\omega(n + \tfrac{1}{2}) - \frac{b^2}{8\alpha^6\hbar\omega}(30n^2 + 30n + 11) \\
&= \hbar\omega\left[n + \tfrac{1}{2} - \frac{b^2}{8\alpha^2 k^2}(30n^2 + 30n + 11)\right] .
\end{aligned}
\tag{5.27}
$$

□

Problem 5.1 *Generalized harmonic oscillator.*

(a) Show that a Hamiltonian operator of the form

$$\hat{\mathcal{H}} = \frac{\hat{p}^2}{2m} + V_0 + \epsilon x + \frac{1}{2}kx^2$$

can be solved exactly as a harmonic oscillator. (*Hint:* Complete the square in x.)

(b) Find the exact energy eigenvalues.

Problem 5.2 *Wave functions.* Write the first-order wave function for the perturbation $\hat{\mathcal{H}}^{(1)} = bx^3$.

Problem 5.3 *First order with* $\hat{\mathcal{H}}^{(1)} = cx^4$.

(a) Find the first-order perturbation energy, $E_n^{(1)}$.

(b) Find the first-order wave function for the harmonic oscillator with $\hat{\mathcal{H}}^{(1)} = cx^4$.

Problem 5.4 *Second order with* $\hat{\mathcal{H}}^{(1)} = cx^4$. Find the second-order energy perturbation energy, $E_n^{(2)}$, and an expression for the total energy that is analogous to Equation (5.27) for $\hat{\mathcal{H}}^{(1)} = cx^4$.

Problem 5.5 *Cross terms.* Show that for a perturbation of the type bx^3+cx^4 that the total energy correction is simply the sum of the perturbations taken separately (i.e., show that the cross terms do not contribute).

5.3 Perturbation Theory for Degenerate States

When two or more eigenstates have the same energy, we need to proceed differently, since the denominator of both the first-order wave functions and the second-order energy corrections are not defined with the usual expressions.

In order to keep the algebra simple while we discuss the method, we will consider a system with only two degenerate states, but the method can easily be extended. Instead of separating the corrections into first-order and higher-order terms, we will deal with the Schrödinger equation directly, so that

$$(\hat{\mathcal{H}}^{(0)} + \hat{\mathcal{H}}^{(1)})\psi_n = E_n\psi_n\,, \tag{5.28}$$

but we will again represent ψ_n in terms of the eigenfunctions of $\hat{\mathcal{H}}^{(0)}$. Hence we write $\psi_n = \sum_i c_{ni}|i\rangle$ and take the scalar product with $\langle k|$ so that Equation (5.28) becomes

$$c_{nk}E_k^{(0)} + \sum_i c_{ni}\mathcal{H}_{ki}^{(1)} = c_{nk}E_n\,, \tag{5.29}$$

where $\mathcal{H}_{ki}^{(1)}$ is defined as in the previous sections, and E_n is the total energy for state n, including the perturbation, and Equation (5.29) may be rearranged to obtain

$$c_{nk}(E_n - E_k^{(0)}) = \sum_i c_{ni}\mathcal{H}_{ki}^{(1)}\,. \tag{5.30}$$

Now with only two degenerate states, ψ_n and ψ_ℓ, we may write

$$\begin{aligned}\psi_n &= c_{nn}|n\rangle + c_{n\ell}|\ell\rangle \\ \psi_\ell &= c_{\ell n}|n\rangle + c_{\ell\ell}|\ell\rangle\,.\end{aligned}$$

Setting $k = n$ first in Equation (5.29), and then $k = \ell$, we obtain the pair of equations:

$$c_{nn}(E_n^{(0)} + \mathcal{H}_{nn}^{(1)} - E_n) + c_{n\ell}\mathcal{H}_{n\ell}^{(1)} = 0 \tag{5.31}$$

$$c_{nn}\mathcal{H}_{\ell n}^{(1)} + c_{n\ell}(E_\ell^{(0)} + \mathcal{H}_{\ell\ell}^{(1)} - E_n) = 0\,. \tag{5.32}$$

The necessary condition for a nontrivial solution is that the determinant of coefficients vanish, or that

$$\begin{vmatrix} E_n^{(0)} + \mathcal{H}_{nn}^{(1)} - E_n & \mathcal{H}_{n\ell}^{(1)} \\ \mathcal{H}_{\ell n}^{(1)} & E_\ell^{(0)} + \mathcal{H}_{\ell\ell}^{(1)} - E_n \end{vmatrix} = 0\,, \tag{5.33}$$

which has solutions

$$E_n = E_n^{(0)} + \tfrac{1}{2}(\mathcal{H}_{\ell\ell}^{(1)} + \mathcal{H}_{nn}^{(1)}) \pm \tfrac{1}{2}\sqrt{(\mathcal{H}_{\ell\ell}^{(1)} - \mathcal{H}_{nn}^{(1)})^2 + 4|\mathcal{H}_{n\ell}^{(1)}|^2}\,, \tag{5.34}$$

where we have used the fact that $E_n^{(0)} = E_\ell^{(0)}$.

Example 5.2
Stark splitting. We will examine Stark splitting in a Hydrogen atom for the $n = 2$ level. This means that a uniform electric field is applied, and we will take it to be in the z-direction for convenience, so the potential energy of an

electron in this external field is $V = \hat{\mathcal{H}}^{(1)} = e\mathcal{E}z = e\mathcal{E}r\cos\theta$. The four states for $n = 2$ are ψ_{200}, ψ_{210}, ψ_{211}, and ψ_{21-1}, and since the perturbation is odd, the first order perturbation, or all the diagonal matrix elements, will vanish for each state. In fact, the only nonzero matrix element is

$$\begin{aligned}\hat{\mathcal{H}}^{(1)}_{200,210} = \hat{\mathcal{H}}^{(1)}_{210,200} &= e\mathcal{E}\langle\psi_{200}|r\cos\theta|\psi_{210}\rangle \\ &= \frac{e\mathcal{E}}{16{a'_0}^4}\int_0^\infty r^4\left(2 - \frac{r}{a'_0}\right)\mathrm{e}^{-r/a'_0}\,\mathrm{d}r\int_0^\pi \sin\theta\cos^2\theta\,\mathrm{d}\theta \\ &= -3a'_0 e\mathcal{E}\,.\end{aligned}$$

The perturbation matrix is then

$$\hat{\mathcal{H}}^{(1)} = \begin{pmatrix} 0 & -3a'_0e\mathcal{E} & 0 & 0 \\ -3a'_0e\mathcal{E} & 0 & 0 & 0 \\ 0 & 0 & 0 & 0 \\ 0 & 0 & 0 & 0 \end{pmatrix}.$$

The degeneracy is still apparent, but only in the upper left submatrix, so we only need to diagonalize the submatrix by setting

$$\begin{vmatrix} E_2^{(0)} - E_2 & -3a'_0e\mathcal{E} \\ -3a'_0e\mathcal{E} & E_2^{(0)} - E_2 \end{vmatrix} = 0\,,$$

whose roots are

$$E_2 = E_2^{(0)} \pm 3a'_0e\mathcal{E}\,.$$

The wave function associated with energy $E_2 = E_2^{(0)} - 3a'_0e\mathcal{E}$ is $\psi = \frac{1}{\sqrt{2}}(\psi_{200} + \psi_{210})$ while the state $\psi = \frac{1}{\sqrt{2}}(\psi_{200} - \psi_{210})$ has energy $E_2 = E_2^{(0)} + 3a'_0e\mathcal{E}$. □

Problem 5.6 *Stark effect for* $n = 3$. Calculate the Stark effect for the $n = 3$ state of Hydrogen. (*Hint*: Two of the energy corrections are $\pm 9a'_0e\mathcal{E}$.)

Problem 5.7 *Degenerate eigenfunctions.* Find the eigenfunctions for Problem 5.6.

Problem 5.8 *2–D harmonic oscillator.* Suppose a two-dimensional harmonic oscillator has the Hamiltonian,

$$\hat{\mathcal{H}} = -\frac{\hbar^2}{2m}\left(\frac{\partial^2}{\partial x^2} + \frac{\partial^2}{\partial y^2}\right) + \tfrac{1}{2}k(1 + bxy)(x^2 + y^2)\,.$$

(a) If $b = 0$, what are the wave functions and energies of the three lowest states?

(b) If b is a small positive number ($0 < b \ll 1$), find from perturbation theory the corrections to the energies of these three lowest states.

Problem 5.9 *Spin Hamiltonian.* Suppose a certain electric potential is described by the spin Hamiltonian,

$$\hat{\mathcal{H}}^{(1)} = D\hat{S}_z^2 + E(\hat{S}_x^2 - \hat{S}_y^2)\,.$$

Treating this as a perturbation of an ion with three degenerate spin-one states, find the energy splitting that results.

5.4 Time-Dependent Perturbation Theory

When the perturbation varies in time, we must use the time-dependent Schrödinger equation,

$$[\hat{\mathcal{H}}^{(0)} + \hat{\mathcal{H}}^{(1)}(\boldsymbol{r},t)]\Psi = \mathrm{i}\hbar\frac{\partial\Psi}{\partial t}\,, \tag{5.35}$$

where

$$\hat{\mathcal{H}}^{(0)}\Psi_i = E_i^{(0)}\Psi_i\,, \qquad \text{and} \qquad \Psi_i = \psi_i \mathrm{e}^{-(\mathrm{i}/\hbar)E_i^{(0)}t}\,.$$

We assume the ψ_i to form a complete orthonormal basis set ($\langle i|j\rangle = \delta_{ij}$), and the expansion coefficients will themselves become time-dependent so that

$$\Psi = \sum_n a_n(t)\Psi_n\,. \tag{5.36}$$

Using Equation (5.36) in Equation (5.35), we find

$$(\hat{\mathcal{H}}^{(0)} + \hat{\mathcal{H}}^{(1)})\sum_n a_n\Psi_n = \sum_n(\mathrm{i}\hbar\dot{a}_n + E_n^{(0)}a_n)\Psi_n\,.$$

We now take the scalar product with $\langle\Psi_k|$, obtaining

$$\sum_n[a_n\langle k|\hat{\mathcal{H}}^{(1)}|n\rangle - \mathrm{i}\hbar\dot{a}_n\delta_{kn}]\mathrm{e}^{(\mathrm{i}/\hbar)(E_k^{(0)}-E_n^{(0)})t} = 0\,.$$

If we define

$$\omega_{kn} \equiv \frac{1}{\hbar}(E_k^{(0)} - E_n^{(0)})\,,$$

then this may be written

$$\mathrm{i}\hbar\dot{a}_k = \sum_n a_n\mathcal{H}_{kn}^{(1)}\mathrm{e}^{\mathrm{i}\omega_{kn}t}\,. \tag{5.37}$$

For a large number of terms in the summation, this is difficult (but not impossible) to solve. We will thus restrict our attention to problems where the initial state is given, so that at $t = 0$, we know that the system is in state j

and $a_j(0) = 1$ and all the rest are initially zero. Then for short times, $a_j(t)$ is still close to unity, and the others are still small, so we may approximate

$$\mathrm{i}\hbar\dot{a}_k \simeq \mathcal{H}_{kj}^{(1)} \mathrm{e}^{\mathrm{i}\omega_{kj}t} .$$

This equation has solutions

$$a_j(t) = 1 - \frac{\mathrm{i}}{\hbar}\int_0^t \mathcal{H}_{jj}^{(1)}(t')\,\mathrm{d}t' , \tag{5.38}$$

$$a_k(t) = -\frac{\mathrm{i}}{\hbar}\int_0^t \mathcal{H}_{kj}^{(1)}(t')\mathrm{e}^{\mathrm{i}\omega_{kj}t'}\,\mathrm{d}t' , \qquad k \neq j . \tag{5.39}$$

Problem 5.10 *Decaying perturbation.* A particle in a one-dimensional box (infinite potential well) is initially in the ground state. At time zero, a weak perturbation of the form $\hat{\mathcal{H}}^{(1)}(x,t) = \epsilon\hat{x}\mathrm{e}^{-t^2}$ is turned on. If the box extends from $x = 0$ to $x = a$, find the probability that the particle will be found in the first excited state as $t \to \infty$.

Problem 5.11 *Perturbed harmonic oscillator.* A particle is initially in the ground state of a one-dimensional harmonic oscillator. At time zero, a perturbation of the form $\hat{\mathcal{H}}^{(1)}(x,t) = V_0(\alpha\hat{x})^3\mathrm{e}^{-t/\tau}$ is turned on ($\alpha x = q$).

(a) Calculate the probability that, after a sufficiently long time (as $t \to \infty$), the system will have made a transition to an excited state. Consider all possible final states.

(b) Find the ratio of the probability that the system will be in a state higher than $n = 1$ to the probability that the system will be in the $n = 1$ state. Evaluate this ratio for $\omega\tau \gg 1$ and for $\omega\tau \ll 1$.

(c) For fixed $V_0/\hbar\omega = \beta$, find the probability that the system remains in the ground state, and evaluate P_0 for $\omega\tau \gg 1$ and for $\omega\tau \ll 1$.

5.4.1 Perturbations That Are Constant in Time

If the perturbation $\hat{\mathcal{H}}^{(1)}$ is constant in time after being turned on at $t = 0$, then the integrals in Equation (5.38) and Equation (5.39) may be evaluated immediately, with the result

$$a_j(t) = 1 - \frac{\mathrm{i}}{\hbar}\mathcal{H}_{jj}^{(1)} t , \tag{5.40}$$

$$a_k(t) = \frac{\mathcal{H}_{kj}^{(1)}}{\hbar\omega_{kj}}(1 - \mathrm{e}^{\mathrm{i}\omega_{kj}t}) , \qquad k \neq j . \tag{5.41}$$

Now the probability that the state Ψ_k is occupied at time t is given by $|a_k(t)|^2$, so it amounts to the probability that the system has made a *transition* from state Ψ_j to Ψ_k. We may thus write from Equation (5.41)

$$P_k(t) = |a_k(t)|^2 = \left| \frac{2\mathcal{H}_{kj}^{(1)}}{E_k^{(0)} - E_j^{(0)}} \right|^2 \sin^2 \frac{(E_k^{(0)} - E_j^{(0)})t}{2\hbar}, \qquad k \neq j\,. \qquad (5.42)$$

We note from this expression that the probability oscillates at frequency ω_{kj} (since $\sin^2 \omega_{kj}t/2 = (1 - \cos\omega_{kj}t)/2$), but that due to its resonant denominator, it is large only where $E_j^{(0)}$ is close to $E_k^{(0)}$. We also note that the assumed smallness of $a_k(t)$ requires either $|\mathcal{H}_{kj}^{(1)}| \ll |E_k^{(0)} - E_j^{(0)}|$ or $\omega_{kj}t \ll 1$.

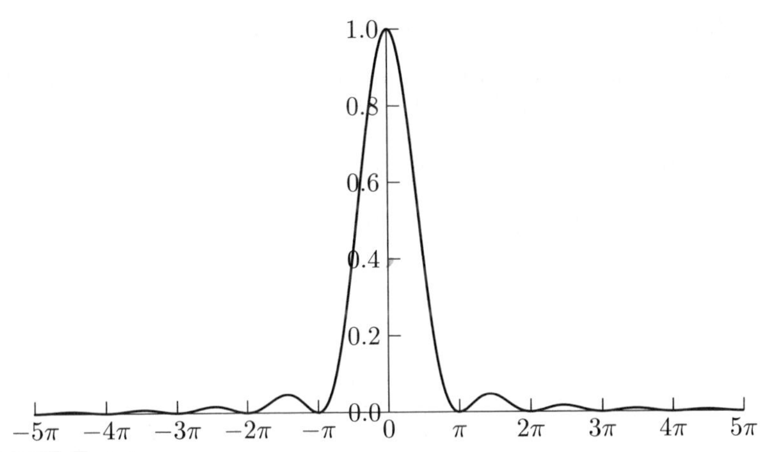

FIGURE 5.1
Plot of $\sin^2 x/x^2$ where $x = Et/2\hbar$, illustrating the δ function character of $\sin^2 x/x^2$.

Now for large t, letting $E \equiv E_k^{(0)} - E_j^{(0)}$, the oscillating term resembles a delta function $\delta(E)$ as illustrated in Figure 5.1 where the only significant contribution is near the origin in E. If we assume there exists a near continuum of final states, characterized by the density of final states $N(E) = \rho(E)\,\mathrm{d}E$, where $N(E)$ is the number of states between E and $E + dE$, then the area under the graph of $|a_k(E,t)|^2$ versus E represents the total transition probability. Hence,

$$\begin{aligned} P(t) = \int_{-\infty}^{\infty} |a_k(E,t)|^2 \rho(E)\,\mathrm{d}E &= |2\mathcal{H}_{kj}^{(1)}|^2 \rho(E) \frac{t}{2\hbar} \int_{-\infty}^{\infty} \frac{\sin^2 x}{x^2}\,\mathrm{d}x \\ &= \frac{2\pi t}{\hbar} |\mathcal{H}_{kj}^{(1)}|^2 \rho(E)\,, \end{aligned} \qquad (5.43)$$

which indicates that the probability of making a transition is linearly increasing with time after the perturbation is turned on. If we look rather at the

transition rate, then the result is

$$R = \frac{2\pi}{\hbar} |\mathcal{H}_{kj}^{(1)}|^2 \rho(E) , \tag{5.44}$$

which is **Fermi's Golden Rule (Number 2)**.

Problem 5.12 *Two-state problem.* If there are only two states, the system oscillates back and forth between them. Assuming only states j (initial state) and k exist (i.e., for which $\mathcal{H}_{kj}^{(1)} \neq 0$), find the oscillation frequency (or frequencies) and $a_j(t)$ and $a_k(t)$.

5.4.2 Perturbations That Are Harmonic in Time

We now assume that the perturbation oscillates in time, again having been turned on at time $t = 0$, and we shall assume the form of the perturbation is

$$\hat{\mathcal{H}}^{(1)}(\boldsymbol{r}, t) = 2\hat{\mathcal{H}}^{(1)}(\boldsymbol{r}) \cos \omega t = \hat{\mathcal{H}}^{(1)}(\mathrm{e}^{\mathrm{i}\omega t} + \mathrm{e}^{-\mathrm{i}\omega t}) . \tag{5.45}$$

Substituting this into Equation (5.39), the integrals are straightforward, so that

$$\begin{aligned} a_k(t) &= -\frac{\mathrm{i}}{\hbar} \mathcal{H}_{kj}^{(1)} \int_0^t [\mathrm{e}^{\mathrm{i}(\omega_{kj}+\omega)t'} + \mathrm{e}^{\mathrm{i}(\omega_{kj}-\omega)t'}] \, \mathrm{d}t' \\ &= -\mathcal{H}_{kj}^{(1)} \left[\frac{\mathrm{e}^{\mathrm{i}(\omega_{kj}+\omega)t} - 1}{\hbar(\omega_{kj}+\omega)} + \frac{\mathrm{e}^{\mathrm{i}(\omega_{kj}-\omega)t} - 1}{\hbar(\omega_{kj}-\omega)} \right] . \end{aligned} \tag{5.46}$$

There are two special cases. When $\omega \sim -\omega_{kj}$, the first term in Equation (5.46) dominates the second term, in which case

$$\hbar\omega \sim E_j^{(0)} - E_k^{(0)} .$$

This corresponds to **stimulated emission** since $E_j^{(0)} > E_k^{(0)}$ and the transition is from an initial state with higher energy to a final state with lower energy. The probability is of course maximum when $\hbar\omega = E_j^{(0)} - E_k^{(0)}$. The other case is where the second term dominates in Equation (5.46), or when $\omega \sim \omega_{kj}$. In this case, we have **resonant absorption**, since

$$\hbar\omega \sim E_k^{(0)} - E_j^{(0)} ,$$

and the transition is from a lower energy state to higher energy state. If the perturbation is due to an electromagnetic wave, the perturbating force is an incident photon with frequency ω. Then the two cases can be visualized in figures (a) and (b) of Figure 5.2. Figure (a) shows one photon coming in and two coming out. The second photon is stimulated and is emitted in phase with the incident photon. Figure (b) shows nothing coming out as the perturbing

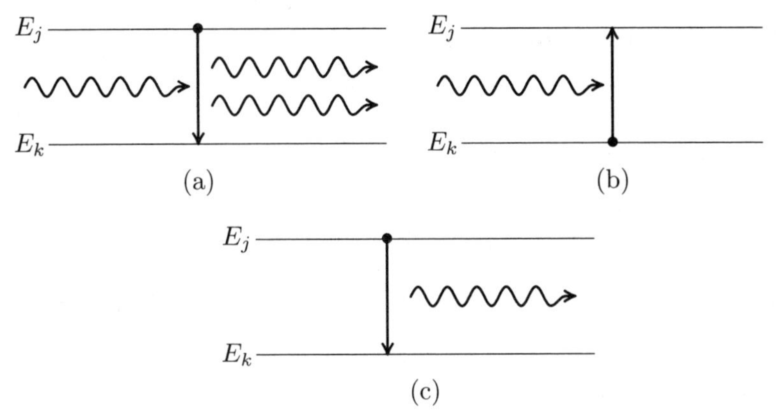

FIGURE 5.2
Three kinds of interaction between photons and atoms. (a) Stimulated emission. (b) Absorption. (c) Spontaneous emission. The first two are resonant interactions while the last is nonresonant

photon is resonantly absorbed. Figure (c) represents spontaneous emission and is a nonresonant process. The relationships between these emission and absorption processes will be treated in more detail in Chapter 9 where their relation to masers and lasers will be discussed.

When neither term is near resonance, a_k is small, so the general condition for a significant transition probability is that $\omega = \pm\omega_{kj}$. Since only one of the terms is important at a time, we may write the transition probability from Equation (5.46) as

$$|a_k(t)|^2 = \left|\frac{2\mathcal{H}^{(1)}_{kj}}{\hbar(\omega \pm \omega_{kj})}\right|^2 \sin^2\left[\frac{(\omega \pm \omega_{kj})t}{2}\right], \tag{5.47}$$

where the minus sign corresponds to absorption and the plus sign to emission. Doing the same integral over energy as in the previous case, the transition rate from state j to state k is given by

$$R_k = \frac{2\pi}{\hbar}|\mathcal{H}^{(1)}_{kj}|^2\rho(E_k)\delta(E_{kj} \mp E), \tag{5.48}$$

where E is the photon energy if the perturbing force is an electromagnetic field.

Problem 5.13 *Dipole matrix elements.* Calculate all of the dipole matrix elements for the transition $n = 2$ to $n = 1$ for the one-electron atom, i.e., find all of the nonzero elements of the form $\langle \boldsymbol{d} \rangle \equiv \langle \psi_{2\ell m}|q\boldsymbol{r}|\psi_{100}\rangle$ for $\langle d_x \rangle$, $\langle d_y \rangle$, and $\langle d_z \rangle$.

5.4.3 Adiabatic Approximation

When the perturbation is turned on *slowly*, so that the characteristic frequency changes little over a period, then we may estimate the probability of a transition from Equation (5.37), where we again assume the initial state to be j and that for nominally short times, only transitions from the initial state need be considered, then

$$\mathrm{i}\hbar\dot{a}_k \simeq \mathcal{H}_{kj}^{(1)} \mathrm{e}^{\mathrm{i}\omega_{kj} t} .$$

Assuming now that $\mathcal{H}_{kj}^{(1)}$ depends on time, we may integrate by parts to obtain

$$\begin{aligned} a_k(t) &= -\frac{\mathrm{i}}{\hbar}\int_0^t \mathcal{H}_{kj}^{(1)}(t')\mathrm{e}^{\mathrm{i}\omega_{kj}t'}\,\mathrm{d}t' \\ &= -\frac{1}{\hbar\omega_{kj}}\mathcal{H}_{kj}^{(1)}(t')\mathrm{e}^{\mathrm{i}\omega_{kj}t'}\Big|_0^t + \frac{1}{\hbar\omega_{kj}}\int_0^t \frac{\mathrm{d}}{\mathrm{d}t'}\mathcal{H}_{kj}^{(1)}(t')\mathrm{e}^{\mathrm{i}\omega_{kj}t'}\,\mathrm{d}t' \\ &= -\frac{1}{\hbar\omega_{kj}}\mathcal{H}_{kj}^{(1)}(t)\mathrm{e}^{\mathrm{i}\omega_{kj}t} + \frac{1}{\hbar\omega_{kj}}\int_0^t \left[\frac{\mathrm{d}}{\mathrm{d}t'}\mathcal{H}_{kj}^{(1)}(t')\right]\mathrm{e}^{\mathrm{i}\omega_{kj}t'}\,\mathrm{d}t' \qquad (5.49) \end{aligned}$$

since $\mathcal{H}_{kj}^{(1)}(0) = 0$ in the adiabatic approximation. Since $\mathcal{H}_{kj}^{(1)}(t)$ is slowly varying, the remaining integral may be made very small, so the adiabatic approximation normally includes only the leading term.

Example 5.3

Equilibrium shift. We consider a harmonic oscillator that undergoes a linear shift of the equilibrium position from 0 to x_1, where x_1 is a constant. The equilibrium position is given by

$$x_0(t) = \begin{cases} x_1 t/t_0 & 0 \le t \le t_0 \\ x_1 & t \ge t_0 \end{cases}$$

We may write the Hamiltonian then as

$$\begin{aligned} \hat{\mathcal{H}}^{(0)} &= \frac{\hat{p}^2}{2m} + \tfrac{1}{2}kx^2 , \qquad t \le 0 \\ \hat{\mathcal{H}} &= \frac{\hat{p}^2}{2m} + \tfrac{1}{2}k[x - x_0(t)]^2 , \qquad t > 0 \\ \hat{\mathcal{H}}^{(1)} &= \hat{\mathcal{H}} - \hat{\mathcal{H}}^{(0)} = \tfrac{1}{2}k\,[x_0(t)]^2 - kxx_0(t) . \end{aligned}$$

We take the system to be initially in the ground state, so $\psi_j = \psi_0(x)$. We wish to find the probability of finding the system in state ψ_1, so $k = 1$. Then the matrix element is given by

$$\begin{aligned} \mathcal{H}_{10}^{(1)} &= \langle\psi_1|\tfrac{1}{2}kx_0(t)^2|\psi_0\rangle - \langle\psi_1|kxx_0(t)|\psi_0\rangle \\ &= 0 - \frac{kx_0(t)}{\alpha}\langle\psi_1|q|\psi_0\rangle \\ &= -\frac{kx_0(t)}{\sqrt{2}\alpha} , \end{aligned}$$

where $q = \alpha x$. Then we have $\omega_{10} = (\frac{3}{2}\hbar\omega - \frac{1}{2}\hbar\omega)/\hbar = \omega$ so

$$a_1(t) = \frac{kx_0(t)}{\hbar\omega}\sqrt{\frac{\hbar}{2m\omega}}\mathrm{e}^{\mathrm{i}\omega t} = \frac{\alpha x_0(t)}{\sqrt{2}}\mathrm{e}^{\mathrm{i}\omega t}.$$

and the probability of finding the system in state ψ_1 is $|a_1(t)|^2 = q_1^2 t^2/2t_0^2$ where $q_1 = \alpha x_1$ for $0 \le t \le t_0$ and $|a_1|^2 = q_1^2/2$ for $t > t_0$. □

Problem 5.14 *The neglected term in the adiabatic approximation.* For the example above for the shifting equilibrium position of a harmonic oscillator,

(a) Denoting the normally kept term as calculated above as $a_1^{(0)}(t)$ and the normally neglected term in Equation (5.49) as $a_1^{(1)}(t)$, find $a_1^{(1)}(t)$.

(b) Show that for short times ($\omega t \ll 1$), $|a_1^{(1)}(t)/a_1^{(0)}(t)| \sim 1$, while for longer times ($\omega t_0 \gg 1$), $|a_1^{(1)}(t_0)/a_1^{(0)}(t_0)| \ll 1$.

5.4.4 Sudden Approximation

For this case, the perturbation is turned on suddenly at $t = 0$, and then held constant. This approximation suits the example above better, so we will compare the results of the two approaches to the same problem. In this case we assume the eigenvalues and eigenfunctions for the initial state are known and the general solution may be written

$$\Psi^{(i)} = \sum_i a_i \psi_i^{(i)} \mathrm{e}^{-\mathrm{i}E_i^{(i)}t/\hbar}, \qquad t < 0,$$

and for the final state, we likewise know the eigenvalues and eigenfunctions, with general solution

$$\Psi^{(f)} = \sum_j b_j \psi_j^{(f)} \mathrm{e}^{-\mathrm{i}E_j^{(f)}t/\hbar}, \qquad t \ge 0.$$

The wave function must be continuous at $t = 0$ (the equilibrium may not be continuous, but the particle probability density must be), so we have at $t = 0$

$$\sum_i a_i \psi_i^{(i)} = \sum_j b_j \psi_j^{(f)},$$

so we take the scalar product with $\langle \psi_k^{(f)}|$, and obtain

$$b_k = \sum_i a_i \langle \psi_k^{(f)} | \psi_i^{(i)} \rangle.$$

For the example above, the initial and final states are harmonic oscillator states, but the initial state has $x = 0$ as the equilibrium position and the final state has x_1 as the equilibrium position. Given the initial state is the ground state and that we wish to have the probability of finding the system in the first excited state of the new system, we calculate

$$b_1 = \langle\psi_1^{(f)}|\psi_0^{(i)}\rangle\,,$$

with

$$\psi_0^{(i)}(q) = \frac{1}{\pi^{1/4}}\mathrm{e}^{-q^2/2}$$
$$\psi_1^{(f)}(q) = \frac{\sqrt{2}}{\pi^{1/4}}\mathrm{e}^{-(q-q_1)^2/2}(q-q_1)\,.$$

Thus,

$$\begin{aligned} b_1 &= \sqrt{\frac{2}{\pi}}\int_{-\infty}^{\infty}\mathrm{e}^{-q^2/2-(q-q_1)^2/2}(q-q_1)\,\mathrm{d}q \\ &= \sqrt{\frac{2}{\pi}}\mathrm{e}^{-q_1^2/4}\int_{-\infty}^{\infty}\mathrm{e}^{-(q-q_1/2)^2}(q-q_1)\,\mathrm{d}q \\ &= \sqrt{\frac{2}{\pi}}\mathrm{e}^{-q_1^2/4}\int_{-\infty}^{\infty}\mathrm{e}^{-u^2}(u-\tfrac{1}{2}q_1)\,\mathrm{d}u \\ &= -\frac{q_1}{\sqrt{2}}\mathrm{e}^{-q_1^2/4}\,. \end{aligned}$$

From this we obtain

$$|b_1|^2 = \frac{q_1^2}{2}\mathrm{e}^{-q_1^2/2}\,.$$

For small q_1, this result is the same as we obtained for the adiabatic approximation.

Problem 5.15 *Spring break.* If one spring of a pair of springs breaks at $t = 0$ such that $k \to \frac{1}{2}k$, find the probability that a system that is initially in the first excited state, ψ_1, will remain in that state, and then calculate the probability that it will be found in the third excited state after the break.

Problem 5.16 *Tritium decay.* Use the sudden approximation to calculate the probability that the 1s electron of tritium will remain in the 1s state (ground state) of $^3\mathrm{He}^+$ when tritium decays by β emission so that Z changes from 1 to 2.

Problem 5.17 *Moving wall.* A particle of mass m is in the ground state of an infinitely deep, one-dimensional, square well.

(a) If the wall separation is suddenly doubled, what is the probability that the particle will remain in the ground state?

(b) If the wall separation is suddenly halved, what is the probability that the particle will remain in the ground state?

5.5 The Variational Method

In many problems, the Hamiltonian is sufficiently complicated that it is difficult to get exact eigenfunctions and eigenvalues for even a good first guess, or in other words, there does not exist an exactly soluble Hamiltonian which is close enough that perturbation theory is reliable. In this case, we may use a method that relies only on a reasonable guess for the wave function, but uses the exact Hamiltonian. In order to see how this works, we will first assume that exact eigenfunctions and eigenvalues are available, but this will be relaxed later after we see the method described.

First, we consider that the eigenvalues of the Hamiltonian may be ordered, so that there is a ground state value, first excited, etc., which we characterize by the inequalities $E_0 < E_1 < E_2 < \cdots$. Each of these values correspond to eigenfunctions such that

$$\hat{\mathcal{H}}\psi_i = E_i\psi_i\,,$$

even though we do not know the eigenfunctions. An arbitrary function may be expressed in terms of these eigenfunctions by superposition as

$$\psi = \sum_i a_i\psi_i\,.$$

If we allow for the moment that these wave functions are *not* normalized, then the energy associated with the arbitrary wave function is

$$\langle\mathcal{H}\rangle = \frac{\langle\psi|\hat{\mathcal{H}}|\psi\rangle}{\langle\psi|\psi\rangle} = \frac{\sum_i |a_i|^2 E_i}{\sum_i |a_i|^2}\,. \tag{5.50}$$

If now we replace each eigenvalue in the sum by the *lowest* eigenvalue, E_0, then we may replace the equality with the inequality

$$\langle\mathcal{H}\rangle \geq \frac{E_0\sum_i |a_i|^2}{\sum_i |a_i|^2} = E_0\,.$$

Equation (5.50) thus provides an *upper bound* on the ground state of the system, although in general it may not be too useful. If now we construct an *arbitrary* wave function to approximate the expected wave function for the ground state, since we do not know what the true eigenfunction is, then we may get a useful bound on the ground state energy. If we, in addition, include one or more adjustable parameters in the trial wave function, we may *minimize* $\langle\mathcal{H}\rangle$ with respect to these parameters, and approach more closely the ground state energy, knowing that we approach it from above. The technique, then, is to choose a trial wave function that

1. Satisfies the boundary conditions the real eigenfunction must satisfy.

2. Resembles as much as possible the true eigenfunction (odd or even, etc.).
3. Is tractable (all integrals may be evaluated).
4. Includes one or more adjustable parameters.

This trial function is then used to evaluate $\langle \mathcal{H} \rangle$ as a function of the parameter or parameters by means of Equation (5.50), and then minimize this function with respect to the various parameters included.

Example 5.4

Hydrogen atom with a Gaussian trial function. We will estimate the ground state energy of the Hydrogen atom with a Gaussian trial function (instead of the simple exponential), of the form $\psi(r) = C\mathrm{e}^{-\alpha r^2}$. Since this is independent of angle, Equation (5.50) then gives the recipe

$$\langle \mathcal{H} \rangle = \langle T \rangle + \langle V \rangle \,,$$

where

$$\begin{aligned}
\langle T \rangle &= -\frac{\hbar^2}{2m} \frac{\int_0^\infty \psi^* \frac{1}{r^2} \frac{\mathrm{d}}{\mathrm{d}r}\left(r^2 \frac{\mathrm{d}\psi}{\mathrm{d}r}\right) 4\pi r^2 \, \mathrm{d}r}{\int_0^\infty |\psi|^2 4\pi r^2 \, \mathrm{d}r} \\
&= -\frac{\hbar^2}{2m} \frac{|C|^2 \int_0^\infty (4\alpha^2 r^4 - 6\alpha r^2) \mathrm{e}^{-2\alpha r^2} \, \mathrm{d}r}{|C|^2 \int_0^\infty r^2 \mathrm{e}^{-2\alpha r^2} \, \mathrm{d}r} \\
&= \frac{3\alpha\hbar^2}{2m} \,, \\
\langle V \rangle &= -\frac{e^2}{4\pi\epsilon_0} \frac{\int_0^\infty r \mathrm{e}^{-2\alpha r^2} \, \mathrm{d}r}{\int_0^\infty r^2 \mathrm{e}^{-2\alpha r^2} \, \mathrm{d}r} \\
&= -\frac{e^2}{4\pi\epsilon_0} \frac{2\sqrt{2\alpha}}{\sqrt{\pi}} \,.
\end{aligned}$$

Combining these, the total energy is

$$\langle \mathcal{H} \rangle = A\alpha - B\sqrt{\alpha} \,,$$

with $A = 3\hbar^2/2m$ and $B = e^2/\pi\epsilon_0\sqrt{2\pi}$. Differentiating, we find

$$\frac{\mathrm{d}\langle \mathcal{H} \rangle}{\mathrm{d}\alpha} = A - \frac{B}{2\sqrt{\alpha}} = 0 \,,$$

so $\sqrt{\alpha} = B/2A$ gives the minimum so that

$$\begin{aligned}
\langle \mathcal{H} \rangle &= \frac{B^2}{4A} - \frac{B^2}{2A} = -\frac{B^2}{4A} \\
&= -\frac{me^4}{12\pi(\pi\epsilon_0)^2\hbar^2} \,.
\end{aligned}$$

Comparing with the exact answer, $\langle \mathcal{H} \rangle / E_0 = 32/12\pi = 8/3\pi$, so the error is approximately 15%. (The approximate energy is *higher* since the energies are both negative.) □

5.5.1 Accuracy

In order to see how the accuracy depends on the trial function, suppose that we write the general wave function in terms of the eigenfunctions, so that $\psi = \sum_i a_i \psi_i$, where we assume that $a_i / a_0 \sim \epsilon$, $i > 0$. Then we calculate

$$\begin{aligned}
\langle \mathcal{H} \rangle &= \frac{\langle \psi | \hat{\mathcal{H}} | \psi \rangle}{\langle \psi | \psi \rangle} \\
&= \frac{\sum_i |a_i|^2 E_i}{\sum_i |a_i|^2} \\
&\approx E_0 \left(1 - \frac{|a_1|^2}{|a_0|^2} + \cdots \right) + \frac{|a_1|^2}{|a_0|^2} E_1 \left(1 - \frac{|a_1|^2}{|a_0|^2} + \cdots \right) + \cdots \\
&= E_0 + \mathcal{O}(\epsilon^2) \,,
\end{aligned}$$

so if ψ is in error by order ϵ, then E_0 is in error by order ϵ^2. This means that a relatively poor guess in the trial functions leads to a better estimate for the energy than one might expect.

Problem 5.18 *Anharmonic oscillator.* Use the variational method to estimate the ground state energy of a particle in an anharmonic potential well whose potential energy is given by $V = \frac{1}{2} m\omega^2 x^2 + bx^4$. For the trial function, take $\psi = \psi_0 + \beta \psi_2$ where ψ_0 and ψ_2 are oscillator wave functions. Let β be the variational parameter.

Problem 5.19 Use the variational method to estimate the ground state energy of a particle of mass m in the potential $V = Cx$ where $x \geq 0$. Take the potential to be infinite for $x < 0$. Try $\psi = x\mathrm{e}^{-ax}$ as the trial function.

Problem 5.20 *Charmonium P states.* Estimate the energy of the Charmonium P-state with $\ell = 1$ by keeping the angular momentum term in the Schrödinger equation and using the trial function, $u(r) = r^2 \mathrm{e}^{-\alpha r/2}$ (since $u \sim r^{\ell+1}$ near $r = 0$). Compare this to the measured value of 3.413 GeV (i.e., find the percentage difference).

Problem 5.21 Use the variational method to estimate the ground state energy of the harmonic oscillator. Take the trial wave function to be $\psi(x) = \operatorname{sech} \alpha x$.

5.6 Wentzel, Kramers, and Brillouin Theory (WKB)

The plane wave described by $\psi = A\mathrm{e}^{\pm ikx}$ that characterizes the wave function for a particle in a uniform potential region is especially simple. If the potential for a more realistic problem changes slowly in x, we might expect that something close to the plane wave solution might be a good first approximation, and the method based on this idea was first put forward by Jeffries [2], and later applied to quantum mechanics by Wentzel [3], Kramers [4], and Brillouin [5]. The method is called either the JWKB method, WKBJ method, or the WKB method (most common in the United States).

5.6.1 WKB Approximation

The method applies to the time-independent Schródinger equation

$$-\frac{\hbar^2}{2m}\frac{\mathrm{d}^2\psi}{\mathrm{d}x^2} + V(x)\psi = E\psi\,,$$

where $V(x)$ is assumed to vary slowly with x so that $V(x)$ changes little over a wavelength. We write the solution as

$$\psi = \mathrm{e}^{\mathrm{i}S(x)/\hbar}\,, \tag{5.51}$$

where

$$S(x) = S_0(x) + \hbar S_1(x) + \tfrac{1}{2}\hbar^2 S_2(x) + \cdots\,, \tag{5.52}$$

and we assume only a few terms will be needed in the expansion. Substituting Equation (5.51) into the Schrödinger equation, it reduces to

$$\frac{1}{2m}(S')^2 - \frac{\mathrm{i}\hbar}{2m}S'' + V = E\,.$$

Using the expansion of Equation (5.52), this becomes

$$\frac{1}{2m}(S_0')^2 + \frac{\hbar^2}{2m}(S_1')^2 + \cdots + \frac{\hbar}{2m}(2S_0'S_1' + \hbar S_0'S_2' + \cdots) + (V - E)$$
$$-\frac{\mathrm{i}\hbar}{2m}(S_0'' + \hbar S_1'' + \cdots) = 0\,.$$

Assuming that $\hbar$ is the smallness expansion parameter, we group terms with the same order in $\hbar$ so that through second order,

$$\frac{1}{2m}(S_0')^2 + (V - E) = 0 \tag{5.53}$$

$$S_0'S_1' - \frac{\mathrm{i}}{2}S_0'' = 0 \tag{5.54}$$

$$S_0'S_2' + (S_1')^2 - \mathrm{i}S_1'' = 0\,. \tag{5.55}$$

We may separate variables in Equation (5.53) so that

$$\mathrm{d}S_0 = \pm\sqrt{2m[E - V(x)]}\,\mathrm{d}x\,,$$

which may be integrated to obtain

$$S_0(x) = \pm \int_{x_0}^{x} \sqrt{2m[E - V(x')]}\,\mathrm{d}x'\,. \tag{5.56}$$

Since S_0 is now determined, we may go to next order and integrate Equation (5.54) once to obtain

$$S_1 = \frac{\mathrm{i}}{2}\ln(S_0')\,,$$

so that

$$\mathrm{e}^{\mathrm{i}S_1} = [2m(E - V)]^{-1/4}\,.$$

If $V(x)$ is sufficiently slowly varying, we can usually get by with only these first two terms and write the solution as

$$\psi(x) = \frac{A\mathrm{e}^{\mathrm{i}S_0/\hbar} + B\mathrm{e}^{-\mathrm{i}S_0/\hbar}}{(E - V)^{1/4}}\,. \tag{5.57}$$

Using the plane wave notation where $k^2 = 2m(E - V)/\hbar^2$, we may define $k(x) = \sqrt{2m[E - V(x)]}/\hbar$ and write the solution as

$$\psi(x) = \frac{A}{\sqrt{k(x)}}\exp\left[\pm\mathrm{i}\int^{x} k(x')\,\mathrm{d}x'\right]. \tag{5.58}$$

In terms of $k(x)$, we may write the validity condition for this solution, which comes from the higher order terms, as

$$\left|\frac{1}{k^2}\frac{\mathrm{d}k}{\mathrm{d}x}\right| \ll 1\,. \tag{5.59}$$

The wave number $k \sim 1/\lambda$, so this condition effectively requires that the change in wavelength should be small over a wavelength, or that $|d\lambda/dx| \ll 1$.

5.6.2 WKB Connection Formulas

From the form of the solution and of $k(x)$ in particular, it is apparent that for $E > V$, we have a propagating wave, since $k(x)$ is real. When $E < V$, however, $k(x)$ is imaginary, and the solution is either growing or decaying exponentially. Furthermore, the region where the transition occurs is a region where the validity condition fails. It is useful to know how to handle such situations, since a propagating wave impinging on a sloping barrier may reflect, or if the barrier turns over again farther along, the wave may *tunnel through* and become propagating again. Since the abrupt barrier we treated in Chapter 2 is frequently not realistic, we address ourselves to the solution of a problem of a sloping barrier. We note that what we seek from this problem is a **connection formula** to connect the solution where $E > V$ with the solution where $E < V$, both of which are valid sufficiently far from the transition point.

To describe this transition region, we assume that close to the transition, the potential varies linearly with x about x_0, the transition point where $E = V(x_0)$. (This amounts to expanding $V(x) = V(x_0) + V'(x_0)(x - x_0) + \cdots$ and keeping only these first two terms.) The Schrödinger equation then becomes

$$-\frac{\hbar^2}{2m}\frac{\mathrm{d}^2\psi}{\mathrm{d}x^2} + V'(x_0)(x - x_0)\psi = 0\,, \tag{5.60}$$

which is the Airy equation (see Section 3.4.2) with $\xi = \alpha(x - x_0)$ and $\alpha = (2mV'(x_0)/\hbar^2)^{1/3}$. Because the Airy equation is still unfamiliar, we will simply list the connection formulas for the two cases, where the potential is increasing to the right (barrier on the right) or decreasing to the right (barrier on the left). In each case k_1 corresponds to a propagating wave and k_2 corresponds to growing or decaying waves, as indicated in Figure 5.3.

1. Barrier on the right.

$$\frac{2}{\sqrt{k_1}}\cos\left(\int_x^a k_1\,\mathrm{d}x' - \frac{\pi}{4}\right) \leftrightarrow \frac{1}{\sqrt{k_2}}\exp\left(-\int_a^x k_2\,\mathrm{d}x'\right)$$
$$\frac{1}{\sqrt{k_1}}\sin\left(\int_x^a k_1\,\mathrm{d}x' - \frac{\pi}{4}\right) \leftrightarrow -\frac{1}{\sqrt{k_2}}\exp\left(\int_a^x k_2\,\mathrm{d}x'\right).$$

2. Barrier on the left.

$$\frac{1}{\sqrt{k_2}}\exp\left(-\int_x^b k_2\,\mathrm{d}x'\right) \leftrightarrow \frac{2}{\sqrt{k_1}}\cos\left(\int_b^x k_1\,\mathrm{d}x' - \frac{\pi}{4}\right)$$
$$\frac{1}{\sqrt{k_2}}\exp\left(\int_x^b k_2\,\mathrm{d}x'\right) \leftrightarrow -\frac{1}{\sqrt{k_1}}\sin\left(\int_b^x k_1\,\mathrm{d}x' - \frac{\pi}{4}\right).$$

Problem 5.22 Using $S_0' = \hbar k(x)$, show that

$$\hbar S_2 = -\frac{1}{2}\frac{k'}{k^2} - \frac{1}{4}\int \frac{k'^2}{k^3}\,\mathrm{d}x\,.$$

Problem 5.23 *WKB solution of the Airy equation.* Find the approximate solution of Equation (5.60), using Equation (5.58).

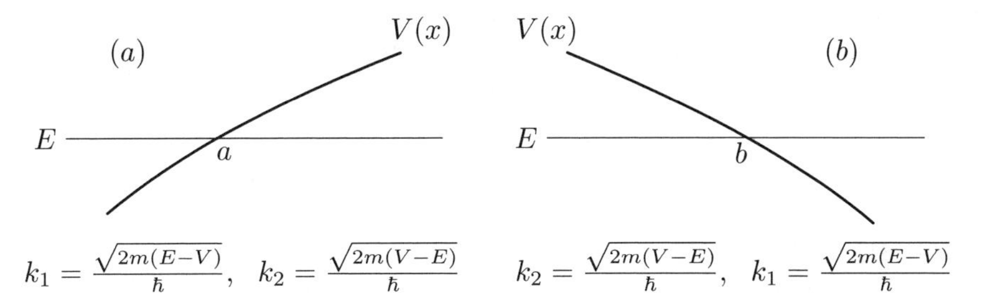

FIGURE 5.3
(a) Sketch of $V(x)$ for a barrier on the right. (b) Sketch of $V(x)$ for a barrier on the left.

5.6.3 Reflection

We may draw some conclusions from the nature of the connection formulas. If we note that the cosine term is connected to a decaying solution on the other side of the barrier, and that the sine solution is connected to a growing solution, it is apparent that any traveling wave ($\sim \exp[\mathrm{i}\int^x k(x')\,\mathrm{d}x']$) must have both kinds of terms since $\mathrm{e}^{ikx} = \cos kx + \mathrm{i}\sin kx$. On the other hand, if the barrier is very thick, the exponentially growing solution cannot be allowed (or must be made extremely small) so the sine term must be excluded. The conclusion is that for a thick barrier, only standing waves are permitted, or saying this another way, the magnitude of the reflection coefficient must be 1. For the phase shift upon reflection, we write the wave function as

$$\begin{aligned}
\psi &\sim A(x)\left(\mathrm{e}^{\mathrm{i}\int^x k\,\mathrm{d}x'} + R\,\mathrm{e}^{-\mathrm{i}\int^x k\,\mathrm{d}x'}\right) \\
&\sim \frac{2A}{\sqrt{k_1}}\cos\left(-\int^x k_1\,\mathrm{d}x' - \frac{\pi}{4}\right) \\
&= \frac{A\mathrm{e}^{\pi\mathrm{i}/4}}{\sqrt{k_1}}\left(\mathrm{e}^{\mathrm{i}\int^x k_1\,\mathrm{d}x'} + \mathrm{e}^{-\mathrm{i}\pi/2}\mathrm{e}^{-\mathrm{i}\int^x k_1\,\mathrm{d}x'}\right),
\end{aligned}$$

from which we conclude that $R = -\mathrm{i}$ or that there is a 90° phase shift upon reflection.

5.6.4 Transmission through a Finite Barrier

When there is a finite width barrier, the connection formulas are useful in constructing the solution. The model is indicated in Figure 5.4 where we will assume an incoming wave from the left, in region I, and only an outgoing wave on the right, in region III. The wave is nonpropagating in region II, and the solution there is made up of both growing and decaying solutions.

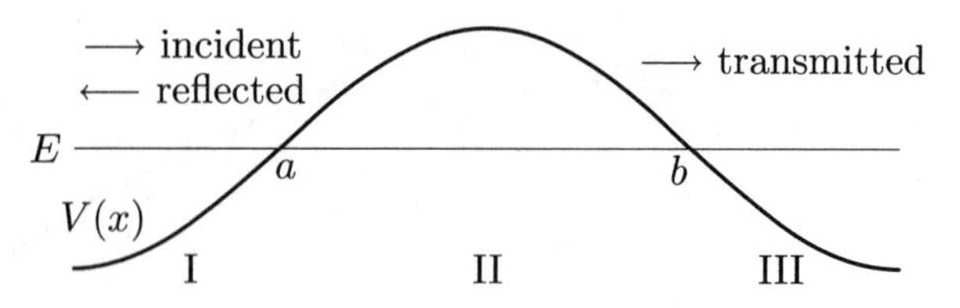

FIGURE 5.4
Barrier transmission with incident wave from the left.

Writing the wave function in each region using the connection formulas, we find

$$
\begin{aligned}
\psi_I &= \frac{A}{\sqrt{k_1}}\cos\left(\int_x^a k_1(x')\,\mathrm{d}x' - \frac{\pi}{4}\right) + \frac{B}{\sqrt{k_1}}\sin\left(\int_x^a k_1(x')\,\mathrm{d}x' \quad \frac{\pi}{4}\right)\\
\psi_{II} &= \frac{A}{2\sqrt{k_2}}\mathrm{e}^{-\int_a^x k_2(x')\,\mathrm{d}x'} - \frac{B}{\sqrt{k_2}}\mathrm{e}^{\int_a^x k_2(x')\,\mathrm{d}x'}\\
&= \frac{A}{2\sqrt{k_2}}\mathrm{e}^{-\int_a^b k_2(x')\,\mathrm{d}x' + \int_x^b k_2(x')\,\mathrm{d}x'} - \frac{B}{\sqrt{k_2}}\mathrm{e}^{\int_a^b k_2(x')\,\mathrm{d}x' - \int_x^b k_2(x')\,\mathrm{d}x'}\\
\psi_{III} &= -\frac{A\mathrm{e}^{-\int_a^b k_2(x')\,\mathrm{d}x'}}{2\sqrt{k_1}}\sin\left(\int_b^x k_1(x')\,\mathrm{d}x' - \frac{\pi}{4}\right)\\
&\quad -\frac{2B\mathrm{e}^{\int_a^b k_2(x')\,\mathrm{d}x'}}{\sqrt{k_1}}\cos\left(\int_b^x k_1(x')\,\mathrm{d}x' - \frac{\pi}{4}\right)\\
&= \left(\frac{\mathrm{i}A\mathrm{e}^{-\eta}}{4} - B\mathrm{e}^{\eta}\right)\frac{\mathrm{e}^{\mathrm{i}\phi_r}}{\sqrt{k_1}} + \left(\frac{A\mathrm{e}^{-\eta}}{4\mathrm{i}} - B\mathrm{e}^{\eta}\right)\frac{\mathrm{e}^{-\mathrm{i}\phi_r}}{\sqrt{k_1}},
\end{aligned}
$$

where $\phi_r(x) = \int_b^x k_1(x')\,\mathrm{d}x' - \pi/4$ and $\eta \equiv \int_a^b k_2(x)\,\mathrm{d}x$, since $\int_a^x = \int_a^b - \int_x^b$. For an outgoing wave in region III, we require $A\mathrm{e}^{-\eta} = 4\mathrm{i}B\mathrm{e}^{\eta}$ to eliminate the incoming $\mathrm{e}^{-\mathrm{i}\phi_r}$ term, so

$$
\psi_{III} = \frac{\mathrm{i}A\mathrm{e}^{-\eta-\mathrm{i}\pi/4}}{2\sqrt{k_1}}\exp\left(\mathrm{i}\int_b^x k_1(x')\,\mathrm{d}x'\right)
$$

On the left, we define

$$
\int_x^a k(x')\,\mathrm{d}x' - \frac{\pi}{4} = -\int_a^x k(x')\,\mathrm{d}x' - \frac{\pi}{4} \equiv -\phi_l(x) - \frac{\pi}{4}
$$

so $\phi_l(x) = \int_a^x k_1(x')\,\mathrm{d}x'$. Using the result for B in region I, representing the solution in terms of incoming and outgoing waves,

$$
\begin{aligned}
\psi_I &= \frac{A}{2\sqrt{k_1}}(\mathrm{e}^{-\mathrm{i}\phi_l-\mathrm{i}\pi/4} + \mathrm{e}^{\mathrm{i}\phi_l+\mathrm{i}\pi/4}) + \frac{B}{2\mathrm{i}\sqrt{k_1}}(\mathrm{e}^{-\mathrm{i}\phi_l-\mathrm{i}\pi/4} - \mathrm{e}^{\mathrm{i}\phi_l+\mathrm{i}\pi/4})\\
&= \frac{A\mathrm{e}^{\mathrm{i}\pi/4}}{2\sqrt{k_1}}\left(1 + \frac{\mathrm{e}^{-2\eta}}{4}\right)\mathrm{e}^{\mathrm{i}\phi_l} + \frac{A\mathrm{e}^{-\mathrm{i}\pi/4}}{2\sqrt{k_1}}\left(1 - \frac{\mathrm{e}^{-2\eta}}{4}\right)\mathrm{e}^{-\mathrm{i}\phi_l},
\end{aligned}
$$

so the *amplitude* transmission (or tunneling) and reflection coefficients [noting that $e^{-i\phi_l}$ is the left-going (reflected) term] are given by

$$t = \frac{e^{-\eta}}{1 + \frac{1}{4}e^{-2\eta}}$$
$$r = -i\left(\frac{1 - \frac{1}{4}e^{-2\eta}}{1 + \frac{1}{4}e^{-2\eta}}\right)$$

where we note that the phase of the reflected wave agrees with the result for the impenetrable barrier and that there is no phase change for the transmitted wave since $ie^{-i\pi/4} = e^{i\pi/4}$. For the magnitudes, we have

$$T = \left|\frac{e^{-\eta}}{1 + \frac{1}{4}e^{-2\eta}}\right|^2 = \frac{1}{(e^{\eta} + \frac{1}{4}e^{-\eta})^2} \tag{5.61}$$

$$R = \left|\frac{-i(1 - \frac{1}{4}e^{-2\eta})}{1 + \frac{1}{4}e^{-2\eta}}\right|^2 = \left(\frac{e^{\eta} - \frac{1}{4}e^{-\eta}}{e^{\eta} + \frac{1}{4}e^{-\eta}}\right)^2 \tag{5.62}$$

For thick enough barriers that $e^{-2\eta} \ll 1$, this simplifies to $T = e^{-2\eta}$ and $R = 1 - T$.

Problem 5.24 *Reflection coefficient.* Show that $R = 1 - T$ for any η.

Problem 5.25 *Transmission coefficient.* Consider a potential barrier described by $V(x) = V_0(1 - x^2/L^2)$ where V_0 is 100 eV and L is 2 Å (1 Å $= 10^{-8}$ m). Find the energy of an electron that will have a 50% probability of transmission. (*Hint:* First use the expression for the transmission coefficient to find η for 50% transmission, then set up the integral and set the limits so that the integrand vanishes at each end.)

6

Atomic Spectroscopy

There are many kinds of effects that resolve the degeneracy in many atomic systems, including external electric and magnetic fields that tend to orient particles or states with electric or magnetic dipole moments, and a number of internal interactions between the particles themselves, either due to their spin or their dipole moments. The agreement between the calculated splittings based on quantum mechanical perturbation calculations and the measured splittings measured in atomic spectroscopy is one of the most impressive accomplishments of quantum mechanics. For the effects of spin–spin correlations, we will examine the phenomenon only qualitatively, but for spin–orbit effects and external magnetic field effects, we will examine the effects quantitatively.

6.1 Effects of Symmetry

6.1.1 Particle Exchange Symmetry

Whereas in classical theory, the exchange operation between two identical particles is relatively straightforward, the further property that these particles are **indistinguishable** adds new features in quantum mechanics. Since classical particles can in principle be "tagged," one can still tell the difference between two states where the difference is that the two particles are simply exchanged. In quantum mechanics, however, two electrons are indistinguishable from one another, so one cannot distinguish which state is which. For example, we can represent a state with two electrons as either $\chi_+(1)\chi_-(2)$ or as $\chi_-(1)\chi_+(2)$. All we really know about this state is that one spin is up and the other down, but since the electrons 1 and 2 are indistinguishable, the two states are likewise indistinguishable. Since we cannot tell which is which, the appropriate representation of the state is a linear combination of the two, with equal weights, that we may write as

$$\chi^{\pm}(1,2) = \frac{1}{\sqrt{2}}[\chi_+(1)\chi_-(2) \pm \chi_-(1)\chi_+(2)]\,. \tag{6.1}$$

It should be noted that $\chi^+(1,2)$ does not change sign when the particles are exchanged but that $\chi^-(1,2)$ does change sign. This does not affect any classical quantity directly, but it does have some indirect effects in quantum mechanics. Because of this effect of interchanging particles, $\chi^+(1,2)$ is said to be **symmetric** and $\chi^-(1,2)$ is said to be **antisymmetric** with respect to particle exchange.

We now generalize these results to more general wave functions that may include both space and spin variables. Denoting a one-electron wave function with a particular set of quantum numbers n, ℓ, m, m_s, for particle 1 by $\psi_a(1)$ and another wave function with quantum numbers n', ℓ', m', m_s', for particle 2 by $\psi_b(2)$, the two-electron wave functions are

$$\psi_{ab}(1,2) = \psi_a(1) \cdot \psi_b(2) \quad \text{and} \quad \psi_{ba}(1,2) = \psi_b(1) \cdot \psi_a(2) \,.$$

We can then construct a set of a symmetric and an antisymmetric wave function such that

$$\psi_S = \tfrac{1}{\sqrt{2}}(\psi_{ab} + \psi_{ba}) \,, \tag{6.2}$$

$$\psi_A = \tfrac{1}{\sqrt{2}}(\psi_{ab} - \psi_{ba}) \,. \tag{6.3}$$

We now define a **particle exchange operator** that exchanges particles 1 and 2, so that

$$\hat{P}_{12}\psi_{ab}(1,2) = \psi_a(2)\psi_b(1) = \psi_{ba}(1,2) \,,$$
$$\hat{P}_{12}\psi_{ba}(1,2) = \psi_b(2)\psi_a(1) = \psi_{ab}(1,2) \,.$$

Using this operator on the wave functions of Equations (6.2) and (6.3), we find

$$\hat{P}_{12}\psi_S = +\psi_S \,,$$
$$\hat{P}_{12}\psi_A = -\psi_A \,.$$

Thus we observe that while neither ψ_{ab} nor ψ_{ba} are eigenfunctions of the exchange operator, both ψ_S and ψ_A are eigenfunctions with eigenvalues ± 1, respectively. We observe further that if the exchange operator (strictly speaking, the *two-particle* exchange operator) commutes with the Hamiltonian for a two-particle system, then the symmetry of the wave function is invariant in time.

We find experimentally that *the wave function for a system of electrons is antisymmetric with respect to the exchange of any two electrons.* This is equivalent to the statement of the **Pauli principle**, which states that no two electrons can occupy the same state. To see that these two statements are equivalent, we suppose that each one-particle state is ψ_a, so that the two particle wave functions are

$$\psi_S = \tfrac{1}{\sqrt{2}}(\psi_{aa} + \psi_{aa}) = \sqrt{2}\psi_{aa} \,,$$
$$\psi_A = \tfrac{1}{\sqrt{2}}(\psi_{aa} - \psi_{aa}) = 0 \,.$$

The antisymmetric wave function for two identical states thus does not exist, and the two statements of the exclusion principle are equivalent.

In order to generalize the two-particle wave functions to N-particle wave functions, we first note that the antisymmetric wave function can be written as

$$\psi_A = \frac{1}{\sqrt{2}} \begin{vmatrix} \psi_a(1) & \psi_a(2) \\ \psi_b(1) & \psi_b(2) \end{vmatrix} = \frac{1}{\sqrt{2}} [\psi_a(1) \cdot \psi_b(2) - \psi_b(1) \cdot \psi_a(2)] .$$

This form can be generalized into what is called the **Slater determinant** given by

$$\psi_A = \frac{1}{\sqrt{N!}} \begin{vmatrix} \psi_a(1) & \psi_a(2) & \cdots & \psi_a(N) \\ \psi_b(1) & \psi_b(2) & \cdots & \psi_b(N) \\ \vdots & \vdots & & \vdots \\ \psi_N(1) & \psi_N(2) & \cdots & \psi_N(N) \end{vmatrix} . \tag{6.4}$$

The complete antisymmetry of this representation is apparent from the rules of determinants whereby the interchange of any two rows or any two columns changes the sign of the determinant. The interchange of any two columns corresponds to the interchange of two particles.

While the examples above described electrons, the symmetry is not restricted to electrons. Experiments show that any identical, indistinguishable particles with half-integral spin ($\pm\frac{1}{2}$, $\pm\frac{3}{2}$, ...) have antisymmetric wave functions, and obey Fermi-Dirac statistics. Furthermore, particles with integral spin (0, ± 1, ± 2, ...) are described by symmetric wave functions and obey Bose-Einstein statistics. The world is thus divided into fermions and bosons. The characteristic states of these two classes of particles are thus constructed in one of the following ways:

1. Fermion states:

$$\psi_A = \begin{cases} \text{symmetric spatial function} \times \text{antisymmetric spin function} \\ \text{antisymmetric spatial function} \times \text{symmetric spin function.} \end{cases}$$

2. Boson states:

$$\psi_S = \begin{cases} \text{symmetric spatial function} \times \text{symmetric spin function} \\ \text{antisymmetric spatial function} \times \text{antisymmetric spin function.} \end{cases}$$

For a specific example, we shall consider a two-particle state where the spatial portions of the wave functions are denoted $\psi_n(1)$ and $\psi_{n'}(2)$ where n and n' represent the spatial quantum numbers and the spin functions by $\chi_\pm(k)$ with $k = 1, 2$. The spatial wave function is then of the form

$$\psi_{\text{spatial}} = \tfrac{1}{\sqrt{2}} [\psi_n(1)\psi_{n'}(2) \pm \psi_{n'}(1)\psi_n(2)] ,$$

where one is symmetric and the other antisymmetric. We represent the four independent spin functions by

$$\begin{gathered}\chi_+(1)\chi_+(2)\\ \tfrac{1}{\sqrt{2}}[\chi_+(1)\chi_-(2)+\chi_-(1)\chi_+(2)]\\ \chi_-(1)\chi_-(2)\\ \tfrac{1}{\sqrt{2}}[\chi_+(1)\chi_-(2)-\chi_-(1)\chi_+(2)]\,,\end{gathered}$$

where the first three are symmetric and the last is antisymmetric. The overall wave functions for two electrons are

$$\frac{1}{\sqrt{2}}[\psi_n(1)\psi_{n'}(2)-\psi_{n'}(1)\psi_n(2)]\cdot\begin{cases}\chi_+(1)\chi_+(2)\\ \tfrac{1}{\sqrt{2}}[\chi_+(1)\chi_-(2)+\chi_-(1)\chi_+(2)]\\ \chi_-(1)\chi_-(2)\end{cases} \tag{6.5}$$

$$\frac{1}{\sqrt{2}}[\psi_n(1)\psi_{n'}(2)+\psi_{n'}(1)\psi_n(2)]\cdot\frac{1}{\sqrt{2}}[\chi_+(1)\chi_-(2)-\chi_-(1)\chi_+(2)]. \tag{6.6}$$

Example 6.1

Exchange force. Consider two particles in the same one-dimensional box, $-a/2 \le x \le a/2$, so the two lowest wave functions for a particle in a box and their energies are

$$\psi_1(x)=\sqrt{\frac{2}{a}}\cos\frac{\pi x}{a}\,,\qquad E_1=\frac{\pi^2\hbar^2}{2ma^2}$$

$$\psi_2(x)=\sqrt{\frac{2}{a}}\sin\frac{2\pi x}{a}\,,\qquad E_2=\frac{4\pi^2\hbar^2}{2ma^2}$$

(a) Suppose the two particles are identical, but distinguishable, such as a neutron and an antineutron. The two-particle states are

$$\begin{aligned}\psi_{11}(x_1,x_2)&=\psi_1(x_1)\psi_1(x_2) \qquad & E&=E_1+E_1\\ \psi_{12}(x_1,x_2)&=\psi_1(x_1)\psi_2(x_2) & E&=E_1+E_2\\ \psi_{21}(x_1,x_2)&=\psi_2(x_1)\psi_1(x_2) & E&=E_2+E_1\\ \psi_{22}(x_1,x_2)&=\psi_2(x_1)\psi_2(x_2) & E&=E_2+E_2\end{aligned}$$

so ψ_{12} and ψ_{21} are degenerate with $E=5\pi^2\hbar^2/2ma^2$. We now wish to calculate the mean separation between the two particles by calculating $(\Delta x_{12})=\sqrt{\langle(x_1-x_2)^2\rangle}$ using $\psi_{12}(x_1,x_2)$. We therefore calculate

$$\begin{aligned}\langle(x_1-x_2)^2\rangle&=\int\!\!\int\psi_1^*(x_1)\psi_2^*(x_2)(x_1-x_2)^2\psi_1(x_1)\psi_2(x_2)\,\mathrm{d}x_1\,\mathrm{d}x_2\\ &=\frac{2}{a}\left(\int_{-\frac{a}{2}}^{\frac{a}{2}}x_1^2\cos^2\frac{\pi x_1}{a}\,\mathrm{d}x_1\right)\left(\int_{-\frac{a}{2}}^{\frac{a}{2}}|\psi_2(x_2)|^2\,\mathrm{d}x_2\right)\end{aligned}$$

$$
\begin{aligned}
&+\frac{2}{a}\left(\int_{-\frac{a}{2}}^{\frac{a}{2}} |\psi_1(x_1)|^2\,\mathrm{d}x_1\right)\left(\int_{-\frac{a}{2}}^{\frac{a}{2}} x_2^2\sin^2\frac{2\pi x_2}{a}\,\mathrm{d}x_2\right)\\
&-\frac{8}{a^2}\left(\int_{-\frac{a}{2}}^{\frac{a}{2}} x_1\cos^2\frac{\pi x_1}{a}\,\mathrm{d}x_1\right)\left(\int_{-\frac{a}{2}}^{\frac{a}{2}} x_2\sin^2\frac{2\pi x_2}{a}\,\mathrm{d}x_2\right)\\
&=\frac{2}{a}\int_{-\frac{a}{2}}^{\frac{a}{2}} x_1^2\cos^2\frac{\pi x_1}{a}\,\mathrm{d}x_1+\frac{2}{a}\int_{-\frac{a}{2}}^{\frac{a}{2}} x_2^2\sin^2\frac{2\pi x_2}{a}\,\mathrm{d}x_2\\
&=a^2\left(\frac{1}{12}-\frac{1}{2\pi^2}\right)+a^2\left(\frac{1}{12}-\frac{1}{8\pi^2}\right)\\
&=a^2\left(\frac{1}{6}-\frac{5}{8\pi^2}\right) \qquad (6.7)
\end{aligned}
$$

where we note that some of the integrals are the normalization integrals, and the cross term involve two integrals, each of which is odd between symmetric limits so that term vanishes, leading to the result that $(\Delta x_{12}) = 0.321a$.

(b) Now we consider two identical, indistinguishable particles, such as two neutrons in a nucleus. Since they both have half-integral spin, they are fermions, so the wave function must be antisymmetric when we exchange particles. Neither $\psi_{12}(x_1, x_2)$ nor $\psi_{21}(x_1, x_2)$ have this property, but an antisymmetric combination does, such as

$$\psi_A(x_1, x_2) = \frac{1}{\sqrt{2}}[\psi_1(x_1)\psi_2(x_2) - \psi_2(x_1)\psi_1(x_2)]$$

This state also has $E = E_1 + E_2$, so we will again calculate (Δx_{12}) for this wave function. We need

$$
\begin{aligned}
\langle(x_1-x_2)^2\rangle &= \int\int \psi_A^*(x_1-x_2)^2\psi_A\,\mathrm{d}x_1\,\mathrm{d}x_2\\
&=\frac{1}{2}\left[\int\int |\psi_1^*(x_1)\psi_2^*(x_2)-\psi_2^*(x_1)\psi_1^*(x_2)|^2\right.\\
&\qquad\left.\times\,(x_1^2+x_2^2-2x_1x_2)\,\mathrm{d}x_1\,\mathrm{d}x_2\right] \qquad (6.8)
\end{aligned}
$$

Eliminating those integrals that are normalization integrals and those that vanish by symmetry, the remaining integrals are

$$
\begin{aligned}
\langle(x_1-x_2)^2\rangle &= \frac{2}{a}\int_{-a/2}^{a/2} x_1^2\cos^2\frac{\pi x_1}{a}\,\mathrm{d}x_1+\frac{2}{a}\int_{-a/2}^{a/2} x_2^2\sin^2\frac{2\pi x_2}{a}\,\mathrm{d}x_2\\
&\quad+2\left(\frac{2}{a}\int_{-a/2}^{a/2} x\psi_2(x)\psi_1(x)\mathrm{d}x\right)^2 \qquad (6.9)
\end{aligned}
$$

We have already evaluated the first two integrals and

$$\int_{-a/2}^{a/2} x\psi_2(x)\psi_1(x)\mathrm{d}x = \frac{16a}{9\pi^2}$$

so

$$\begin{aligned}\langle(x_1 - x_2)^2\rangle &= a^2\left(\frac{1}{12} - \frac{1}{2\pi^2}\right) + a^2\left(\frac{1}{12} - \frac{1}{8\pi^2}\right) + 2\left(\frac{16a}{9\pi^2}\right)^2 \\ &= 0.1682a^2 \qquad (6.10)\end{aligned}$$

so $(\Delta x_{12}) = 0.410a$. Even though there is no force between the particles, the exchange symmetry keeps the particles farther apart than in case (a). This effective "repulsion" is referred to as the **exchange force** even though no force is involved.

(c) For the final case, we assume the particles are identical and indistinguishable except that they have integral spin, so they are bosons. The calculation is similar except that the wave function is symmetric and of the form

$$\psi_S(x_1, x_2) = \frac{1}{\sqrt{2}}[\psi_1(x_1)\psi_2(x_2) + \psi_2(x_1)\psi_1(x_2)]$$

The nonvanishing integrals for this case are the same as the previous case, but the result is

$$\begin{aligned}\langle(x_1 - x_2)^2\rangle &= a^2\left(\frac{1}{12} - \frac{1}{2\pi^2}\right) + a^2\left(\frac{1}{12} - \frac{1}{8\pi^2}\right) - 2\left(\frac{16a}{9\pi^2}\right)^2 \\ &= 0.03845a^2 \qquad (6.11)\end{aligned}$$

so $(\Delta x_{12}) = 0.196a$. The correlation in this case tends to bring them closer together.

□

Problem 6.1 Fill in the missing steps and evaluate the integrals leading to Equations (6.10) and (6.11).

Problem 6.2 Write an antisymmetric wave function for three electrons and show that it is equivalent to that from Equation (6.4).

Problem 6.3 *Particle exchange operator.*

(a) Apply the particle exchange operator $\hat{P}_{12}$ to each of the probability densities $\psi_{ab}^*\psi_{ab}$, and $\psi_{ba}^*\psi_{ba}$. Do they remain unchanged?

(b) Operate on $\psi_S^*\psi_S$ and $\psi_A^*\psi_A$ with $\hat{P}_{12}$. Do they remain unchanged?

(c) Do the above results suggest a physical reason for choosing ψ_S and ψ_A as wave functions rather than ψ_{ab} and ψ_{ba}? If so, why?

Problem 6.4 *Triplets and singlets.*

(a) Show that the spin portion of the wave functions of Equation (6.5) correspond to $S = 1$ with $M_S = 1, 0, -1$ by showing that they are eigenfunctions of $\hat{S}^2(1,2)$ and $\hat{S}_z(1,2)$ with the appropriate eigenvalues.

(b) Show that the spin portion of the wave function of Equation (6.6) corresponds to $S = 0$ with $M_S = 0$ by showing that it is an eigenfunction of $\hat{S}^2(1,2)$ and $\hat{S}_z(1,2)$ with the appropriate eigenvalues.

6.1.2 Exchange Degeneracy and Exchange Energy

We now wish to examine the effects these exchange symmetries may have on a quantum mechanical system. We begin again with a two-particle system, and consider the Hamiltonian for two *noninteracting* particles:

$$\hat{\mathcal{H}}(1,2) = \left[-\frac{\hbar^2}{2m}\nabla_1^2 + V_1\right] + \left[-\frac{\hbar^2}{2m}\nabla_2^2 + V_2\right] = \hat{\mathcal{H}}_1 + \hat{\mathcal{H}}_2\,, \tag{6.12}$$

$$[\hat{P}_{12}, \hat{\mathcal{H}}(1,2)] = 0. \tag{6.13}$$

Because of the result of Equation (6.13), the wave function symmetry will be invariant in time. Even if the Hamiltonians were interacting through symmetric terms such as the Coulomb force, the symmetries would be unchanged. Now this noninteracting system is degenerate, as we can see by examining the individual one electron wave functions, where

$$\begin{aligned}
\hat{\mathcal{H}}(1)\psi_n(1) &= E_n\psi_n(1)\\
\hat{\mathcal{H}}(2)\psi_n(2) &= E_n\psi_n(2)\\
\hat{\mathcal{H}}(1)\psi_{n'}(1) &= E_{n'}\psi_{n'}(1)\\
\hat{\mathcal{H}}(2)\psi_{n'}(2) &= E_{n'}\psi_{n'}(2)\,,
\end{aligned}$$

so that

$$\hat{\mathcal{H}}(1,2)\psi_A = (E_n + E_{n'})\psi_A\,,$$

where ψ_A is one of the four antisymmetric wave functions given by Equations (6.5) and (6.6).

While these states are degenerate with noninteracting particles, the degeneracy is resolved when interactions are included because the spins are correlated. For the solution of Equation (6.6), where the spins are antiparallel so that $S = 0$, which we call the **singlet** state, the electrons are allowed to come very close to one another. In fact, the probability of finding them close to one another is maximized when the two sets of coordinates overlap, since then the spatial portion of the wave function is doubled. For the parallel spin case, however, where $S = 1$, which we call the **triplet** state (all three of the solutions of Equation (6.5) have $S = 1$, but $M_S = 1, 0, -1$ for the three states), the electrons are forbidden to approach too close, since the spatial portion of the wave function vanishes as the coordinates of the two particles overlap. This means the electrons effectively avoid each other while the singlet electrons effectively attract each other. This tendency to avoid or attract is euphemistically described as due to the "exchange force," which is not a force at all, but simply an effect due to spin correlations. There is no classical analog for this "force" nor is there any operator that properly describes the effect. It is a purely quantum mechanical effect due to the correlations between the spin alignment of particles.

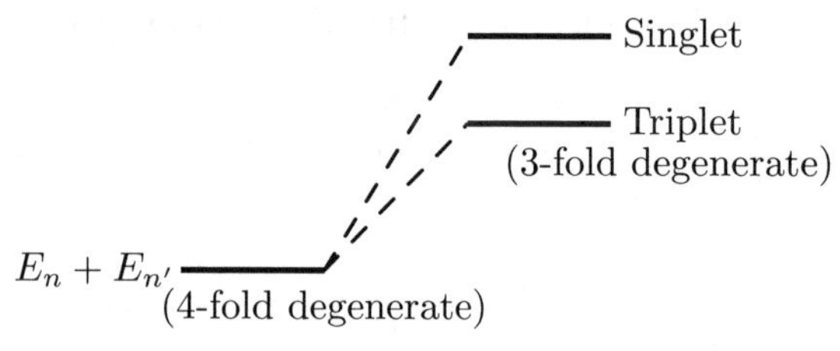

FIGURE 6.1
The exchange energy due to the Coulomb repulsion for two electrons.

With these correlation effects in mind, we can now see how the inclusion of interactions among the electrons will break the degeneracy. If, for example, we consider the Coulomb repulsion between the electron pair, it is obvious that electrons which are on the average closer to one another will have a larger average Coulomb potential energy term, so that the singlet solution will have a larger energy perturbation than the triplet cases, since in those cases, the electrons are farther apart on the average. This is depicted in Figure 6.1 where the triplet states are still degenerate with respect to one another (since all three have the same spatial wave function), but lie lower than the singlet state. Since electrons tend to occupy the lowest energy states, the triplet states tend to be occupied first, so that the spins tend to be aligned. It is in this sense that it is often said that the exchange force tends to align the spins. This effect is extremely important in the case of ferromagnetism, where in a closely coupled system of a crystal lattice, all of the spins in an entire macroscopic region, which is called a domain, are aligned in the same direction. Since the magnetic moment for an electron is $\boldsymbol{\mu} = -(e/2m)\boldsymbol{S}$, the net magnetic moment is extremely strong. In an unmagnetized sample of a ferromagnetic material, each domain is strongly magnetized but the domains are randomly aligned, while a magnetized sample has the domains at least partially aligned.

6.2 Spin–Orbit Coupling in Multielectron Atoms

6.2.1 Spin–Orbit Interaction

The spin–orbit interaction, which was introduced as Equation (5.5) in the list of perturbations in Section 5.1, derives from the energy associated with the magnetic moment of the electron spin interacting with the magnetic field due to the orbital motion of the electron. We usually think of the electron circulating about the nucleus, but it is just as valid to think, from the point of view of the electron, of the motion of the nucleus about the electron. Seen this way, there is current and hence a magnetic field at the electron's position, and its magnetic moment tends to align with this field.

From the coordinate frame of the electron, the magnetic field may be obtained from the relativistic transformation,

$$\boldsymbol{B}'_{\perp} = \gamma\left(\boldsymbol{B}_{\perp} - \frac{\boldsymbol{v}\times\boldsymbol{E}}{c^2}\right) \simeq -\frac{\boldsymbol{v}\times\boldsymbol{E}}{c^2}\,,$$

since there is no external magnetic field and $\gamma \simeq 1$. The electric field is simply the Coulomb field, which may be expressed in terms of the Coulomb potential energy as

$$\boldsymbol{E} = -\nabla\Phi = \frac{1}{e}\frac{\mathrm{d}V}{\mathrm{d}r}\hat{e}_r\,,$$

where $V = q\Phi = -e\Phi$. With $\boldsymbol{v} = v\hat{e}_\phi$, this gives

$$\boldsymbol{B}' = \frac{v}{ec^2}\frac{\mathrm{d}V}{\mathrm{d}r}\hat{e}_z\,.$$

Now the angular momentum is given by $\boldsymbol{\ell} = \boldsymbol{r}\times\boldsymbol{p} = rmv\hat{e}_z$, so that the field can be expressed in terms of the orbital angular momentum, such that

$$\boldsymbol{B}' = \frac{1}{emc^2}\frac{1}{r}\frac{\mathrm{d}V}{\mathrm{d}r}\boldsymbol{\ell}\,. \tag{6.14}$$

We found in Section 4.4.1 that the magnetic moment of the electron is related to its spin angular momentum. For the orbital magnetic moment we found that

$$|\mu| = \frac{e}{2m}\ell\,,$$

where ℓ is the orbital angular momentum. Keeping track of the directions and signs, the general relationship between angular momentum and magnetic moment is

$$\boldsymbol{\mu} = -\frac{e}{2m}\boldsymbol{\ell}\,. \tag{6.15}$$

Unfortunately, the relationship between **spin angular momentum** and the magnetic moment of an electron is **anomalous**. This means it does not follow the general relationship above, but is off by a factor of two, so that

$$\boldsymbol{\mu}_e = -\frac{e}{m}\boldsymbol{s}\,. \tag{6.16}$$

The reason for this anomaly is to be discovered in relativistic quantum mechanics. Using this relationship, however, the spin–orbit energy is obtained from

$$W = -\boldsymbol{\mu}\cdot\boldsymbol{B} = \frac{\boldsymbol{s}\cdot\boldsymbol{\ell}}{m^2c^2r}\frac{\mathrm{d}V}{\mathrm{d}r}\,. \tag{6.17}$$

6.2.2 The Thomas Precession

While the spin–orbit perturbation is the principal correction after including the exchange interaction and the electrostatic mutual repulsion term, one must check to find if any other terms are of similar order to the spin–orbit term. While not listed separately in some references, there is one weak relativistic effect that is not only similar in magnitude to the spin–orbit correction but scales exactly the same so that it is directly proportional to it and half as large. This term arises from the different reference frames of the electron and the nucleus that leads to the two separate "clocks" being slightly out of synchronization. If an observer traveling with the electron sees a time interval T, an observer on the nucleus will see an interval $T' = \gamma T$, where the speed is the orbital speed of the electron (we may assume circular orbits for simplicity). The orbital angular velocities measured by these two observers are thus $2\pi/T$ and $2\pi/T'$, respectively. In the reference frame of the electron, its spin vector continues to point in the same direction in space, but from the point of view of the other observer, the spin vector precesses at a rate given by the difference between the two frequencies. The kinematic precessional angular frequency is then given by

$$\Omega_k = \frac{2\pi}{T'}\left[\left(1-\frac{v^2}{c^2}\right)^{-1/2} - 1\right] \simeq \frac{2\pi}{T'}\frac{v^2}{2c^2}\,.$$

But from the basic definitions and Newton's law,

$$\frac{2\pi}{T'} = \omega = \frac{|\boldsymbol{\ell}|}{mr^2}\,, \quad \text{and} \quad \frac{mv^2}{r} = -\frac{\mathrm{d}V}{\mathrm{d}r}\,,$$

so that

$$\Omega_k = -\frac{|\boldsymbol{\ell}|}{2m^2c^2r}\frac{\mathrm{d}V}{\mathrm{d}r}\,. \tag{6.18}$$

We must also consider the Larmor precession, however, since a magnetic dipole in a magnetic field $\boldsymbol{B}'$ will precess about the field, which in this case is $\boldsymbol{B}'$. This precession is obtained by setting the torque equal to the rate of change of the angular momentum (in this case the **spin** angular momentum), so that

$$\boldsymbol{\mu}\times\boldsymbol{B}' = \frac{\mathrm{d}\boldsymbol{s}}{\mathrm{d}t} = \boldsymbol{\Omega}_\ell \times \boldsymbol{s}\,,$$

so that

$$\Omega_\ell = \frac{eB'}{m} = \frac{|\boldsymbol{\ell}|}{m^2c^2r}\frac{\mathrm{d}V}{\mathrm{d}r}\,. \tag{6.19}$$

It is apparent that this classical precession frequency is exactly twice the kinematic precessional frequency given by Equation (6.18) and of the opposite sign. If this classical result corresponds to our estimate of the spin–orbit energy, and then the Thomas precession term is added, it results in halving the spin–orbit coupling term, so that we replace Equation (6.17) by half its value,

$$W = -\boldsymbol{\mu}\cdot\boldsymbol{B} = \frac{\boldsymbol{s}\cdot\boldsymbol{\ell}}{2m^2c^2r}\frac{\mathrm{d}V}{\mathrm{d}r}\,, \tag{6.20}$$

for our perturbation estimate. Integrating over the radial coordinate in order to obtain the first order perturbation theory result, we may write

$$\langle W\rangle = \frac{1}{2}\frac{Ze^2}{4\pi\epsilon_0 m^2c^2}\left\langle\frac{1}{r^3}\right\rangle \boldsymbol{s}\cdot\boldsymbol{\ell}\,, \tag{6.21}$$

where the result from Equation (B.68) in Appendix B is

$$\left\langle\frac{1}{r^3}\right\rangle = \frac{Z^3}{a_0^3n^3\ell(\ell+\frac{1}{2})(\ell+1)}\,. \tag{6.22}$$

Problem 6.5 Evaluate $\langle r^{-3}\rangle$ for cases (a) $n=2,\ \ell=1$, (b) $n=3,\ \ell=1$, and (c) $n=3,\ \ell=2$, and show that they each satisfy Equation (6.22).

6.2.3 *LS* Coupling, or Russell–Saunders Coupling

When the energy shift due to spin–orbit effects occurs in a many-electron atom, the radial part of the wave functions prevents us from making accurate evaluations of the potential function due to the complexity of the multielectron system. The angular part, however, is still amenable to the same type of analysis we did for the one-electron atom, except that we now deal with the total orbital angular momentum and the total spin angular momentum, so that we have

$$\begin{aligned} \boldsymbol{J} &= \boldsymbol{L}+\boldsymbol{S} \\ |\boldsymbol{J}|^2 &= |\boldsymbol{L}|^2+|\boldsymbol{S}|^2+2\boldsymbol{L}\cdot\boldsymbol{S} \\ \text{so that}\qquad \boldsymbol{L}\cdot\boldsymbol{S} &= \tfrac{1}{2}(|\boldsymbol{J}|^2-|\boldsymbol{L}|^2-|\boldsymbol{S}|^2)\,. \end{aligned}$$

Representing the contribution due to the radial part by a constant $C' = C\hbar^2$, where C is obtained from Equation (6.21), we may represent the perturbation by

$$\langle\boldsymbol{L}\cdot\boldsymbol{S}\rangle = \tfrac{1}{2}C'[J(J+1)-L(L+1)-S(S+1)]\,,$$

since $|\boldsymbol{J}|^2$, $|\boldsymbol{L}|^2$, and $|\boldsymbol{S}|^2$ are quantized constants of the motion. For a group of energy levels with a given L and S, the *spacing* is then proportional to

$J(J+1)$, and given by

$$E_{J+1} - E_J = \tfrac{1}{2}C'[(J+1)(J+2) - J(J+1)] = C'(J+1)\,,$$

so *the spacing between levels of a multiplet with given L and S is proportional to the larger of the two J values.* This is called the **Landé interval rule**.

For an example, we consider an atom with two valence electrons and suppose one is in a $4p$ state ($n=4$, $\ell=1$) and the other in a $4d$ state ($n=4$, $\ell=2$) using the spectroscopic notation that $\ell = 0,1,2,3,4,5,\ldots$ is represented by the designation $s,p,d,f,g,\ldots$ respectively, where the letters stand for **sharp, principal, diffuse, fuzzy,** and then in alphabetical order (g,h,i,...). The exchange energy or spin–spin correlation energy will first split the states with $S=0$ being higher in energy than the $S=1$ state. Then the mutual electrostatic interaction will further split the states, and this is governed principally by the orbital angular momentum, since states with higher angular momentum will generally be more distant on the average than those with lower angular momentum, and the larger values of L will lie lowest. Finally, the spin–orbit coupling will split these into even finer multiplets as indicated in Figure 6.2. The spacing illustrates the Landé interval rule for a case with one electron in a $4p$ state and the other in a $4d$ state.

The first splitting is due to the exchange, or spin–spin correlations. To see how this goes for any number of electrons, we need to examine how each additional electron affects the splitting due to this effect.

For a specific example, let us consider a case with three electrons, and to be explicit, we shall assume they are in the $2p3p4d$ state. We first add the two p electrons together, and then add the third electron to the combination. For the two electrons in our example, the two spins can combine to an $S=0$ state, which we call a **Singlet** since there are $2S+1=1$ state. They may also combine to form a **Triplet** $S=1$ state, since there are $2S+1=3$ states for this case. Adding more electrons adds new levels according to Figure 6.3, where not all of the levels may actually occur, since the exclusion principle (no two electrons may have the same quantum numbers) may exclude some possibilities if n and ℓ are the same for both electrons. With n different for all three electrons, the exclusion principle will not affect this example. When we add the third electron, we see from Figure 6.3 that there will be further splitting into two doublet states ($S=\frac{1}{2}$) and one quartet state ($S=\frac{3}{2}$). Each additional electron splits the states further until the shell is half full (if all have the same n and ℓ), then the exclusion principal reduces the number of levels until with all but one state in a shell filled, we are back to a singlet state.

		$^2D_{3/2,5/2}$		
	$^2P_{1/2,3/2}$	$^2D_{3/2,5/2}$	$^2F_{5/2,7/2}$	
$^2S_{1/2}$	$^2P_{1/2,3/2}$	$^2D_{3/2,5/2}$	$^2F_{5/2,7/2}$	$^2G_{7/2,9/2}$

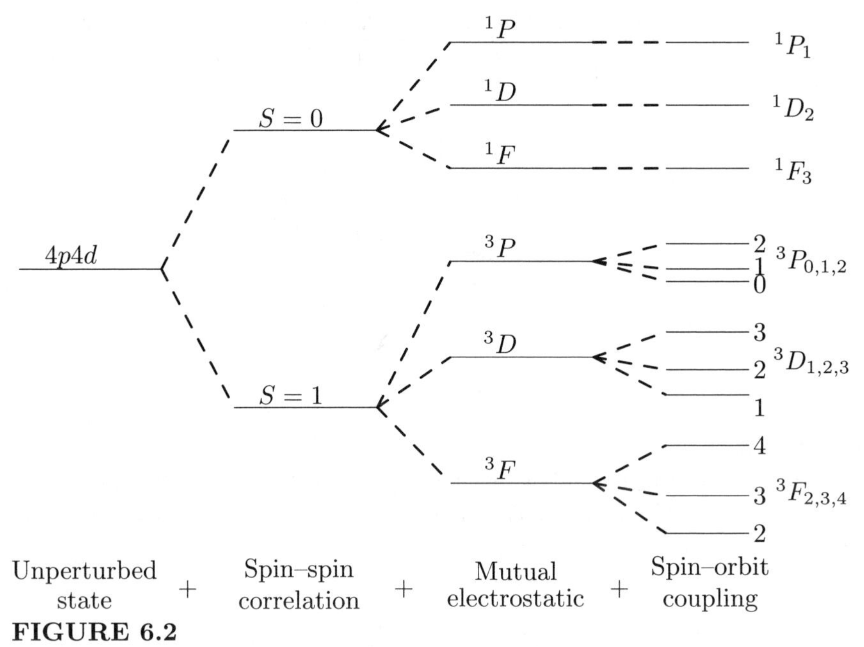

FIGURE 6.2
LS coupling diagram for a 4*p* and a 4*d* electron.

We next combine the orbital angular momenta to find what combinations may occur. It is apparent, for example, that if we combine the two p states ($\ell = 1$), we can have an S state ($L = 0$), a P state ($L = 1$), and a D state ($L = 2$). The maxima and minima are determined from the sum and difference of the two ℓ values, and all integral values in between can occur as they do not need to be parallel or antiparallel. Adding the third d electron then leads to the cases

$$\begin{aligned} S + d &\longrightarrow D \\ P + d &\longrightarrow P, D, F \\ D + d &\longrightarrow S, P, D, F, G \end{aligned}$$

At the last stage, we add the $\boldsymbol{L} \cdot \boldsymbol{S}$ coupling, which will separate each of these into multiplets according to their respective J values. Then each of the two doublet sets will have and the quartet will have

$$\begin{array}{ccccc} & & {}^4D_{1/2,3/2,5/2,7/2} & & \\ & {}^4P_{1/2,3/2,5/2} & {}^4D_{1/2,3/2,5/2,7/2} & {}^4F_{3/2,5/2,7/2,9/2} & \\ {}^4S_{3/2} & {}^4P_{1/2,3/2,5/2} & {}^4D_{1/2,3/2,5/2,7/2} & {}^4F_{3/2,5/2,7/2,9/2} & {}^4G_{5/2,7/2,9/2,11/2}, \end{array}$$

which adds up to a total of *65 distinct levels* for this case. It is apparent that spectroscopy can be very complicated.

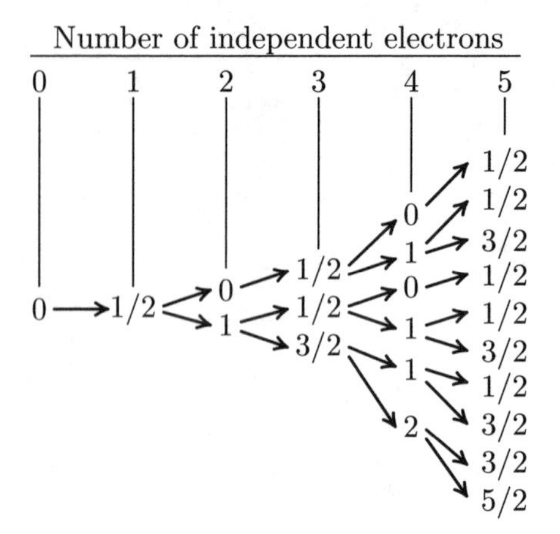

FIGURE 6.3
Resultant spin quantum numbers from combining 1, 2, 3, 4, and 5 independent electrons. For over half-full shells, the exclusion principle may exclude some cases.

In Figures 6.2 and 6.4 and in the list of the doublet and quartet states above, the designation of each state after all of the effects are included is called its **Spectral term**. This gives the S-value in the leading superscript, the L-value through the principal letter, and the J-value through the subscript where the recipe for the notation is ${}^{2S+1}(L)_J$ where the numbers for the superscript and subscript are used, but the letter corresponding to the L-value is used. The superscript therefore gives the number of states due to the spin, so that, for example, an $S = 1$ state has $2S + 1 = 3$ for the superscript and indicates a *triplet*.

In order to examine the effects of the exclusion principle, we examine another case, this time with only two electrons, but both with at least one and possibly two identical quantum numbers. To keep it simple, we consider two p electrons in shells n and n', and we will examine both $n \neq n'$ and $n = n'$. The splitting is illustrated in Figure 6.4, where the states indicated by the dashed lines are forbidden by the exclusion principle if $n = n'$.

In order to understand which states will be excluded, group theory is the best way to calculate the general case, but that is beyond the scope of our study, so we rather tabulate the various quantum numbers for the case $n = n'$, keeping track of the various possible values of $m_{\ell 1}$, $m_{\ell 2}$, m_{s1}, and m_{s2} in Table 6.1. We first strike out the states with both spins up or both spins down with $m_{\ell 1} = m_{\ell 2}$, since they are indistinguishable. Furthermore, since electrons may be interchanged without effect, a simultaneous switch of $m_{\ell 1} \leftrightarrow m_{\ell 2}$ and $m_{s1} \leftrightarrow m_{s2}$ is identical, so we do not duplicate states. This reduces the number of possible states to 15. We now must match these with the states in Figure 6.4.

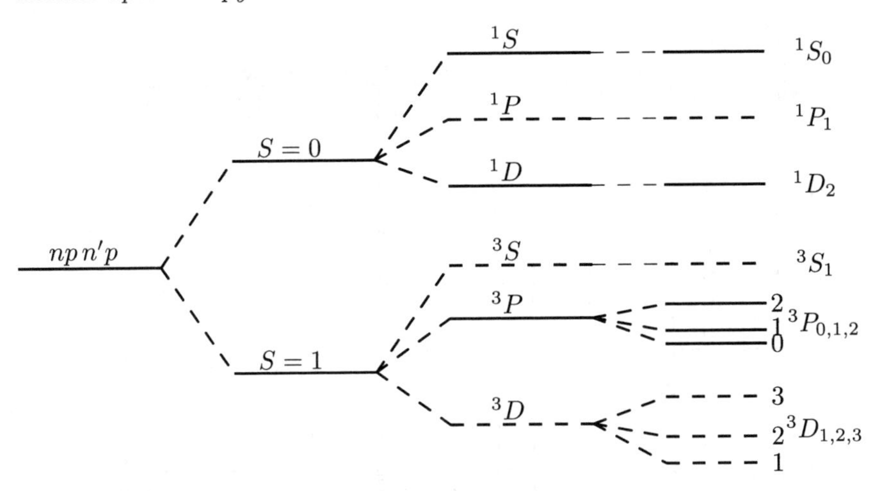

FIGURE 6.4
LS coupling diagram for two p electrons. If $n = n'$, then the exclusion principle excludes states with dashed lines.

1. First, there are no $m_J = 3$ values in Table 6.1 since that would require both $m_{\ell 1} = m_{\ell 2}$ and $m_{s1} = m_{s2}$ and this is excluded by the exclusion principle. This rules out the $^3\mathrm{D}_{1,2,3}$ state, since if one is missing, all are missing.

2. Next, the state labeled 2 requires a state with $S = 1$ (triplet) and $L \geq 1$, so we must have the $^3\mathrm{P}_{0,1,2}$ with $1 + 3 + 5 = 9$ M_J states.

3. Then, the state labeled 1 must have $L = 2$, so we must have the $^1\mathrm{D}_2$ with 5 states for a total of 14 out of the 15 states.

4. There is only one state left, so this requires the $^1\mathrm{S}_0$ state, and hence the $^1\mathrm{P}_1$ and the $^3\mathrm{S}_1$ states are missing.

On the basis of the considerations above, we can formulate some rules (commonly known as Hund's rules) to predict the spectroscopic character for the ground state of an atom, since the ground state is the lowest lying state:

> The ground state should correspond to the *highest* values of L and S, and if the state is less than half full, the *smallest* value of J, that are possible under the exclusion principle. (If the state is more than half full, the *largest* value of J will lie lowest.)

When there are more than two equivalent electrons, it is very difficult to follow the procedure above. Using group theory methods, the states that will survive are shown in Table 6.2, where for each case only unclosed shell electron configurations are shown. The example, which had two p states in the same

TABLE 6.1
Values of m_ℓ and m_s for two equivalent p electrons.

$m_{\ell 1}$	$m_{\ell 2}$	m_{s1}	m_{s2}	label	$m_{\ell 1}$	$m_{\ell 2}$	m_{s1}	m_{s2}	label
1	1	+	+	out	0	−1	+	+	11
1	1	+	−	1	0	−1	+	−	12
1	1	−	+	1	0	−1	−	+	13
1	1	−	−	out	0	−1	−	−	14
1	0	+	+	2	−1	1	+	+	6
1	0	+	−	3	−1	1	+	−	8
1	0	−	+	4	−1	1	−	+	7
1	0	−	−	5	−1	1	−	−	9
1	−1	+	+	6	−1	0	+	+	11
1	−1	+	−	7	−1	0	+	−	13
1	−1	−	+	8	−1	0	−	+	12
1	−1	−	−	9	−1	0	−	−	14
0	1	+	+	2	−1	−1	+	+	out
0	1	+	−	4	−1	−1	+	−	15
0	1	−	+	3	−1	−1	−	+	15
0	1	−	−	5	−1	−1	−	−	out
0	0	+	+	out					
0	0	+	−	10					
0	0	−	+	10					
0	0	−	−	out					

shell is on the third line, which indicates that remaining states are ^{1}S, ^{1}D, and a ^{3}P, and that this is the same for either two p electrons or four (6-2=4) p electrons, since a filled p-shell has six electrons. All of the corresponding J states are implied for each term.

Problem 6.6 *Two equivalent d electrons.* Work out the fine-structure splitting for two electrons in $nd\,n'd$ states, and sketch the figure analogous to Figure 6.4 and the table corresponding to Table 6.1.

Problem 6.7 What spectral terms result from an electron configuration $3d4f$ assuming LS coupling? Indicate on a sketch the expected spacings.

Problem 6.8 Show that the following ground state terms satisfy the rules above: B($^2P_{1/2}$), Sc($^2D_{3/2}$), Se($^3P_2$), Zr($^3F_2$), Nb($^6D_{1/2}$), Pr($^4I_{9/2}$), Ta($^4F_{3/2}$).

6.2.4 Selection Rules for LS Coupling

In order to fully understand the spectral lines from the splittings above, it is not sufficient to know only the energy levels for each state. Since allowed transitions occur *between* two states by dipole radiation (a single photon emitted or absorbed), there are rules that govern which transitions may occur and which may not occur. When spin effects are included, the simple rules we

TABLE 6.2
LS terms arising from equivalent electrons. In the first column are the configurations of the (unclosed shell) electrons. In the second column are the possible *LS* values and their multiplicities. For some of the *f*-electrons, the exponents inside the parentheses indicate the number of distinct terms having the multiplicity given (e.g., for f^5, f^9, there are 4 ^{2}P states, 5 ^{2}D states, etc.).

Configuration	*LS* terms
s^2	^{1}S
p^1, p^5	^{2}P
p^2, p^4	^{1}(SD) ^{3}P
p^3	^{2}(PD) ^{4}S
d^1, d^9	^{2}D
d^2, d^8	^{1}(SDG) ^{3}(PF)
d^3, d^7	^{2}D ^{2}(PDFGH) ^{4}(PF)
d^4, d^6	^{1}(SDG) ^{3}(PF) ^{1}(SDFGI) ^{3}(PDFGH) ^{5}D
d^5	^{2}D ^{2}(PDFGH) ^{4}(PF) ^{2}(SDFGI) ^{4}(DG) ^{6}S
f^1, f^{13}	^{2}F
f^2, f^{12}	^{1}(SDGI) ^{3}(PFH)
f^3, f^{11}	^{2}(PD^2 F^2 G^2 H^2 IKL) ^{4}(SDFGI)
f^4, f^{10}	^{1}(S^2 D^4 FG^4 H^2 I^3 IKL^2 N) ^{3}(P^{3}D^2 F^4 G^3 H^4 I^2 K^2 LM) ^{5}(SDFGI)
f^5, f^9	^{2}(P^4 D^5 F^7 G^6 H^7 I^5 K^5 L^3 M^2 NO) ^{4}(SP^2 D^3 F^4 G^4 H^3 I^3 K^2 LM) ^{6}(PFH)
f^6, f^8	^{1}(S^4 PD^6 F^4 G^8 H^4 I^7 K^3 L^4 M^2 N^2 Q) ^{3}(P^6 D^5 F^9 G^7 H^9 I^6 K^6 L^3 M^3 NO) ^{5}(SPD^3 F^2 G^3 H^2 I^2 KL) ^{7}F
f^7	^{2}(S^2 P^5 D^7 F^{10} G^{10} H^9 I^9 K^7 L^5 M^4 N^2 OQ) ^{4}(S^2 P^2 D^6 F^5 G^7 H^5 I^5 K^3 L^3 MN) ^{6}(PDFGHI) ^{8}S

obtained from calculating which dipole moment terms were nonzero simply from the spatial portion of the wave function are not adequate. The more general rules for LS coupling, for the vast majority of states, are:

1. Transitions occur only between configurations in which *one electron* changes its state. (Only one electron "jumps" at a time.)

2. The ℓ-value of the jumping electron must change by one unit

$$\Delta\ell = \pm 1 .$$

3. For the atom as a whole, the quantum numbers L, S, J, and M_J must change as follows:

$$\Delta S = 0,$$

$$
\begin{aligned}
\Delta L &= 0,\ \pm 1, \\
\Delta J &= 0,\ \pm 1, \qquad \text{but } J=0 \to J=0 \text{ forbidden} \\
\Delta M_J &= 0,\ \pm 1, \qquad \text{but } M_J=0 \to M_J=0 \text{ forbidden if } \Delta J = 0\,.
\end{aligned}
$$

6.2.5 Zeeman Effect

Another aspect of the interactions between the orbital angular momentum and spin angular momentum is observable under the influence of an external magnetic field. The coupling is again due to the magnetic moments, and the energy associated with the coupling is given by

$$W = -\boldsymbol{\mu}\cdot\boldsymbol{B}\,.$$

Since this is typically even smaller than the fine structure splitting, we may again treat this as a perturbation. Adding the individual orbital magnetic moments and spin magnetic moments for each electron, we may write

$$
\begin{aligned}
\boldsymbol{\mu} &= -\sum_{i=1}^{N}\left[\frac{1}{2}\left(\frac{e}{m}\right)\boldsymbol{l}_i + \left(\frac{e}{m}\right)\boldsymbol{s}_i\right] \\
&= -\frac{1}{2}\left(\frac{e}{m}\right)(\boldsymbol{L}+2\boldsymbol{S}) \qquad (6.23) \\
&= -\frac{1}{2}\left(\frac{e}{m}\right)(\boldsymbol{J}+\boldsymbol{S})\,. \qquad (6.24)
\end{aligned}
$$

From this, it is apparent that the magnetic moment will not be parallel to the total angular momentum $\boldsymbol{J}$ unless $\boldsymbol{S} = 0$. This is due to the fact that the proportionality constants for orbital and spin magnetic moments are different. It follows that we do not have a simple constant of the motion here to discuss the precession of the magnetic moment in the external field. In order to make a good estimate of the effect, we note that in the absence of the magnetic field, $\boldsymbol{J}$ is a constant of the motion, and $\boldsymbol{L}$ and $\boldsymbol{S}$ precess about $\boldsymbol{J}$ at a rate depending on the strength of the LS coupling. This precession frequency is just the energy from the LS coupling divided by $\hbar$.

When the magnetic field is applied, however, the total angular momentum is no longer a constant of the motion, since it will precess about the direction of the applied magnetic field, as illustrated in Figure 6.5. For a weak field, this precession will be much slower than the $\boldsymbol{L}\cdot\boldsymbol{S}$ precession (in the upper half of the figure), so the time average of the magnetic moment will be well approximated by its component along $\boldsymbol{J}$, multiplied by the component of $\boldsymbol{J}$ along $\boldsymbol{B}$. The idea is that $\boldsymbol{\mu}$ precesses relatively rapidly about $\boldsymbol{J}$ (in the lower half of the figure) while $\boldsymbol{J}$ precesses comparatively slowly about $\boldsymbol{B}$. The energy is then given by

$$W = \frac{1}{2}\left(\frac{e}{m}\right)\frac{[(\boldsymbol{J}+\boldsymbol{S})\cdot\boldsymbol{J}](\boldsymbol{J}\cdot\boldsymbol{B})}{|\boldsymbol{J}|^2}$$

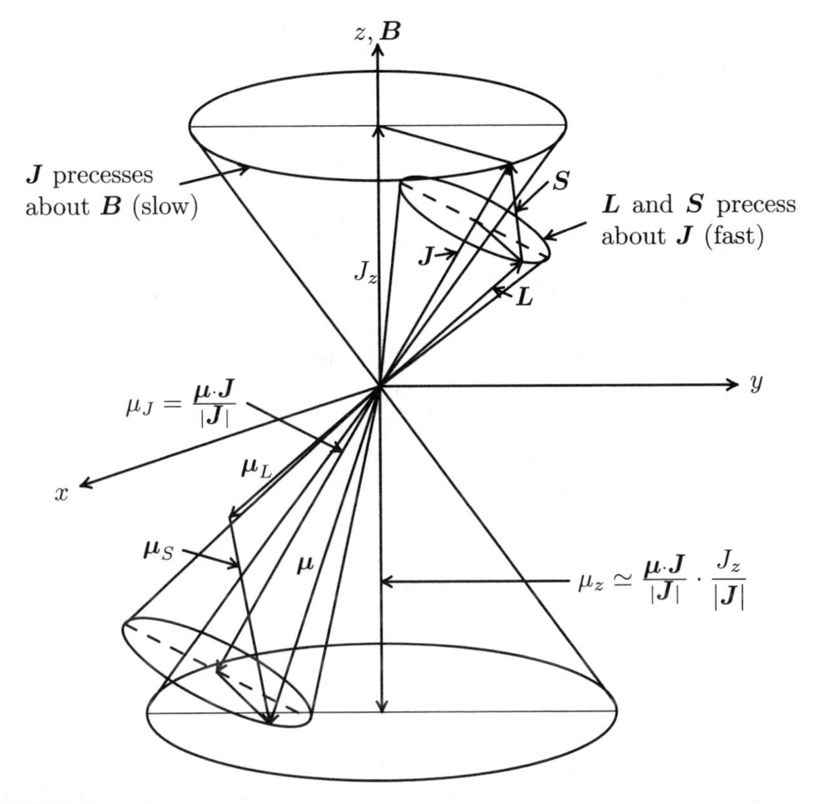

FIGURE 6.5
Illustration of precessions leading to approximations for the Zeeman effect for weak fields ($B < 1$ T).

$$= \frac{1}{2}\left(\frac{eB}{m}\right)\frac{(|\boldsymbol{J}|^2 + \boldsymbol{J}\cdot\boldsymbol{S})J_z}{|\boldsymbol{J}|^2} . \tag{6.25}$$

We may evaluate $\boldsymbol{J}\cdot\boldsymbol{S}$ the way we evaluated $\boldsymbol{L}\cdot\boldsymbol{S}$, so that

$$\boldsymbol{J}\cdot\boldsymbol{S} = \tfrac{1}{2}(|\boldsymbol{J}|^2 + |\boldsymbol{S}|^2 - |\boldsymbol{L}|^2) ,$$

so $\hat{W}$ becomes

$$\hat{W} = \frac{1}{2}\left(\frac{eB}{m}\right)\frac{[|\hat{\boldsymbol{J}}|^2 + \frac{1}{2}(|\hat{\boldsymbol{J}}|^2 + |\hat{\boldsymbol{S}}|^2 - |\hat{\boldsymbol{L}}|^2)]\hat{J}_z}{|\hat{\boldsymbol{J}}|^2} .$$

Since all the operators for $\boldsymbol{J}$, $\boldsymbol{L}$, and $\boldsymbol{S}$ commute with one another, the operator $\hat{W}$ is straightforward, and we may replace the operators by their corresponding eigenvalues, with the result

$$\langle W\rangle = \frac{1}{2}\left(\frac{eB}{m}\right)\frac{\{J(J+1) + \frac{1}{2}[J(J+1) + S(S+1) - L(L+1)]\}M_J\hbar}{J(J+1)}$$

$$= \frac{1}{2}\left(\frac{e\hbar B}{m}\right)\left[1 + \frac{J(J+1) + S(S+1) - L(L+1)}{2J(J+1)}\right] M_J$$
$$= \frac{1}{2}\left(\frac{e\hbar B}{m}\right) g M_j \tag{6.26}$$

where

$$g \equiv 1 + \frac{J(J+1) + S(S+1) - L(L+1)}{2J(J+1)} \tag{6.27}$$

is called the **Landé g-factor**. Because the splitting is now proportional to the M_J value, the degeneracy of the spin–orbit coupling is now resolved, and forms a new multiplet with $2J+1$ members, running from $-J$ to J. In classical theory, each level characterized by a single value of L, J, and S should have only three levels because of the selection rule $\Delta M_J = 0, \pm 1$. We can see how this could occur if the g-values for each state were the same, since the energy shift for each level is given by

$$\Delta E_1 = \mu_B B g_1 M_{J_1}\,, \qquad \text{and} \qquad \Delta E_2 = \mu_B B g_2 M_{J_2}\,,$$

and the frequency of the spectral transition would be

$$\Delta\omega_{12} = \frac{\Delta E_1 - \Delta E_2}{\hbar}$$
$$\Delta E_{12} = \mu_B B(g_1 M_{J_1} - g_2 M_{J_2})\,, \tag{6.28}$$

where $\mu_B = e\hbar/2m$ (called the **Bohr magneton**), so that if $g_1 = g_2$, then $\Delta E_{12} = \mu_B B g_1 \Delta M_J$, giving just the three components. This is in *qualitative* agreement with the classical theory. The so-called "anomalous" Zeeman patterns are then primarily due to the fact that the *Landé g-factor* varies from level to level, and not that the selection rule is violated.

Example 6.2

Zeeman structure. For an example with half-integral spin, we consider the levels corresponding to a transition from a $^{10}H_{3/2}$ state to a $^{10}G_{1/2}$ state. For the upper state, we have $S = 9/2$, $L = 5$, and $J = 3/2$. The g-factor for this state is then

$$g_1 = 1 + \frac{\frac{3}{2}(\frac{3}{2}+1) + \frac{9}{2}(\frac{9}{2}+1) - 5(5+1)}{2\frac{3}{2}(\frac{3}{2}+1)} = \frac{4}{5}\,.$$

For the lower state, we have $S = 9/2$, $L = 4$, and $J = 1/2$. The g-factor for this state is then

$$g_2 = 1 + \frac{\frac{1}{2}(\frac{1}{2}+1) + \frac{9}{2}(\frac{9}{2}+1) - 4(4+1)}{2\frac{1}{2}(\frac{1}{2}+1)} = \frac{14}{3}\,.$$

The initial state has four M_J states $(\frac{3}{2}, \frac{1}{2}, -\frac{1}{2}, -\frac{3}{2})$ and the final state has two M_J states $\pm\frac{1}{2}$, so the possible transitions are illustrated in Figure 6.6.

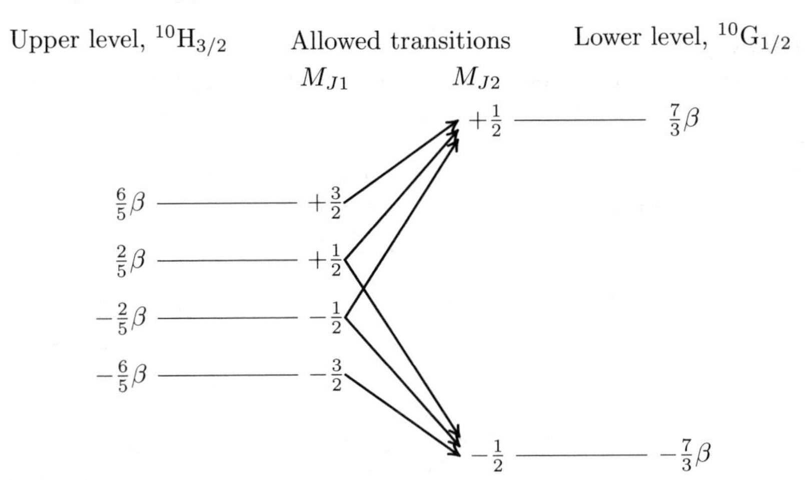

FIGURE 6.6
Allowed transitions for the transition $^{10}H_{3/2} \rightarrow {}^{10}G_{1/2}$ with $\beta = \mu_B B$.

TABLE 6.3
M_J-values and spectral frequencies for the transition $^{10}H_{3/2} \rightarrow {}^{10}G_{1/2}$.

	$\Delta M_J = +1$		$\Delta M_J = 0$		$\Delta M_J = -1$	
M_{J1}	$-\frac{1}{2}$	$-\frac{3}{2}$	$+\frac{1}{2}$	$-\frac{1}{2}$	$+\frac{3}{2}$	$+\frac{1}{2}$
M_{J2}	$+\frac{1}{2}$	$-\frac{1}{2}$	$+\frac{1}{2}$	$-\frac{1}{2}$	$+\frac{1}{2}$	$-\frac{1}{2}$
ΔE_{12}	$(-\frac{2}{5} - \frac{7}{3})\beta$	$(-\frac{6}{5} + \frac{7}{3})\beta$	$(\frac{2}{5} - \frac{7}{3})\beta$	$(-\frac{2}{5} + \frac{7}{3})\beta$	$(\frac{6}{5} - \frac{7}{3})\beta$	$(\frac{2}{5} + \frac{7}{3})\beta$

We may then determine the energies associated with these transitions by grouping them according to the values of ΔM_J as in Table 6.3.

The resulting energies are $\pm 41\beta/15$, $\pm 29\beta/15$, and $\pm 17\beta/15$ where $\beta = \mu_B B$. The spectroscopist sees the pattern of these six states as six lines as illustrated in Figure 6.7 (to scale).

6	6	17	6	6

FIGURE 6.7
Zeeman splittings in units of $2\beta/15$ for the transition $^{10}H_{3/2} \rightarrow {}^{10}G_{1/2}$.

□

Example 6.3
$^3P_1 \rightarrow {}^3D_2$ *transitions.* As another example with integral J and S, we consider transitions from $^3\mathrm{P}_1 \rightarrow {}^3\mathrm{D}_2$. Figure 6.8 shows the g-values for the upper and lower levels and illustrates the allowed transitions. Then Table 6.4 shows how the energies for each transition are found, and finally, Figure 6.9 shows graphically the splittings (to scale). □

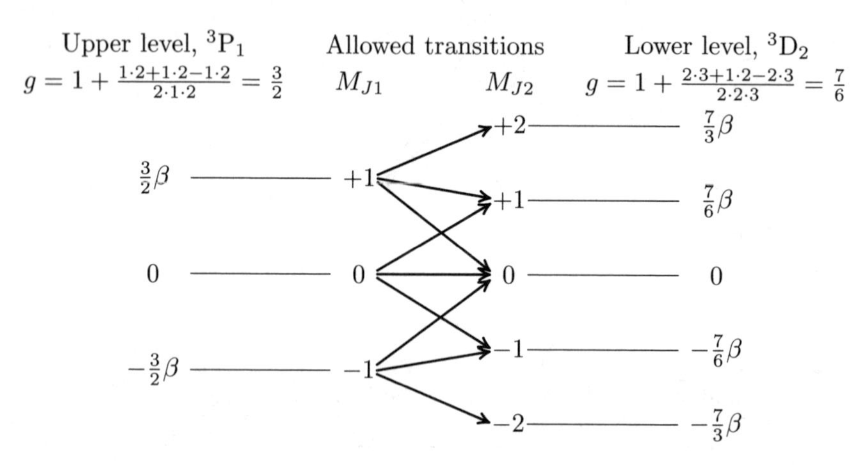

FIGURE 6.8
The g-values and allowed transitions for the transition $^3\mathrm{P}_1 \rightarrow {}^3\mathrm{D}_2$ with $\beta = \mu_B B$.

TABLE 6.4
M_J-values and spectral frequencies for the transition $^3\mathrm{P}_1 - {}^3\mathrm{D}_2$.

	$\Delta M_J = +1$			$\Delta M_J = 0$			$\Delta M_J = -1$		
M_{J1}	+1	0	−1	+1	0	−1	+1	0	−1
M_{J2}	+2	+1	0	+1	0	−1	0	−1	−2
ΔE_{12}	$(\frac{3}{2} - \frac{7}{3})\beta$	$-\frac{7}{6}\beta$	$-\frac{3}{2}\beta$	$(\frac{3}{2} - \frac{7}{6})\beta$	0	$-(\frac{3}{2} - \frac{7}{6})\beta$	$\frac{3}{2}\beta$	$\frac{7}{6}\beta$	$-(\frac{3}{2} - \frac{7}{3})\beta$

Problem 6.9 Find the Zeeman structure of a spectral line that results from the transition $^4\mathrm{F}_{3/2} \rightarrow {}^4\mathrm{D}_{5/2}$.

Problem 6.10 *Zeeman width.*

(a) By what factor will the total spread of the Zeeman pattern of the transition $^{10}\mathrm{H}_{5/2} \rightarrow {}^{10}\mathrm{G}_{3/2}$ exceed the classical value?

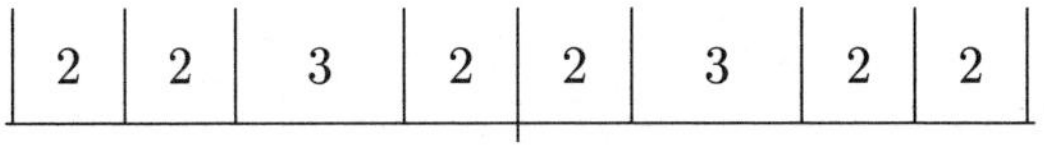

2	2	3	2	2	3	2	2

FIGURE 6.9
Zeeman splittings in units of $\beta/6$ for the transition $^3\mathrm{P}_1$ – $^3\mathrm{D}_2$.

(b) How many lines will appear in the Zeeman pattern of this transition?

Problem 6.11 A certain spectral line is known to result from a transition from a ^{3}D level to another level whose LS designation is unknown. The Zeeman pattern of the line is shown (to scale) in Figure 6.10.

(a) List all the possible spectral terms for the upper state and the corresponding lower states that satisfy the two selection rules: $\Delta L = 0, \pm 1$ and $\Delta J = 0, \pm 1$.

(b) Reduce this set using the full set of selection rules.

(c) Estimate the number of lines for each remaining set of upper and lower states.

(d) Work out the Zeeman splittings by doing the figures corresponding to Figures 6.8 and 6.9 and the table corresponding to Table 6.4 for those that have the correct number of lines.

(e) Determine the spectral terms for both the upper and lower state for the case that corresponds to Figure 6.10.

FIGURE 6.10
Zeeman pattern for $^3\mathrm{D}_? \rightarrow ?$.

7

Quantum Statistics

When we consider large numbers of particles, it becomes advantageous to talk of their properties in an average sense rather than describing each particle individually. The function that gives this description is the **distribution function**, which describes how many particles have energy between E and $E+dE$ if there are N total particles. The most general form of the distribution might be $f(x, y, z, v_x, v_y, v_z, t)$, but if the system is time-independent and spatially uniform, it simplifies. Furthermore, if the distribution is isotropic, then we can express the distribution function as a function of speed, but for our purposes, the energy distribution is preferred. In formulating the distribution function, we will need to know how many kinds of particles there are and how they interact with one another. We shall deal first only with one kind of particle at a time and assume that they interact with one another only weakly, and we shall furthermore assume that the distribution is stationary in time.

The fundamental technique in determining the distribution function is based on counting and the calculus of variations in order to find the most probable distribution. We shall consider three kinds of particles: (a) **Identical, distinguishable particles**, (b) **Identical, indistinguishable particles** that obey the **Pauli exclusion principle**, (*or particles with half-integral spin*), and (c) **Identical, indistinguishable particles** that do not obey the Pauli exclusion principle, (*or particles with integral spin*). The first group is composed of classical particles (they have the same charge and mass, but we can imagine painting them different colors, or writing a number on them to tell them apart) and they obey **Maxwell–Boltzmann** statistics. The second group is composed of **fermions** and they obey **Fermi–Dirac** statistics. The third group is composed of **bosons** and they obey **Bose–Einstein** statistics.

7.1 Derivation of the Three Quantum Distribution Laws

We consider now an assembly of N particles, all of which belong to only one of the three groups defined above. We assume them to be noninteracting, and move under the influence of some potential $V(\boldsymbol{r})$, and occupy specific energy states ψ_i with energy ϵ_i, $i = 0, 1, 2, \ldots$ For simplicity, we shall assume all of these states are nondegenerate, so that the states can be ordered sequentially according to their energy. We can treat degenerate states by assigning statistical weights to the degenerate levels, and still leave the energy levels distinct.

One of the obvious difficulties we encounter in this description is the enormous number of particles we may have to consider. For this reason, we will not consider each state individually, but since they are presumed to be so close together, we will consider dividing the energy into "cells" so that each cell has many energy states, but the energy width of each cell is too small for us to distinguish. This means that we cannot measure the energy to any greater precision than a cell width. We shall then take every particle in cell s to have the mean energy of that cell, ϵ_s, since we are unable to make any finer distinctions. In addition to the energy levels, we need to specify how many particles occupy each cell, and we shall let n_s be the number of particles in cell s with with energy ϵ_s. We shall also indicate the number of *states* per cell by g_s and we assume $g_s \gg 1$ for all s. The relationships are clearly governed by the total number of particles and total energy such that

$$\sum_{s=1}^{\infty} n_s = N \tag{7.1}$$

$$\sum_{s=1}^{\infty} n_s \epsilon_s = E\,. \tag{7.2}$$

It is clear that for all s greater than some s_{max}, we must have $n_s = 0$. Also, if we are dealing with fermions, we also need to specify that only one particle can occupy each of the g_s states because of the Pauli exclusion principle.

7.1.1 The Density of States

The quantity g_s, which represents the number of states between E and $E+\delta E$ where δE is the width of a cell, is sometimes determined by the specifics of a problem, but is often given by the density of states in phase space, which is a universal quantity. In order to calculate this quantity, we imagine an arbitrarily deep three-dimensional square well, and examine how the spacing between the states varies with the energy, assuming that we are far from the bottom of the well so that the number of states per cell varies slowly enough to be represented by a continuous function. The energy states of a particle in a box, as given by Equation (3.6), enable us to calculate this spacing between energy states for very general systems if they are large and the quantum numbers are large.We begin by noting that each state may be regarded as occupying a rectangular box whose size (in *inverse* coordinate space) is represented by incrementing each of the integers by one, so the volume of this "box" is $(1/a)(1/b)(1/c) = 1/V$ where $V = abc$ is the volume of the box that contains the particle. If we imagine a coordinate system in this

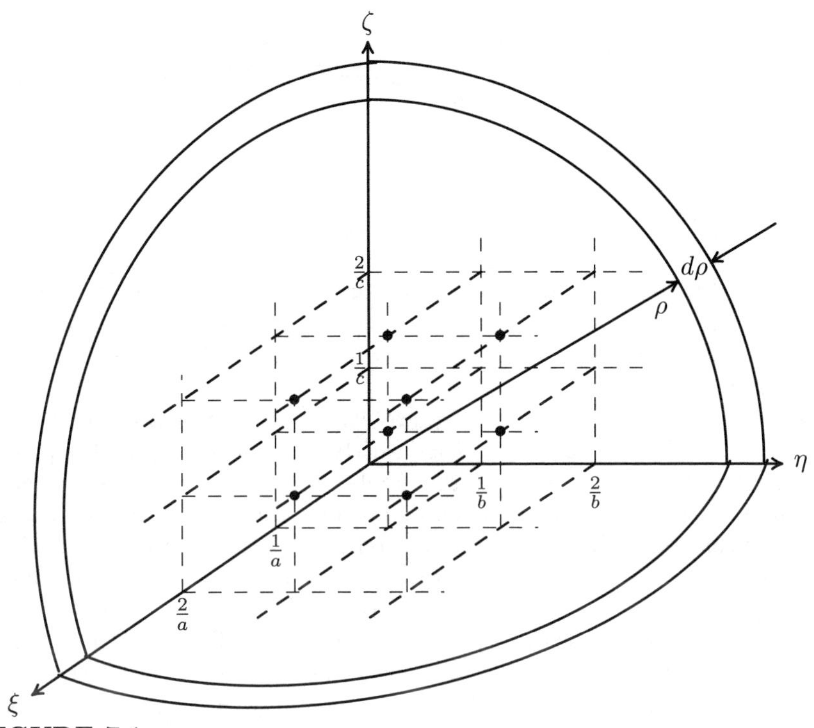

FIGURE 7.1
Energy states in inverse coordinates.

inverse space, illustrated in Figure 7.1, where we define $\xi = \ell/a$, $\eta = m/b$, and $\zeta = n/c$, where now ℓ, m, and n are so large that we may consider them to be continuous variables, then $E = (h^2/8m)(\xi^2 + \eta^2 + \zeta^2)$. The radial coordinate in this space is ρ where $\rho^2 = \xi^2 + \eta^2 + \zeta^2$ so $E = (h^2/8m)\rho^2$. Then the volume in this space between E and $E + dE$ is proportional to the volume between ρ and $\rho + d\rho$, which from the figure is the surface area of a sphere (in the first octant only, since the integers are nonnegative), given by $4\pi\rho^2/8$ times the thickness of the shell, $d\rho$, so for the number of *states* between ρ and $\rho + d\rho$, we find

$$N(\rho)\,\mathrm{d}\rho = \frac{\text{volume of shell}}{\text{volume/state}} = \frac{4\pi\rho^2\,\mathrm{d}\rho}{8(1/V)} = \frac{\pi V \rho^2\,\mathrm{d}\rho}{2}\,.$$

What we wish to find is the number of states between E and $E + dE$, so using $\rho = \sqrt{8mE}/h$ and $d\rho = \sqrt{2m}/hE^{-1/2}\,\mathrm{d}E$, we find that

$$N(E)\,\mathrm{d}E = \frac{4\pi V\sqrt{2m^3}E^{1/2}\,\mathrm{d}E}{h^3}\,. \tag{7.3}$$

This result is the **density of states** per energy interval. Another important way of representing this result is to find the density of states in coordinate-momentum space, which we call **phase space**. Using $E = p^2/2m$, we can express the result as

$$N\,\mathrm{d}x\,\mathrm{d}y\,\mathrm{d}z\,\mathrm{d}p_x\,\mathrm{d}p_y\,\mathrm{d}p_z = \frac{4\pi\sqrt{2m^3}\,\mathrm{d}V}{h^3}\frac{p}{\sqrt{2m}}\frac{p\,\mathrm{d}p}{m}$$
$$= \frac{dV}{h^3}4\pi p^2\,\mathrm{d}p$$
$$= \frac{1}{h^3}\,\mathrm{d}x\,\mathrm{d}y\,\mathrm{d}z\,\mathrm{d}p_x\,\mathrm{d}p_y\,\mathrm{d}p_z\,. \tag{7.4}$$

Thus, **the density of states in phase space** is h^{-3}. We will later see that when spin is included, the density of states for electrons will be twice this value, since there are two electron states for every energy state, one with spin up and one with spin down.

This still leaves us with a massive problem, since the number of arrangements among the various states that satisfy Equation (7.1) and Equation (7.2) is typically even larger than the number of particles. In order to tackle this problem, we distinguish between a **macroscopic distribution** ("course-grained") that gives the cell populations only, and a **microscopic distribution** ("fine-grained") which describes the arrangements of the particles among the levels within the cells as well as the cell populations. We then want to know which of the various possible macroscopic distributions is *most likely* to occur, and we will take this most likely distribution as our distribution function. In order to estimate the likelihood of any particular macroscopic distribution we take the following postulate

> **Postulate:** *Every physically distinct microscopic distribution* of the N particles among the various energy levels ϵ_i that satisfies both the condition that the total energy be $E + \delta E$ and the requirements of the exclusion principle, if it applies, *is equally likely to occur.*

By physically distinct, we mean that the distribution is distinguishable. The quantity δE is the uncertainty level and corresponds approximately to the width of the cells. From the postulate, it is apparent that the probability of finding a particular macroscopic distribution is proportional to the number of distinguishable microscopic distributions from which it may be constructed. Our problem is first to count these possible ways, and then to determine which has the largest probability. We thus need to determine $P(n_s)$ for each type of particle, where $P(n_s)$ is the number of microscopic distributions for the population distribution characterized by a particular set of the n_s that satisfies Equations (7.1) and (7.2), which are taken to be given.

Problem 7.1 *Density of states in two dimensions.*

(a) Consider a two-dimensional flat space where particles can only move in a plane. Repeat the derivation for the density of states beginning with

$$E = \frac{h^2}{8m}\left[\left(\frac{n_x}{a}\right)^2 + \left(\frac{n_y}{b}\right)^2\right]$$

and find $N(E)$ for this case.

(b) Find N for the density of states in the four-dimensional phase space, $\mathrm{d}x\,\mathrm{d}y\,\mathrm{d}p_x\,\mathrm{d}p_y$.

7.1.2 Identical, Distinguishable Particles

For this case, P is determined from the product of the number of ways we can choose the N particles to be placed in the various cells times the number of distinguishable ways these particles within each cell may be arranged. We begin with cell 1, and note that we can choose the first of the n_1 particles N ways, the second $N-1$ ways, etc., so that

$$P_1' = N(N-1)(N-2)\cdots(N-n_1+1) = \frac{N!}{(N-n_1)!},$$

where P_s' is the number of ways n_s particles may chosen to be put in cell s. Here we have counted each sequence separately, where we wish to know only *which* particles are in cell 1. We must therefore divide by the number of sequences of n_1 particles, and this is just $n_1!$ (since we may choose the first one n_1 ways, the second n_1-1 ways, etc.). Thus out of the total distribution, the number of ways we can choose to fill the first cell is

$$P_1 = \frac{P_1'}{n_1!} = \frac{N!}{n_1!(N-n_1)!}.$$

For the second quota, we have removed n_1 particles from N, otherwise we proceed as before, so we may write

$$P_2 = \frac{(N-n_1)!}{n_2!(N-n_1-n_2)!}.$$

Similarly,

$$P_3 = \frac{(N-n_1-n_2)!}{n_3!(N-n_1-n_2-n_3)!},$$

and so forth. Thus we have for the total P

$$P(n_1,\ldots,n_s,\ldots) = P_1\cdots P_s\cdots \tag{7.5}$$

$$= \frac{N!(N-n_1)!\cdots(N-n_1\cdots-n_{s-1})!\cdots}{n_1!(N-n_1)!\cdots n_s!(N-n_1\cdots-n_s)!\cdots} \tag{7.6}$$

$$= \frac{N!}{n_1!\cdots n_s!\cdots} = N!\prod_{s=1}^{\infty}\frac{1}{n_s!}. \tag{7.7}$$

We now must calculate the number of distinguishable ways the n_s particles in cell s may be placed among the g_s states. Since these particles are distinguishable, every arrangement is distinguishable, and each of the n_s particles is equally likely to be in any one of the g_s states, so there are g_s ways to put the first particle in, g_s ways to put the second particle in, etc. Thus, the number of ways for the n_s particles too be placed is $g_s^{n_s}$, and we then have for the total number of distinct microscopic distributions

$$P_C(n_s) = N! \prod_{s=1}^{\infty} \frac{g_s^{n_s}}{n_s!} \qquad \text{Classical Particles} \tag{7.8}$$

7.1.3 Identical, Indistinguishable Particles with Half-Integral Spin (Fermions)

For this case we cannot tell which particles are in which box, so we need only count which of the g_s *states* are occupied by the n_s particles in each cell. Thus we may put the first particle in any of the g_s states, the second particle in any of the remaining $g_s - 1$ states (since no more than one particle per state), etc. Thus the number of ways they may be put into the states is given by

$$P_s' = g_s(g_s - 1)\cdots(g_s - n_s + 1) = \frac{g_s!}{(g_s - n_s)!}\,.$$

This expression counts each sequence separately, but since they are indistinguishable we must divide by the number of possible sequences, which is $n_s!$, so for each cell we have

$$P_s = \frac{P_s'}{n_s!} = \frac{g_s!}{n_s!(g_s - n_s)!}\,.$$

Putting this together for all the cells, noting that all of the cells are independent, we obtain

$$P_F(n_s) = \prod_{s=1}^{\infty} P_s = \prod_{s=1}^{\infty} \frac{g_s!}{n_s!(g_s - n_s)!} \qquad \text{Fermions} \tag{7.9}$$

7.1.4 Identical, Indistinguishable Particles with Integral Spin (Bosons)

The indistinguishability of the particles again prevents us from knowing which particles are in which cells, but now the exclusion principle does not limit the number of particles that can go into each state. We thus may find the number of arrangements of the n_s particles among the g_s states (note that here, $n_s > g_s$ is possible, but not necessary) from a pictorial scheme. We imagine that cell s is laid out with $n_s + g_s - 1$ holes into which either white pegs (particles) or black pegs (partitions) may be put. An example is shown in Figure 7.2 with $g_s = 16$ and $n_s = 24$. We note that $g_s - 1$ partitions are just sufficient to divide the cell into g_s intervals (states), and that the remaining holes, which are separated into groups by these partitions, then represent the

○○●○○○○●●○●○○●●●○○○○●○●○●●●○○○○○○○●●○○●

2 4 0 1 2 0 0 4 1 1 0 0 7 0 2 0

FIGURE 7.2
Illustration of method for finding the number of ways of placing $n_s = 24$ bosons among $g_s = 16$ states, where any number of particles are allowed in each state. (From Ref. [6])

distribution of the remaining particles among the states. Thus the number of distinct permutations of the black and white pegs among the holes is just the same as the number of distinct arrangements of our n_s indistinguishable particles among the g_s states. Now the number of permutations of $n_s + g_s - 1$ *distinguishable* objects is just $(n_s + g_s - 1)!$, but the sequences of the n_s particles or the $g_s - 1$ partitions are themselves indistinguishable, so we must divide by the number of sequences of each of these, leading to

$$P_s = \frac{(n_s + g_s - 1)!}{n_s!(g_s - 1)!} ,$$

so for all the cells, we have for this case

$$P_B(n_s) = \prod_{s=1}^{\infty} \frac{(n_s + g_s - 1)!}{n_s!(g_s - 1)!} \qquad \text{Bosons.} \tag{7.10}$$

Problem 7.2 *Boson states.*

(a) For the example in Figure 7.2, find P_s. You may wish to use the approximate formula for the factorial in Problem 7.3a.

(b) For a simpler case with $n_s = 4$ and $g_s = 2$, find P_s and sketch each of the P_s distinguishable arrangements (show with open circles and solid circles as in Figure 7.2).

7.1.5 The Distribution Laws

We have now established the number of microscopic distributions for a given macroscopic distribution, and according to our postulate, $P(n_s)$ is proportional to the probability of that particular macroscopic distribution being observed. To find the most probable distribution, which we shall denote as $n_s(\epsilon_s)$, we must maximize P for each case subject to the auxiliary conditions that the total number N is fixed and the total energy E is fixed.

Each of the distributions is discrete, or a function of discrete variables, but we have required at each step that n_s and g_s be large (except for the very high lying states that are sparsely filled), so we will *assume* that these variables are continuous. Next, since the logarithm of a function is a monotonic function of its argument, we will maximize the logarithm of P rather than P directly. Then the three cases become

$$\ln P_C(n_s) = \ln N! + \sum_{s=1}^{\infty}(n_s \ln g_s - \ln n_s!) \qquad \text{Classical} \tag{7.11}$$

$$\ln P_F(n_s) = \sum_{s=1}^{\infty}[\ln g_s! - \ln n_s! - \ln(g_s - n_s)!] \qquad \text{Fermions} \tag{7.12}$$

$$\ln P_B(n_s) = \sum_{s=1}^{\infty}[\ln(n_s + g_s - 1)! - \ln n_s! - \ln(g_s - 1)!] \quad \text{Bosons}\,. \tag{7.13}$$

We may write the conditions we wish to impose as

$$\begin{aligned} \delta(\ln P) &= 0\,, \\ \delta N &= 0\,, \\ \delta E &= 0\,, \end{aligned}$$

where the first maximizes the probability, and the other two indicate the constraints that we keep the number of particles and the energy fixed. In principle, we apply each of these separately, but using the method of undetermined multipliers, we may combine these three conditions into one condition,

$$\delta(\ln P) - \alpha\delta N - \beta\delta E = 0\,, \tag{7.14}$$

where α and β are unknown constants. At this point, they are chosen for convenience, but we shall see later that they have important physical significance.

For the first case of classical particles, Equation (7.14) becomes

$$\sum_{s=1}^{\infty}(\ln g_s - \ln n_s - \alpha - \beta\epsilon_s)\delta n_s = 0\,. \tag{7.15}$$

We have used here the approximation that $\delta \ln n! \simeq \ln n \delta n$, which comes from the Stirling approximation to the factorial function for large values.

Now in Equation (7.15), the δn_s are nearly free parameters. They are subject to only two constraints, namely N and E must remain fixed. This means all but two of the δn_s may be chosen arbitrarily, and the two constraints can always be satisfied by the proper choice of the final two values of δn_s. If we choose δn_1 and δn_2 to be those two that are not arbitrary and correspondingly choose α and β such that

$$\begin{aligned} \ln g_1 - \ln n_1 - \alpha - \beta\epsilon_1 &= 0 \\ \ln g_2 - \ln n_2 - \alpha - \beta\epsilon_2 &= 0\,, \end{aligned}$$

then Equation (7.15) must always be satisfied for arbitrary δn_s, $s \neq 1, 2$. Now if the quantity inside the parentheses of Equation (7.15) should differ from zero for $s = \ell$, $\ell > 2$, then by the proper choice of δn_ℓ, the sum can always be made to be *different from* 0. Since this is unacceptable, it follows that

$$\ln g_s - \ln n_s - \alpha - \beta \epsilon_s = 0 \tag{7.16}$$

for *all* s, since we have forced this condition for the first two by the choice of the undetermined constants, and shown that all others must vanish because of the arbitrariness of the δn_s for $s > 2$. At this point, Equation (7.16) is the solution of the problem we set out to solve, since we may rearrange it to obtain

$$n_s(\epsilon_s) = \frac{g_s}{\mathrm{e}^{\alpha + \beta \epsilon_s}} \,. \tag{7.17}$$

Problem 7.3 *The Stirling approximation to the factorial function.*

(a) Compare the % error in Stirling's formula,

$$n! \simeq \mathrm{e}^{-n} n^n \sqrt{2\pi n}\,,$$

with the factorial function for $n = 5, 20, 400$.

(b) Show that the approximation for the variation of the factorial, $\delta \ln n! \simeq \ln n \, \delta n$, follows from Stirling's formula.

(c) Check the accuracy (% error) for the approximate relation $\delta \ln n! \approx \ln n \, \delta n$ for $n = 5, 20, 400$ and $\delta n = \pm 1, \pm 2, \pm 5$. For the derivative of $\ln n!$, use the general expression for the derivative: $\mathrm{d}f(x)/\mathrm{d}x = [f(x + \delta x) - f(x - \delta x)]/2\,\delta x$ (which approaches $f'(x)$ as $\delta x \to 0$).

Problem 7.4 *Fermi–Dirac and Bose–Einstein distribution functions.* Using the same methods, find the corresponding distribution functions for the other two cases, and show that all three may be written in the form

$$n_s(\epsilon_s) = \frac{g_s}{\mathrm{e}^{\alpha + \beta \epsilon_s} \pm 1, 0}\,, \tag{7.18}$$

where we choose $+1$ for the Fermi–Dirac distribution, -1 for the Bose–Einstein distribution, and 0 for the classical (Maxwellian) distribution. (Assume $g_s \gg 1$.)

7.1.6 Evaluation of the Constant Multipliers

In principle, the constant multipliers can be obtained by using the two constraints and the distribution functions, so that

$$\sum_{s=1}^{\infty} \frac{g_s}{\mathrm{e}^{\alpha + \beta \epsilon_s} \pm 1, 0} = N, \qquad \sum_{s=1}^{\infty} \frac{g_s \epsilon_s}{\mathrm{e}^{\alpha + \beta \epsilon_s} \pm 1, 0} = E\,,$$

but this is generally not solvable in closed form for any distribution. In order to understand the significance of the constant multipliers, we consider a mixture of two classes of particles and maximize the product of the two probabilities. If we consider particles of type s and ℓ, which may have the same or different statistics (both fermions, both bosons, or a combination of both), but are distinguishable from each other, then we must maximize the joint probability

$$P(n_s, n_\ell) = P(n_s)P(n_\ell)$$

subject to the constraints

$$\sum_{s=1}^{\infty} n_s = N_1 \tag{7.19}$$

$$\sum_{\ell=1}^{\infty} n_\ell = N_2 \tag{7.20}$$

$$\sum_{s=1}^{\infty} n_s \epsilon_s + \sum_{\ell=1}^{\infty} n_\ell \epsilon_\ell = E\,. \tag{7.21}$$

We will now clearly need *three* undetermined multipliers, which we may call α_1, α_2, and β. The same process we developed before will lead to the same distribution functions as before for each type of particle, *but they will both have the same* β. Generalizing, this means that for an arbitrary mixture of any number of types of particles, there is one **universal constant** β that characterizes all particles in equilibrium.

Problem 7.5 *Multiple species.* Carry out the analysis above for a mixture of fermions and bosons, and verify the above conclusions.

Since there is only one value of β, then it must be so that each species has its own total energy, so that $E = E_1 + E_2$ in the example above, and this is the statistical analog of Dalton's Law of Partial Pressures from thermodynamics. For Maxwell–Boltzmann statistics for the classical particles, we have

$$n_s(\epsilon_s) = \frac{g_s}{\mathrm{e}^{\alpha+\beta\epsilon_s}} \tag{7.22}$$

but we need to know g_s to proceed. Here we use the density of states that represents the number of states with energy between ϵ_s and $\epsilon_s + \Delta\epsilon_s$, which we found in Equation (7.3) to be

$$g_s = \frac{4\pi V\sqrt{2m^3}}{h^3}\sqrt{\epsilon_s}\Delta\epsilon_s\,,$$

where $V = abc$ is the volume of the box. Equation (7.22) then becomes

$$\begin{aligned} N &= \frac{4\pi V\sqrt{2m^3}}{h^3}\mathrm{e}^{-\alpha}\int_0^\infty \epsilon^{1/2}\mathrm{e}^{-\beta\epsilon}\,\mathrm{d}\epsilon \\ &= \frac{4\pi V\sqrt{2m^3}}{h^3}\mathrm{e}^{-\alpha}\beta^{-3/2}\frac{\sqrt{\pi}}{2}\,, \end{aligned} \tag{7.23}$$

$$E = \frac{4\pi V\sqrt{2m^3}}{h^3} \mathrm{e}^{-\alpha} \int_0^\infty \epsilon^{3/2} \mathrm{e}^{-\beta\epsilon}\, \mathrm{d}\epsilon$$
$$= \frac{4\pi V\sqrt{2m^3}}{h^3} \mathrm{e}^{-\alpha} \beta^{-5/2} \frac{3\sqrt{\pi}}{4} . \tag{7.24}$$

Problem 7.6 Supply the missing steps in Equation (7.23) and Equation (7.24).

Problem 7.7 Evaluate α and β from Equations (7.23) and (7.24).
Partial ans.: $\beta = 3N/2E$.

We can now complete the derivation of the quantum distribution laws by noting that

$$\frac{1}{\beta} = \frac{2E}{3N} = \frac{2}{3}\langle\epsilon\rangle .$$

From an ideal gas thermometer, however, the average kinetic energy in a dilute gas is

$$\langle\epsilon\rangle = \frac{3\kappa T}{2}$$

where T is the thermodynamic temperature, and κ is Boltzmann's constant. Thus we conclude that

$$\beta = \frac{1}{\kappa T}$$

and the three distribution laws take the form

$$\begin{aligned} n_s &= g_s \mathrm{e}^{-\alpha-\epsilon_s/\kappa T} && \text{Maxwell–Boltzmann} \\ n_s &= \frac{g_s}{\mathrm{e}^{\alpha+\epsilon_s/\kappa T}+1} && \text{Fermi–Dirac} \\ n_s &= \frac{g_s}{\mathrm{e}^{\alpha+\epsilon_s/\kappa T}-1} && \text{Bose–Einstein} \end{aligned} \tag{7.25}$$

where g_s will depend on the system and α is a normalization constant.

7.2 Applications of the Quantum Distribution Laws

7.2.1 General Features

It is often convenient to consider the quantity

$$n(\epsilon) \equiv \frac{n_s}{g_s} ,$$

which is the probability that a state is occupied and is called the **occupation index**. The similarities and differences can then be compared by examining a plot of the occupation index for each species.

1. From Figure 7.3a, The Maxwell–Boltzmann distribution is exponential at all energies and temperatures.

2. From Figure 7.3b, we can see that the Fermi–Dirac occupation index can never exceed unity, which it approaches at low temperatures and low energies. This is an obvious requirement of the exclusion principle that we used to construct the distribution function. In order for this to happen, α must become large and negative, since $\epsilon/\kappa T$ becomes large and positive. This may be achieved by letting $\alpha \to -\epsilon_F/\kappa T$, so that for all $\epsilon < \epsilon_F$, the occupation index approaches unity as $T \to 0$ while $n(\epsilon) \to 0$ for $\epsilon > \epsilon_F$. In general, ϵ_F varies weakly with temperature, but it is defined to be that energy where $n(\epsilon_F) = 1/2$. ϵ_F is the Fermi energy (denoted ϵ_m in the figure. For $\epsilon \gg \epsilon_F$, the distribution is again exponential and behaves like a Maxwell–Boltzmann case that is offset by energy ϵ_F.

3. The Bose–Einstein distribution is illustrated in Figure 7.3c where it differs from the Maxwell–Boltzmann distribution only for low energies where $n(\epsilon) \gg 1$, since only if there are many particles per state do the quantum effects of bosons exhibit themselves. For this case α is typically large and positive unless one is describing photons, were it vanishes since the number of photons is not generally conserved.

A property shared by all three of these distribution functions is that at sufficiently high temperatures (or at sufficiently low occupation index), all are essentially Maxwellian because the exponential factor greatly exceeds the factor of unity in the denominator that distinguishes them from one another. Hence we may write for sparsely occupied states, *independently of the type of particle*, that

$$\frac{n(\epsilon_1)}{n(\epsilon_2)} = \frac{\mathrm{e}^{\epsilon_2/\kappa T}}{\mathrm{e}^{\epsilon_1/\kappa T}} = \mathrm{e}^{(\epsilon_2-\epsilon_1)/\kappa T} \tag{7.26}$$

so that the *relative* occupation index between two states depends only on the energy difference between the two states and the temperature and *not on the energy.* This ratio is known as the **Boltzmann factor**. For a system near equilibrium, those states a few κT above the ground state energy are sparsely occupied, so their occupation index is very nearly given by $\mathrm{e}^{-\epsilon'/\kappa T}$ with $\epsilon' = \epsilon - \epsilon_0$ where ϵ_0 is the ground state energy. We list several examples of the use of the Boltzmann factor:.

1. **Excited States of Atoms.** If we consider a gas of atoms that are excited by collisions among one another, we can use the Boltzmann factor to estimate the equilibrium populations of the various atomic excited states as a function of temperature, and use these to estimate the relative intensity of the various lines radiated by the gas. Conversely, we may use the measured radiation intensities and the Boltzmann factors to estimate the temperature.

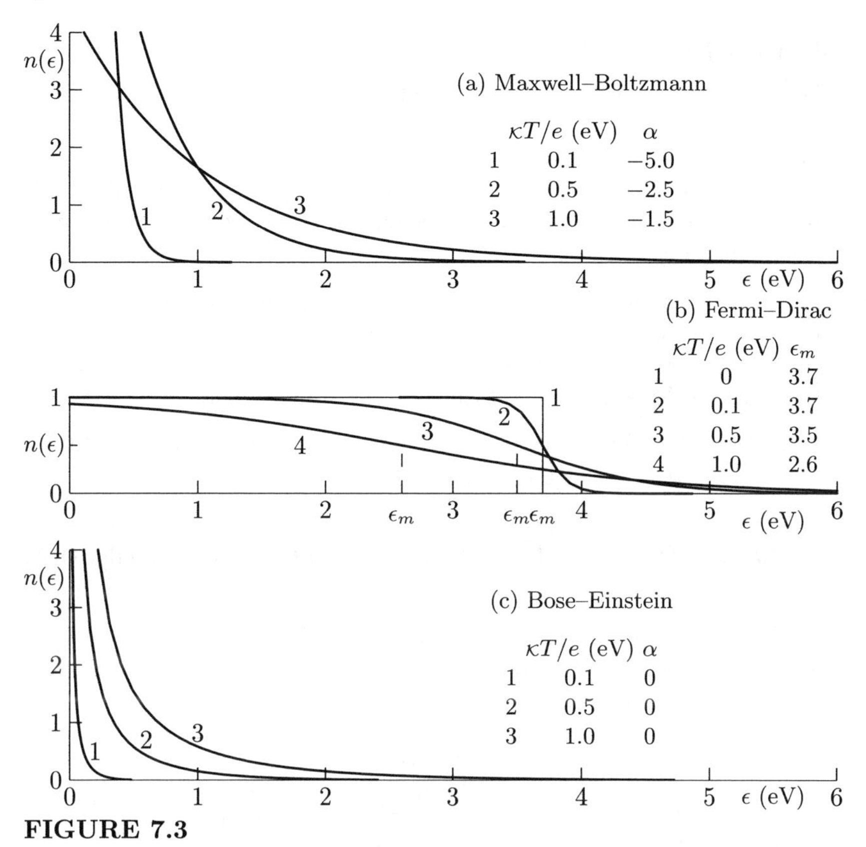

FIGURE 7.3
Illustration of the three distribution laws. (*a*) The Maxwell–Boltzmann distribution. (*b*) The Fermi–Dirac distribution. (*c*) The Bose–Einstein distribution. All energies and temperatures expressed in eV where 1 eV$\leftrightarrow$ 11,604 K.

2. **Excited States of Molecules.** The same kinds of things as we noted for atoms can also be applied to molecules. Thus we can determine the excitation of the various vibrational and rotational states of molecules.

3. **Ionization or Dissociation.** The energy levels required for either ionization of atoms or dissociation of molecules may also be represented by Boltzmann factors, so the ionization density or density of dissociated atoms or molecules may be estimated.

4. **Orientational Distributions.** In many cases in either a magnetic field or an electric field, electric or magnetic dipoles will try to align themselves with the field and in the process change their energy. The relative populations among the various orientational states can often be estimated from the Boltzmann factor.

Problem 7.8 *Ionization of hydrogen.*

(a) When states are degenerate, we need to include some information about the degree of degeneracy in order to properly assess the occupation index, since each of the degenerate states permits additional microscopic distributions. If we take the degeneracy of state s to be d_s, what will be the Boltzmann factor relating two states with energies ϵ_1 and ϵ_2 and degeneracies d_1 and d_2?

(b) If the degeneracy of the states of the Hydrogen atom is $d_n = 2n^2$, $n = 1, 2, \ldots$, find the relative number of atoms in the solar chromosphere that are in the ground state and the first, second, and third excited states ($n = 2, 3, 4$). Take the temperature of the chromosphere to be 5,000 K.

(c) Why does the Balmer series ($n > 2$ to $n = 2$) appear prominently in the *absorption* spectrum of the sun?

7.2.2 Applications of the Maxwell–Boltzmann Distribution Law

We consider first a thermal distribution of harmonic oscillators at the same frequency that are taken to be identical *distinguishable* elements whose possible energy states are $(n + \frac{1}{2})\hbar\omega$. If there are N total oscillators and ΔN_s of them in state s, then

$$\frac{\Delta N_s}{N} = A\mathrm{e}^{-\epsilon_s/\kappa T} = A\exp\left[-\left(s + \frac{1}{2}\right)\frac{\hbar\omega}{\kappa T}\right] .$$

We evaluate A from

$$1 = \sum_{s=0}^{\infty} \frac{\Delta N_s}{N} = A\mathrm{e}^{-\frac{1}{2}\hbar\omega/\kappa T} \sum_{s=0}^{\infty} \mathrm{e}^{-s\hbar\omega/\kappa T} .$$

The sum here is the familiar geometric series, since if we let $x = \exp(-\hbar\omega/\kappa T)$, then the sum is $1 + x + x^2 + \cdots = 1/(1-x)$, so we have

$$A = \mathrm{e}^{\frac{1}{2}\hbar\omega/\kappa T}[1 - \mathrm{e}^{-\hbar\omega/\kappa T}] .$$

The average energy per oscillator is thus

$$\begin{aligned}
\langle E \rangle &= \sum_{s=0}^{\infty} \epsilon_s \frac{\Delta N_s}{N} \\
&= A \sum_{s=0}^{\infty} \epsilon_s \mathrm{e}^{-\epsilon_s/\kappa T} \\
&= A\kappa T^2 \frac{\partial}{\partial T} \sum_{s=0}^{\infty} \mathrm{e}^{-\epsilon_s/\kappa T} \\
&= \frac{1}{2}\hbar\omega + \frac{\hbar\omega}{\mathrm{e}^{\hbar\omega/\kappa T} - 1} .
\end{aligned} \tag{7.27}$$

Problem 7.9 Supply the missing steps in Equation (7.27).

Problem 7.10 Show that, as κT becomes very large compared to the interval $\hbar\omega$, $\langle E\rangle \to \kappa T$.

Example 7.1

Specific Heat of Crystalline Solids. Applying this distribution to crystalline solids gives a good account for the observed relation between the specific heat and the temperature. From thermodynamics and the law of Dulong and Petit, the specific heat depends weakly on temperature and is roughly the same from element to element. The value of the specific heat is close to $3R$ where $R = N_0\kappa$ is the gas constant per mole and N_0 is Avogadro's number. This comes from classical thermodynamics, since if there are N_0 atoms per mole, and each atom has three degrees of freedom, then the total number of degrees of freedom is $3N_0$. Then the law of equipartition of energy requires each degree of freedom to have energy κT, so the total energy is $3N_0\kappa T$ per mole. Then the specific heat should be

$$C_v = \frac{\mathrm{d}E}{\mathrm{d}T} = 3N_0\kappa = 3R\,.$$

Near zero temperature, however, significant variations were observed as the specific heat dropped rapidly toward zero. Einstein first saw this as a quantum mechanical problem in 1911 and assumed that the $3N_0$ degrees of freedom could be described by $3N_0$ harmonic oscillators of the same frequency and derived Equation (7.27) as their distribution. He assumed that the derivative of $3N_0$ times this expression would give the proper result, which did fall to zero rapidly, but not with the observed temperature dependence. Debye in 1912 solved the problem by assuming the distribution was that of an **elastic continuum**, where

$$\mathrm{d}N = N(\omega)\,\mathrm{d}\omega = A\omega^2\,\mathrm{d}\omega\,,$$

where $N(\omega)$ is the number of modes whose vibration frequencies lie between ω and $\omega + d\omega$, and that there was a maximum energy or frequency that corresponded to adjacent atoms moving in opposite directions. This maximum frequency corresponds to the value such that

$$3N_0 = \int_0^{\omega_m} \mathrm{d}N = A\int_0^{\omega_m} \omega^2\,\mathrm{d}\omega\,,$$

which gives $A = 9N_0/\omega_m^3$. If we combine this with the average energy of Equation (7.27), we find for the total energy

$$E = \frac{9N_0\hbar\omega_m}{8} + \frac{9N_0\hbar}{\omega_m^3}\int_0^{\omega_m} \frac{\omega^3\,\mathrm{d}\omega}{\mathrm{e}^{\hbar\omega/\kappa T} - 1}\,.$$

Differentiating with respect to temperature leads to the specific heat

$$C_v = \frac{\partial E}{\partial T} = \frac{9N_0\hbar^2}{\omega_m^3 \kappa T^2} \int_0^{\omega_m} \frac{\omega^4 e^{\hbar\omega/\kappa T}\, d\omega}{(e^{\hbar\omega/\kappa T} - 1)^2} . \tag{7.28}$$

If we now change variables to $x = \hbar\omega/\kappa T$, this becomes

$$\begin{aligned} C_v &= \frac{9N_0\kappa^4 T^3}{\hbar^3\omega_m^3} \int_0^{\hbar\omega_m/\kappa T} \frac{x^4 e^x\, dx}{(e^x - 1)^2} \\ &= 9R\left(\frac{T}{\theta}\right)^3 \int_0^{\theta/T} \frac{x^4 e^x\, dx}{(e^x - 1)^2} , \end{aligned} \tag{7.29}$$

where $\theta = \hbar\omega_m/\kappa$ and is called the **Debye characteristic temperature**. The Debye theory of specific heats is thus a one-parameter theory, and agrees well with experiment as shown in Figure 7.4. □

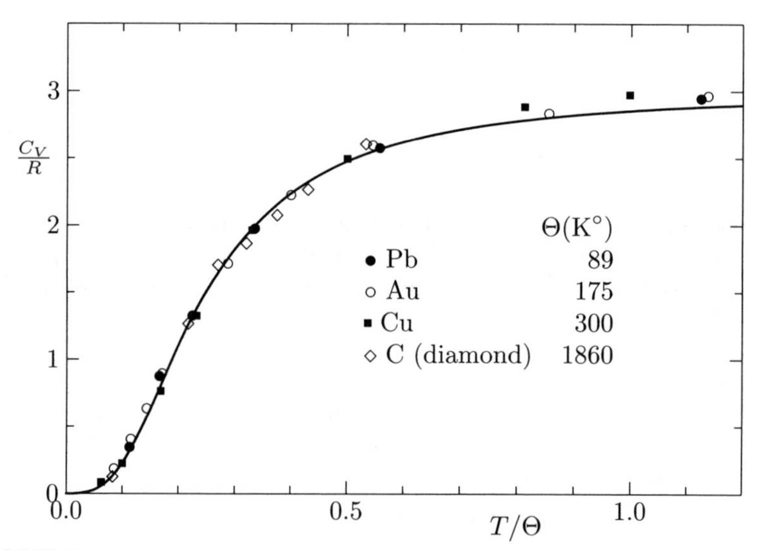

FIGURE 7.4
Debye specific heat compared with observed data for several substances.

Problem 7.11 Fill in the missing steps in the derivation of Equation (7.29).

Problem 7.12 Show that for very low temperatures, the specific heat varies as T^3, and at very high temperatures approaches $3R$.

Example 7.2
Black-body Radiation Law. Our other example will be to derive the Planck black-body radiation law. For this case we follow a similar treatment for

the energy density as we did in the specific heat case, except that for the distribution function we use the result from the Rayleigh–Jeans law that came from classical thermodynamics, which was that

$$N(\lambda) = \frac{8\pi}{\lambda^4}$$

per unit volume. Using this along with the mean energy per oscillator, we have

$$\mathrm{d}U_\lambda = \frac{4}{c} I(\lambda, T)\,\mathrm{d}\lambda = \left[\frac{1}{2}\frac{hc}{\lambda} + \frac{hc}{\lambda(\mathrm{e}^{hc/\lambda\kappa T} - 1)}\right] N(\lambda)\,\mathrm{d}\lambda \tag{7.30}$$

where the relation between the energy density, U_λ, and the radiation intensity, $I(\lambda, T)$, come from electromagnetic theory. Then the spectral intensity is given by

$$I(\lambda, T) = \frac{\pi hc^2}{\lambda^5} + \frac{2\pi hc^2}{\lambda^5(\mathrm{e}^{hc/\lambda\kappa T} - 1)} . \tag{7.31}$$

This is the same as the Planck distribution except for the zero-point term, which cannot be included since it gives an infinite result when integrated over all wavelengths (the ultraviolet catastrophe of the original Rayleigh–Jeans result). We can justify ignoring this term either by considering a slightly different model for the oscillators where this term does not appear, or note that zero-point oscillations neither radiate nor exchange energy with the wall of the cavity, so they cannot enter into the radiation spectrum. □

7.2.3 Applications of the Fermi–Dirac Distribution Law

We have already considered the Fermi energy and the density of states, but now we can combine the density of states (including the twofold degeneracy for electrons due to their spin)

$$g_s = \frac{8\pi V(2m^3)^{1/2}}{h^3}\epsilon_s^{1/2}\,\Delta\epsilon_s$$

with the distribution function to obtain

$$n_s = \frac{8\pi V(2m^3)^{1/2}\epsilon_s^{1/2}\,\Delta\epsilon_s}{h^3[\mathrm{e}^{(\epsilon_s - \epsilon_F)/\kappa T} + 1]} . \tag{7.32}$$

If we take the limit where $T \to 0$, then all states with $\epsilon < \epsilon_F$ will be filled and all states with $\epsilon > \epsilon_F$ will be empty. Then the total number of particles must be

$$N_0 = \sum_{\epsilon_s = 0}^{\epsilon = \epsilon_F} g_s = \frac{8\pi V(2m^3)^{1/2}}{h^3}\int_0^{\epsilon_F} \epsilon^{1/2}\,\mathrm{d}\epsilon$$

which we use to determine the Fermi energy

$$\epsilon_F = \frac{h^2}{2m}\left(\frac{3N_0}{8\pi V}\right)^{2/3} . \tag{7.33}$$

Example 7.3
Semiconductors. Now when we consider electrons and holes in a semiconductor, which is characterized by a band of energies where states are forbidden and the Fermi energy lies inside that band, we wish to estimate the electron density in the conduction band, which lies above the forbidden band, and the hole density in the valence band, which lies below the forbidden band. The energy level in this estimate is relative, so we define the zero of energy to be at the valence band edge as in Figure 7.5, and we take the Fermi level to be nearly midway between the valence band edge and the conduction band edge. The width of the forbidden band is then denoted as ϵ_g and is called the band gap.

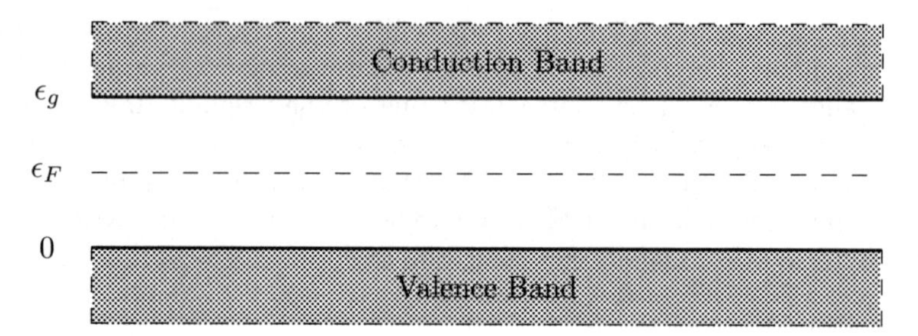

FIGURE 7.5
Energy levels in the valence and conduction bands of a semiconductor.

Denoting the electron density in the conduction band by n and the hole density in the valence band by p, then for electrons we have

$$n = \frac{N_n}{V} = \frac{8\pi(2m_e^{*3})^{1/2}}{h^3}\int_{\epsilon_g}^{\infty}\frac{(\epsilon-\epsilon_g)^{1/2}\,\mathrm{d}\epsilon}{\mathrm{e}^{(\epsilon-\epsilon_F)/\kappa T}+1} . \tag{7.34}$$

The lower limit on the integral is ϵ_g because free electrons exist only for $\epsilon > \epsilon_g$ and m_e^* is the effective electron mass. Since we assume that $\epsilon_g \gg \kappa T$ and that $\epsilon_F \simeq \epsilon_g/2$, then for all $\epsilon > \epsilon_g$, the exponential is large and we can neglect the $+1$ in the denominator and write

$$\begin{aligned} n &= \frac{8\pi(2m_e^{*3})^{1/2}}{h^3}\mathrm{e}^{\epsilon_F/\kappa T}\int_{\epsilon_g}^{\infty}(\epsilon-\epsilon_g)^{1/2}\mathrm{e}^{-\epsilon/\kappa T}\,\mathrm{d}\epsilon \\ &= \frac{8\pi(2m_e^{*3})^{1/2}(\kappa T)^{3/2}}{h^3}\mathrm{e}^{-(\epsilon_g-\epsilon_F)/\kappa T}\int_0^{\infty}u^{1/2}\mathrm{e}^{-u}\,\mathrm{d}u \end{aligned} \tag{7.35}$$

where we have used $u = (\epsilon - \epsilon_g)/\kappa T$. This is a tabulated integral, so

$$n = N_c \mathrm{e}^{-(\epsilon_g - \epsilon_F)/\kappa T}\,, \qquad N_c = \frac{2}{h^3}(2\pi m_e^* \kappa T)^{3/2}\,. \tag{7.36}$$

For holes, the probability that a hole exists is the same as the probability that an electron does *not* exist in that state, so for holes, the occupation index is given by

$$\begin{aligned} n_h(\epsilon) &= 1 - n_e(\epsilon) \\ &= 1 - \frac{1}{\mathrm{e}^{(\epsilon - \epsilon_F)/\kappa T} + 1} \\ &= \frac{1}{\mathrm{e}^{(\epsilon_F - \epsilon)/\kappa T} + 1}\,. \end{aligned} \tag{7.37}$$

Using this occupation index for holes, we may write for the hole density

$$p = \frac{N_p}{V} = \frac{8\pi(2m_h^{*3})^{1/2}}{h^3} \int_{-\infty}^{0} \frac{(-\epsilon)^{1/2}\,\mathrm{d}\epsilon}{\mathrm{e}^{(\epsilon_F - \epsilon)/\kappa T} + 1}\,.$$

The lower limit on the integral is not truly infinite, since the band below has finite width, but it is many κT wide, and the only significant contribution is from the layer a few κT wide, so the infinite limit is used. m_h^* is the effective hole mass. We may again neglect the $+1$ since the exponent is always large over the entire range of the integral. Proceeding as in the electron case, only with with $u = -\epsilon/kT$, the result is

$$p = N_v \mathrm{e}^{-\epsilon_F/\kappa T}\,, \qquad N_v = \frac{2}{h^3}(2\pi m_h^* \kappa T)^{3/2}\,, \tag{7.38}$$

where we use p to represent the hole density (since holes behave as if they have a positive charge) just as we used n to represent the electron density.

For an intrinsic semiconductor, where there are no impurities, then the only way to get an electron in the conduction band is to take one from the valence band, so we must have

$$n = p = n_i(T) \tag{7.39}$$

where $n_i(T)$ is the intrinsic carrier concentration which is given by

$$np = n_i^2 = N_c N_v \mathrm{e}^{-\epsilon_g/\kappa T} \tag{7.40}$$

so that

$$n_i(T) = \frac{2}{h^3}(2\pi m_e \kappa T)^{3/2} \left(\frac{m_e^* m_h^*}{m_e^2}\right)^{3/4} \mathrm{e}^{-\epsilon_g/2\kappa T}\,. \tag{7.41}$$

We can also solve Equation (7.39) for the Fermi level so that

$$\epsilon_F = \frac{1}{2}\epsilon_g + \frac{3\kappa T}{4} \ln\left(\frac{m_h^*}{m_e^*}\right)\,. \tag{7.42}$$

□

TABLE 7.1
Semiconductor Properties

Parameter	Symbol	Ge	Si	InSb	GaAs
Electron effective mass	m_e^*	$0.22\ m_e$	$1.08\ m_e$	$0.013 m_e$	$0.072 m_e$
Hole effective mass	m_h^*	$0.37\ m_e$	$0.59\ m_e$	$0.52\ m_e$	$0.51\ m_e$
Electron mobility	μ_e	0.38	0.14	7.80	.65
Hole mobility	μ_h	0.20	0.05	1.00	.044
Band Gap (eV)(77°K)	ϵ_g	0.73	1.147	0.22	1.4
Band Gap (eV)(300°K)	ϵ_g	0.66	1.1	0.16	1.4
Intrinsic density(300°K)	n_i	$2.37 \cdot 10^{19}$	$1.0 \cdot 10^{16}$	$1.35 \cdot 10^{22}$	$8.3 \cdot 10^{15}$
Rel. dielectric constant	$\epsilon_r = \kappa$	16	11.8	16.8	11.6

Problem 7.13 Fill in the steps leading to Equations (7.38), (7.41), and (7.42).

Problem 7.14 From Table 7.1, find the Fermi energy relative to the valence band edge and the intrinsic density for both silicon and InSb at 300 K. Compare with the measured intrinsic densities (i.e., find the percentage differences).

Example 7.4

Specific Heat of Electrons. From classical thermodynamics, the specific heat of electrons in a solid should have been very large, because of the large number of electrons, and should have contributed substantially to the specific heat of the solid material, whereas it was found to be nowhere near as large as expected. This is because only those electrons near the Fermi energy contribute. We can calculate the total energy using Equation (7.32) as

$$E = \frac{3N_0}{2\epsilon_F^{3/2}} \int_0^\infty \frac{\epsilon^{3/2}\,\mathrm{d}\epsilon}{\mathrm{e}^{(\epsilon-\epsilon_F)/\kappa T}+1}\,. \tag{7.43}$$

This is difficult to evaluate, but for low temperatures we can break the integral into a piece that includes most of the electrons below the Fermi level, and is easy to calculate, and some pieces that are important near the Fermi level. Breaking the integral into two pieces below and above the Fermi level, we first integrate up to the Fermi level, and write

$$\begin{aligned} \int_0^{\epsilon_F} \frac{\epsilon^{3/2}\,\mathrm{d}\epsilon}{\mathrm{e}^{(\epsilon-\epsilon_F)/\kappa T}+1} &= \int_0^{\epsilon_F} \epsilon^{3/2}\,\mathrm{d}\epsilon - \int_0^{\epsilon_F} \frac{\epsilon^{3/2}\,\mathrm{d}\epsilon}{\mathrm{e}^{(\epsilon_F-\epsilon)/\kappa T}+1} \\ &= \frac{2}{5}\epsilon_F^{5/2} - \kappa T \int_0^{\epsilon_F/\kappa T} \frac{(\epsilon_F - u\kappa T)^{3/2}\,\mathrm{d}u}{\mathrm{e}^u+1}\,, \end{aligned} \tag{7.44}$$

where we used the substitution $\epsilon = \epsilon_F - u\kappa T$ and the upper limit may be replaced by infinity since the limit is much larger than unity and there is no significant contribution to the integral over the additional range. The integral beyond the Fermi level gives (with $\epsilon = \epsilon_F + u\kappa T$)

$$\int_{\epsilon_F}^{\infty} \frac{\epsilon^{3/2}\,\mathrm{d}\epsilon}{\mathrm{e}^{(\epsilon-\epsilon_F)/\kappa T}+1} = \kappa T \int_0^{\infty} \frac{(\epsilon_F + u\kappa T)^{3/2}\,\mathrm{d}u}{\mathrm{e}^u + 1}\,. \tag{7.45}$$

If now we recognize that the principal contribution of the remaining integrals is principally about $u = 0$ or from the neighborhood of the Fermi level, then we may expand $(\epsilon_F \pm u\kappa T)^{3/2} = \epsilon_F^{3/2} \pm \frac{3}{2}u\kappa T\epsilon_F^{1/2} + \cdots$ and combine the integrals to obtain

$$\int_0^{\infty} \frac{\epsilon^{3/2}\,\mathrm{d}\epsilon}{\mathrm{e}^{(\epsilon-\epsilon_F)/\kappa T}+1} = \frac{2}{5}\epsilon_F^{5/2} + 3(\kappa T)^2\epsilon_F^{1/2}\int_0^{\infty}\frac{u\,\mathrm{d}u}{\mathrm{e}^u+1} + \cdots\,. \tag{7.46}$$

Then using

$$\int_0^{\infty}\frac{u\,\mathrm{d}u}{\mathrm{e}^u+1} = \frac{\pi^2}{12}\,,$$

the specific heat is found to be

$$C_e = \frac{3\pi^2 N_0 \kappa^2 T}{4\epsilon_F} = 3R\left(\frac{\pi^2\kappa T}{4\epsilon_F}\right)\,, \tag{7.47}$$

which agrees rather well with measurements for low temperature metals and exhibits the linear temperature dependence observed. The observations must be carried out at low temperature because the specific heat of a conductor is due to a combination of lattice vibrations and electron motions, and only when the contribution from the lattice vibrations begins to vary as T^3 can the linear dependence due to electrons be observed. □

Problem 7.15 *Specific heat of electrons.* Fill in the missing steps leading to Equation (7.47).

Problem 7.16 *Specific heat of electrons versus specific heat of lattice vibrations.*

(a) Calculate the Fermi energy for copper in electron Volts (eV). The density of copper is $8.9 \cdot 10^3$ kg/m^3 and the Atomic mass number is 63.54. Assume one electron/atom.

(b) Estimate the temperature where the electron specific heat is equal to the specific heat due to lattice vibrations for copper. Use Equation (7.29) (or Figure 7.4) for the lattice vibrations.

Example 7.5

Degeneracy Pressure. One effect of the Pauli Exclusion Principle is a pressure which is not related to temperature. From the calculation of the density of states, we found that various states in a three-dimensional box are closely packed, and when one adds the exclusion principle which permits only two particles per state (for fermions), the particles fill all of the states starting at the bottom up to Fermi level. At absolute zero, they would lie as close to the bottom as possible, but even at high temperatures, few will lie higher in energy than a few kT above the highest energy at zero temperature. In a way similar to the calculation of the Fermi energy of Equation (7.33), the total energy is found from

$$\begin{aligned} E_{\text{total}} &= \int_0^{\epsilon_F} n(\epsilon)\epsilon \, d\epsilon \\ &= \frac{8\pi V\sqrt{2m^3}}{h^3} \int_0^{\epsilon_F} \epsilon^{3/2} \, d\epsilon \\ &= \frac{8\pi V\sqrt{2m^3}}{h^3} \frac{2}{5} \epsilon_F^{5/2} \\ &= \frac{\hbar^2 \pi^3}{10m} \left(\frac{3N}{\pi}\right)^{5/3} \frac{1}{V^{2/3}} . \end{aligned}$$

From this result, we find the mean energy $\langle E \rangle = E_{\text{total}}/N = 3\epsilon_F/5$.

If one were now to try to decrease the volume, the energy would have to increase, so there is an effective pressure resisting this compression, given by

$$p_{\text{deg}} = -\frac{\partial E_{\text{total}}}{\partial V} .$$

We call this the **degeneracy pressure**, denoted by p_{deg}. We can easily calculate this pressure from

$$\begin{aligned} p_{\text{deg}} &= -\frac{\partial E_{\text{total}}}{\partial V} \\ &= \frac{\hbar^2 \pi^3}{10m} \left(\frac{3N}{\pi}\right)^{5/3} \frac{2}{3V^{5/3}} \\ &= \frac{\hbar^2 \pi^3}{15m} \left(\frac{3N}{\pi V}\right)^{5/3} . \end{aligned} \tag{7.48}$$

□

Example 7.6

Bulk Modulus. The bulk modulus B of any material (the inverse of the compressibility) is defined as

$$B = -V\frac{\partial p}{\partial V} .$$

For a degenerate material, this quantity is given by

$$B = -V\frac{\partial p_{\text{deg}}}{\partial V}$$

$$= -V\left[-\frac{\hbar^2\pi^3}{15m}\left(\frac{3N}{\pi}\right)^{5/3}\frac{5}{3V^{8/3}}\right]$$

$$= \frac{\hbar^2\pi^3}{9m}\left(\frac{3n}{\pi}\right)^{5/3},$$

where $n = N/V$. □

Problem 7.17 *Bulk modulus for metals.* For sodium, $n = 2.65 \cdot 10^{28}\,/\mathrm{m}^3$. Compare (percent error) the calculated value of the bulk modulus with the tabulated value, $6.4 \cdot 10^{10}$ dynes/cm^2 (note units in tabulated value are not SI units!).

7.2.4 Stellar Evolution

In a main sequence star, the pressure supporting the star against gravity derives from the nuclear furnace inside keeping the star hot. In older stars, however, when the hydrogen begins to run out, having converted most of its hydrogen to helium, the thermal pressure is no longer able to support the star, so it begins to collapse. In the process of collapsing, however, the adiabatic compression heats the He to such an extent that a new kind of nuclear burning begins. Two of the processes that occur are

$$^4\mathrm{He} + {}^4\mathrm{He} \to {}^8\mathrm{Be} + \gamma - 95\ \mathrm{keV}\,,$$
$$^4\mathrm{He} + {}^8\mathrm{Be} \to {}^{12}\mathrm{C} + \gamma + 7.4\ \mathrm{keV}\,,$$

so the first requires high temperatures to start, while the second begins to heat the star, along with other processes that involve N and O. Eventually, even the He is used up, and as heavier elements are formed, the process stops around Iron, since near Iron, no more exothermic nuclear processes exist. As this stage is approached, there is no longer any barrier to collapse except degeneracy pressure. We can estimate this degeneracy pressure if we assume that the mass density is constant (independent of radius) and if we assume the star is spherical. In this case we estimate the gravitational potential energy to be

$$dV_g = -G\frac{(\frac{4}{3}\pi r^3\rho)(4\pi r^2\rho\,\mathrm{d}r)}{r}$$

$$= -\frac{(4\pi)^2 G\rho^2 r^4\,\mathrm{d}r}{3}$$

where the first mass on the top line represents the mass inside the radius r, while the second mass represents the mass in a shell of thickness dr at radius r. Integrating, we find

$$V_g = -\frac{(4\pi\rho)^2 G}{3}\int_0^R r^4\,\mathrm{d}r = -\frac{(4\pi\rho)^2 G R^5}{15}\,.$$

From our definition of the density, we have $4\pi R^3 \rho/3 = M = N m_n$ where m_n is the mass of a nucleon. Expressing both the density and the potential in terms of V using $\rho = N m_n / V$ and $R = (3V/4\pi)^{1/3}$, we find

$$V_g = -\frac{3G}{5}(Nm_n)^2 \left(\frac{4\pi}{3V}\right)^{1/3} .$$

Now we must balance the gravitational pressure (negative pressure, tending to collapse the star),

$$p_g = -\frac{\partial V_g}{\partial V} = -\frac{1}{5}\left(\frac{4\pi}{3}\right)^{1/3} G(Nm_n)^2 V^{-4/3} , \tag{7.49}$$

against the degeneracy pressure (tending to resist the collapse),

$$p_{\text{deg}} = \frac{\hbar^2 \pi^3}{15 m_e}\left(\frac{3n}{\pi}\right)^{5/3} = \frac{\hbar^2 \pi^3}{15 m_e}\left(\frac{3N_e}{\pi}\right)^{5/3} V^{-5/3} , \tag{7.50}$$

where N_e is the number of electrons which is taken to be half the number of nucleons, or $N_e \sim N/2$. Balance occurs when

$$V^{1/3} = \frac{3\pi\hbar^2}{4Gm_n^2 m_e (2N)^{1/3}} ,$$

or when the radius is

$$R^* = \left(\frac{3V}{4\pi}\right)^{1/3} = \frac{3(3\pi^2)^{1/3}\hbar^2}{8Gm_n^2 m_e N^{1/3}} . \tag{7.51}$$

Problem 7.18 *Degenerate stars.* For our sun, with mass $M_\odot = 2 \cdot 10^{30}$ kg,

(a) Show that the degenerate radius is approximately $7 \cdot 10^3$ km.

(b) Estimate the mean kinetic energy of an electron and a neutron in the *degenerate* sun and compare this energy with the corresponding rest energies.

7.2.4.1 Relativistic Effects

While for our sun, the relativistic effects are not strong, for more massive stars, the relativistic effects increase to the point that the total energy per particle exceeds the rest energy by a sufficient amount that the rest energy may be neglected in the first approximation. In such cases, we must reexamine the density of states, since now we have

$$\begin{aligned} \mathrm{d}N &= \frac{2V}{h^3}\mathrm{d}p_x\,\mathrm{d}p_y\,\mathrm{d}p_z = \frac{8\pi V}{h^3}p^2\,\mathrm{d}p \\ &\approx \frac{8\pi V}{h^3}\frac{\epsilon^2}{c^3}\,\mathrm{d}\epsilon \end{aligned}$$

since $E = \sqrt{p^2c^2 + m^2c^4} \sim pc$. Hence, for N, we find

$$N = \frac{8\pi V}{h^3c^3} \int_0^{\epsilon_F} \epsilon^2 \, \mathrm{d}\epsilon = \frac{8\pi V}{h^3c^3} \frac{\epsilon_F^3}{3} \, .$$

We therefore have $\epsilon_F = \pi\hbar c(3n/\pi)^{1/3}$. For the total energy, we find

$$\begin{aligned} E_{\text{total}} &= \int_0^{\epsilon_F} \epsilon N(\epsilon) \, \mathrm{d}\epsilon \\ &= \frac{8\pi V}{h^3c^3} \int_0^{\epsilon_F} \epsilon^3 \, \mathrm{d}\epsilon \\ &= \frac{2\pi V}{h^3c^3} \epsilon_F^4 \\ &= \frac{3\hbar c N^{4/3}}{4} \left(\frac{3\pi^2}{V}\right)^{1/3} \, . \end{aligned} \tag{7.52}$$

For this case, the mean energy is $\langle E \rangle_{\text{rel}} = E_{\text{total}}/N = 3\epsilon_F/4$. From the total energy, we obtain the degeneracy pressure from Equation (7.48) to be

$$p_{\text{deg}} = -\frac{\partial E_{\text{total}}}{\partial V} = \frac{(3\pi^2)^{1/3}\hbar c}{4} \left(\frac{N}{V}\right)^{4/3} \, . \tag{7.53}$$

We now note that both the gravitational pressure p_g from Equation (7.49) and the degeneracy pressure p_{deg} from Equation (7.53) scale as $V^{-4/3}$, so there is no fixed radius (except through the neglected finite mass corrections). For large enough N, the star will continue to collapse. To find the critical number where gravitation wins and the collapse begins, we balance the relativistic degeneracy pressure from Equation (7.53) with the gravitational pressure from Equation (7.49),

$$\frac{(3\pi^2)^{1/3}\hbar c}{4} \left(\frac{N}{V}\right)^{4/3} - \frac{1}{5} \left(\frac{4\pi}{3}\right)^{1/3} G(Nm_n)^2 V^{-4/3} = 0 \, ,$$

and solve for N_{crit} to find

$$N_{\text{crit}} = \frac{15\sqrt{5\pi}}{16} \left(\frac{\hbar c}{Gm_n^2}\right)^{3/2} \, . \tag{7.54}$$

Problem 7.19 *Critical mass.*

(a) Calculate the value of N_{crit}.

(b) Compare the critical mass with the solar mass, $M_{\text{crit}}/M_\odot = ?$

(c) Estimate the mean kinetic energy of an electron and a neutron in this critical star and compare this energy with the corresponding rest energies.

7.2.4.2 Neutron Stars

If the mass exceeds this critical mass, it will again collapse until the density is so high that inverse neutron decay begins through the reaction

$$e^- + p \rightarrow n + \nu$$

until all of the electrons and protons are converted to neutrons. The neutrons are still not relativistic, so we can once again estimate the radius where the degeneracy pressure balances the gravitational pressure by simply replacing the electron mass by the neutron mass in Equation (7.51) but letting $N_e = N$ in Equation (7.50) instead of $N_e \sim N/2$, resulting in

$$R_n^* = \left(\frac{3V}{4\pi}\right)^{1/3} = \left(\frac{81\pi^2}{16}\right)^{1/3} \frac{\hbar^2}{Gm_n^3 N^{1/3}} . \tag{7.55}$$

For a neutron star with $M_n/M_\odot = 7$, we find the approximate radius is $R_n^* = 6.5$ km!

Problem 7.20 *Neutron stars.*

(a) Estimate the mean energy/neutron in a neutron star with $N = 7N_\odot$ and compare this with the rest energy of a neutron (i.e., find $\langle E\rangle/m_n c^2$).

(b) We found in Problem 7.19 that the ratio of mean energy to rest energy for electrons at the critical mass was about 2.9. Estimate the mass of a star (in solar masses) which would give the mean energy per particle equal to three times the rest energy for a neutron. This gives a rough estimate when a neutron star may collapse (eventually into a black hole).

7.2.5 Applications of the Bose–Einstein Distribution Law

Here we consider two examples: the black-body distribution, of course, since photons are bosons, having spin 1; and superfluids. We will make comments about superconductors, but not in detail.

Example 7.7

Black-body radiation law. In order to use the Bose–Einstein statistics, we need to know the appropriate density of states for photons, since the expression we have been using explicitly includes a mass term and is thus unacceptable in its present form. When we recall the derivation of the density of states in phase space, however, we have

$$\begin{aligned} g_s &= \frac{V}{h^3}\,\mathrm{d}p_x\,\mathrm{d}p_y\,\mathrm{d}p_z \\ &= \frac{4\pi V}{h^3} p^2\,\mathrm{d}p \\ &= \frac{4\pi V}{h^3 c^3}\epsilon^2\,\mathrm{d}\epsilon \end{aligned}$$

since for a photon $\epsilon = pc$, and when we include the twofold degeneracy for photons due their spin equal to ± 1, along with $\alpha = 0$ since photons are not in general conserved, we have

$$n_s = \frac{8\pi V}{h^3 c^3} \frac{\epsilon_s^2 \, \mathrm{d}\epsilon_s}{\mathrm{e}^{\epsilon_s/\kappa T} - 1} . \tag{7.56}$$

The energy density is therefore given by

$$\mathrm{d}U = \frac{n_s \epsilon_s}{V} = \frac{8\pi}{h^3 c^3} \frac{\epsilon_s^3 \, \mathrm{d}\epsilon_s}{\mathrm{e}^{\epsilon_s/\kappa T} - 1} . \tag{7.57}$$

This result, when converted to a function of λ through $\epsilon_s = hc/\lambda$, is equivalent to the Planck distribution, and we have no zero-point term to deal with. □

Problem 7.21 *Planck distribution.* Show that Equation (7.57) leads to the Planck distribution [Equation (7.31) without the zero-point term].

Example 7.8

Superfluids. Superfluids are bizarre substances. They exhibit zero viscosity and infinite thermal conductivity, and even support a unique type of wave called "second sound." The sudden appearance of the superfluid state in helium is dramatically evident when lowering the temperature by pumping on the liquid helium to cause it to boil. The boiling is evident until the temperature drops to a point where the boiling suddenly stops and the liquid is quiescent as the temperature continues to fall. The boiling, of course, is because the heat in the fluid cannot escape to the surface fast enough by heat conduction, so bubbles form deep inside the liquid and float to the surface. As soon as a minute portion of the fluid makes the transition to the superfluid state, however, that minute portion with infinite thermal conductivity is capable of carrying all of the heat directly to the surface, so the boiling stops. This and other bizarre features will be described as we come to understand some of the peculiarities of this phenomenon.

We begin the description with the conservation of particles for bosons, since helium has four nucleons and two electrons, so the spin is necessarily integral. We thus write

$$\begin{aligned} N = \sum_{s=1}^{\infty} n_s &= \frac{4\pi V (2M^3)^{1/2}}{h^3} \int_0^\infty \frac{\epsilon^{1/2} \, \mathrm{d}\epsilon}{\mathrm{e}^{\alpha + \beta\epsilon} - 1} \\ &= \frac{4\pi V (2M^3)^{1/2} (\kappa T)^{3/2}}{h^3} \int_0^\infty \left[u^{1/2} \mathrm{e}^{-\alpha - u} \sum_{r=0}^{\infty} \mathrm{e}^{-r(\alpha + u)} \right] \mathrm{d}u \text{ with } u = \beta\epsilon \\ &= \frac{4\pi V (2M^3)^{1/2} (\kappa T)^{3/2}}{h^3} \Gamma\left(\frac{3}{2}\right) \sum_{p=1}^{\infty} \left(\frac{\mathrm{e}^{-p\alpha}}{p^{3/2}} \right) . \end{aligned} \tag{7.58}$$

There are three features of Equation (7.58) that we call attention to:

1. On the right, we have a product of a monotonically increasing function of T and a monotonically decreasing function of α. Thus we know that *α is a monotonically increasing function of T.*

2. The function of T has no lower bound. The function of α, on the other hand, has an upper bound when $\alpha \to 0$, *yet the product is constant.*

3. The conclusion from these two observations is that physically acceptable values of α can be obtained *only for temperatures exceeding some critical value.*

$\square$

Problem 7.22 Supply the missing steps in Equation (7.58).

Problem 7.23 *The λ Point.* The specific gravity of liquid helium is 0.15. Use this to find at what temperature Equation (7.58) would give $\alpha = 0$ for helium. First evaluate the sum to *three* significant figures. *Partial ans.* $\sum_{p=1}^{\infty} \frac{1}{p^{3/2}} \simeq 2.6$.

Using a straightforward application of the Bose–Einstein statistics, we find that it does not give a fully acceptable answer. Either we have made an invalid assumption in applying the principles, or the theory is inadequate. In either case, it is not surprising that something strange happens at the point where theory appears to break down.

Actually, we have made one assumption that is not appropriate at very low temperatures. This was the replacement of the sum by an integral. This replacement is valid any time all the appropriate numbers are large, but the replacement of $\Delta\epsilon_s$ by a $\mathrm{d}\epsilon_s$ for the *very first state*, or the zero point motion (not zero energy, but the ground state as in the harmonic oscillator) is not uniformly valid. We thus separate the ground state from all the rest, and use the expression

$$\begin{aligned} N &= \frac{1}{\mathrm{e}^{\alpha} - 1} + \frac{4\pi V (2M^3)^{1/2}}{h^3} \int_0^{\infty} \frac{\epsilon^{1/2}\,\mathrm{d}\epsilon}{\mathrm{e}^{\alpha+\beta\epsilon} - 1} \\ &= \frac{1}{\mathrm{e}^{\alpha} - 1} + \left(\frac{T}{T_c}\right)^{3/2} f(\alpha) N\,. \end{aligned} \tag{7.59}$$

Clearly $f(0) = 1$. The second term is of course the only important term for $T > T_c$, while the ground state term is comparable to the second term for $T < T_c$ and eventually dominates as $T \to 0$. What happens, then, below T_c, is that there is a type of condensation called a **Bose–Einstein Condensate** at low temperatures where the helium atoms fall into the ground state in enormous numbers. It is the properties of these ground state atoms that produce all the bizarre effects. The population of these ground state atoms can be seen easily from Equation (7.59) if we write the first term as n_1 to represent the number in the ground state. We have then that

$$n_1 = N\left[1 - \left(\frac{T}{T_c}\right)^{3/2}\right] \tag{7.60}$$

and we estimate the magnitude of α from

$$n_1 = \frac{1}{e^{\alpha} - 1} \quad \rightarrow \quad \alpha = \ln\left(1 + \frac{1}{n_1}\right) \simeq \frac{1}{n_1}.$$

This behavior is illustrated in Figure 7.6.

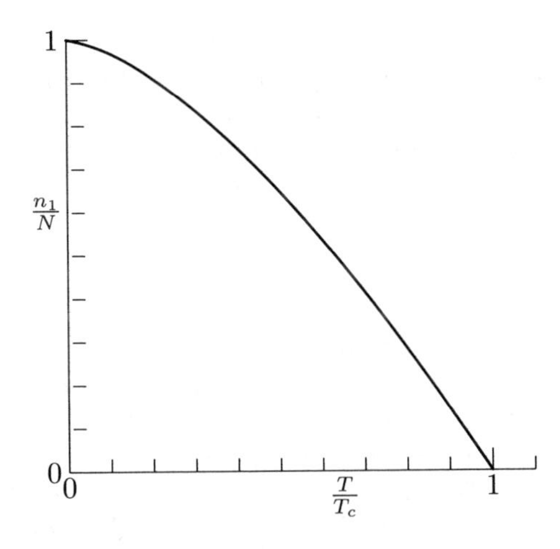

FIGURE 7.6
Population of the ground state of a Bose–Einstein system of free particles.

The lack of viscosity and the effectively infinite thermal conductivity are due to the fact that an enormous number of the helium atoms are in the ground state. Taking this group as a whole, the uncertainty in momentum is effectively zero, since the ground state energy is well-defined. The uncertainty principle then indicates that the uncertainty in location is essentially unbounded. It is no longer possible to say where any of these atoms that make up the condensate are located. Since thermal conductivity measures how fast heat moves from one place to another, it involves particles with kinetic energy moving from one place to another, but these particles are not here one moment and elsewhere the next, since they are effectively everywhere. They form what is sometimes called a giant quantum state. Heat is essentially moving at the speed of light as all atoms are connected by this giant state. Viscosity also involves the drag due to collisions between particles when trying to create a velocity (or momentum) gradient in space. But once again, the nonlocalization of the atoms prevents gradients, so no viscosity is observed. When there is both normal fluid and superfluid together (which is always the case below the λ point), only the normal fluid exhibits viscosity.

Another bizarre effect is that if superfluid is in an open dewar (no cork on top), liquid helium is observed to drip from the bottom of the dewar. The helium does not pass through the dewar, nor is it apparent that it flows over the top, but the nonlocalization means that there is a certain probability that the atoms will be found outside, where they may condense and drip off the bottom. The phenomenon of "second sound" is due to a wave passing through the mixture of normal and superfluid where the densities of the two fluids oscillate as the wave passes, shifting the positive energy in the wave back and forth between the two fluids.

In relatively recent times, another type of Bose–Einstein condensate has been achieved by lowering a group of particles in a trap to temperatures of μK where they condense into the ground state. This technique has allowed several species of atoms to experience this state. The number of atoms in the trap is very small compared to that of a sample of liquid helium, but large enough to exhibit the signature of the condensate.

7.2.5.1 Superconductors

Since the phenomenon of Bose–Einstein condensation which gives rise to the phenomenon of superfluidity is dependent upon the particles being bosons, one would not expect any kind of similar behavior with electrons since they are fermions. If, however, there were some way to pair up electrons, one with spin up and one with spin down, and they more or less traveled together, the pair would behave like a boson. In fact, at low enough temperatures, there are interactions between electron pairs through phonons, which are quantized acoustic waves, which does correlate electron pairs called Cooper pairs after Leon Cooper who first postulated their existence. This led to the Bardeen–Cooper–Schrieffer (BCS) theory, which explained for the first time the phenomenon of superconductivity, which was observed experimentally near the beginning of the 20th century. At higher temperatures, lattice vibrations generally decorrelate the electrons and the effect vanishes, but many materials, both pure and compound, exhibit superconductivity. The recent discovery of high temperature superconductors (above liquid nitrogen temperature) is still not fully understood, but must in some way be related to Bose–Einstein statistics and paired electrons.

8

Band Theory of Solids

8.1 Periodic Potentials

One of the most basic features of solid materials is that at the microscopic level, the atoms are characteristically lined up in an orderly fashion. When there is no order, materials are called amorphous, but even in a great many materials we do not think of as crystalline, there exist small domains within which the atoms are orderly but the adjacent domains have no systematic order with respect to one another. The fundamental effect of the order within a domain or within a larger crystal is to produce a periodic potential well instead of the isolated atomic potentials we have considered previously. This chapter will examine the effects of this periodicity in an idealized way.

If one first considers a series of atomic potentials, each at a fixed spacing a from one another, the superposition of the overlapping Coulomb potentials will produce a net potential as pictured in Figure 8.1. At the edge of the material, the potential will of course rise to zero, but the overlapping potentials inside will produce a lowered maximum potential and far from the edge the potential will be periodic. In our idealized case, we will basically ignore the deviations from periodicity at the edge and take the potential to be exactly periodic.

In addition to the periodicity of the well due to the overlapping potentials, the proximity of the nearest neighbors causes the energy states to be perturbed. Since the potential barriers are not infinite, there is tunneling between atoms, and we need to examine the effect of coupled wells. Considering two wells only with a barrier between them, as in Figure 8.2, the three regions are characterized by the left, center, and right wave functions

$$\psi_\ell(x) = A\mathrm{e}^{\mathrm{i}\alpha x} + B\mathrm{e}^{-\mathrm{i}\alpha x}\,, \qquad -a \le x \le -b\,, \tag{8.1}$$

$$\psi_c(x) = C\mathrm{e}^{\beta x} + D\mathrm{e}^{-\beta x}\,, \qquad -b \le x \le b\,, \tag{8.2}$$

$$\psi_r(x) = F\mathrm{e}^{\mathrm{i}\alpha x} + G\mathrm{e}^{-\mathrm{i}\alpha x}\,, \qquad b \le x \le a\,, \tag{8.3}$$

where $E < V$ so that

$$\alpha = \sqrt{2mE}/\hbar\,,$$
$$\beta = \sqrt{2m(V-E)}/\hbar\,.$$

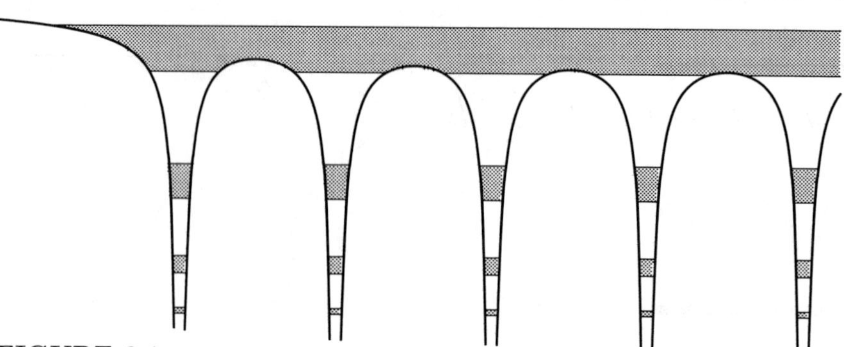

FIGURE 8.1
Periodic potential just inside edge of lattice. Shaded areas represent bands of allowed energy states.

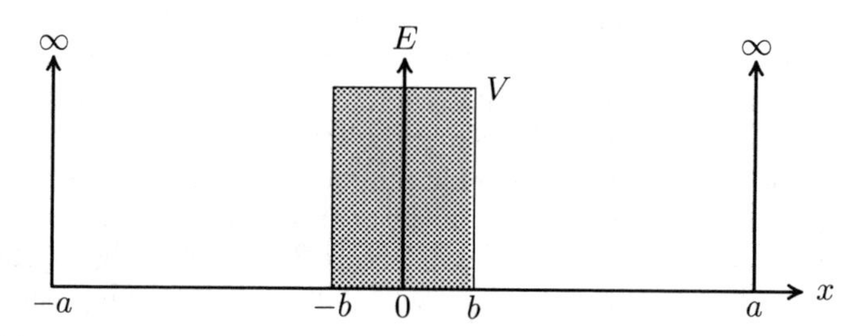

FIGURE 8.2
Two rectangular potential wells separated by a barrier.

The matching conditions require $\psi_\ell(-a) = 0$, $\psi_r(a) = 0$ since the walls are infinite, and then $\psi_\ell(-b) = \psi_c(-b)$, $\psi'_\ell(-b) = \psi'_c(-b)$, $\psi_c(b) = \psi_r(b)$, and $\psi'_c(b) = \psi'_r(b)$. There are two classes of solutions for this problem, one symmetric (even about $x = 0$) and one antisymmetric (odd about $x = 0$). For the even solution, we find that $A = G$, $B = F$, $C = D$, which reduces the problem to the condition

$$\tan\alpha(a-b) = -\frac{\alpha}{\beta\tanh\beta b}\,. \tag{8.4}$$

For the odd solution, we find that $A = -G$, $B = -F$, $C = -D$, which leads to the condition

$$\tan\alpha(a-b) = -\frac{\alpha}{\beta}\tanh\beta b\,. \tag{8.5}$$

If the barrier is high, so that the coupling is weak, $\beta b \gg 1$, so that we may approximate $\tanh\beta b \simeq 1 - 2\mathrm{e}^{-2\beta b}$. For each case we assume that $\alpha(a-b) =$

$n\pi + \epsilon$ with $\epsilon \ll 1$, and hence $\tan \alpha(a-b) \simeq \epsilon$. Then the approximate energies for the two cases given by Equations (8.4) and (8.5) may be obtained from

1. for the symmetric case:

$$\epsilon \simeq -\frac{\alpha}{\beta \tanh \beta b} \simeq -\frac{\alpha}{\beta}(1 + 2\mathrm{e}^{-2\beta b}),$$

   so that

$$\alpha_1 = \sqrt{2mE}/\hbar \simeq \frac{n\pi}{a-b} - \frac{\alpha(1 + 2\mathrm{e}^{-2\beta b})}{\beta(a-b)},$$

2. and for the antisymmetric case:

$$\epsilon \simeq -\frac{\alpha}{\beta}\tanh \beta b \simeq -\frac{\alpha}{\beta}(1 - 2\mathrm{e}^{-2\beta b}),$$

   so that

$$\alpha_1 = \sqrt{2mE}/\hbar \simeq \frac{n\pi}{a-b} - \frac{\alpha(1 - 2\mathrm{e}^{-2\beta b})}{\beta(a-b)}.$$

Combining these results, the coupled energies are perturbed from the infinite barrier case with $\alpha_0 = n\pi/(a-b)$ (uncoupled energies) and have been split into two distinct energies,

$$\alpha = \alpha_0 - \frac{\alpha}{\beta(a-b)} \mp \frac{2\alpha}{\beta(a-b)}\mathrm{e}^{-2\beta b}. \tag{8.6}$$

In general, if there had been three wells, the energy would be split into three levels, and for N wells there would be N levels. For macroscopic crystals that may have 10^{10} wells lined up in each of three dimensions, the original one atom level would be split into 10^{30} levels. These are called energy **bands**, and are shaded in Figure 8.1. From the nature of the splitting term, it is clear that states lying higher (closer to the top of the barrier) will be split into wider bands as illustrated in the figure. In order to understand better the nature of these bands, we must first examine the effects of periodicity on both the levels and widths of these bands, and then examine their effects on the motion of electrons in these bands.

Problem 8.1 *Two coupled wells.* Show that even and odd solutions for the two coupled well problem illustrated in Figure 8.2 lead to Equations (8.4) and (8.5). Then fill in the details leading to Equation (8.6).

8.2 Periodic Potential — Kronig–Penney Model

For the effects of the periodicity of the potential well, which is periodic in the lattice spacing of the crystal which we denote a, the wave function is presumed to satisfy the periodic boundary condition,

$$\psi(x + Na) = \psi(x)\,, \tag{8.7}$$

where $Na = L_x$, the macroscopic length of the sample. The Hamiltonian is presumed to be microscopically periodic, however, so that

$$\hat{\mathcal{H}}(x) = -\frac{\hbar^2}{2m}\frac{\mathrm{d}^2}{\mathrm{d}x^2} + \hat{V}(x)\,,$$

with the periodic potential well

$$\hat{V}(x + a) = \hat{V}(x)\,,$$

so the Hamiltonian must have the same periodicity such that

$$\hat{\mathcal{H}}(x + a) = \hat{\mathcal{H}}(x)\,.$$

In order to explore the effects of periodicity, we first examine the translation operator $\hat{T}$ that has the property

$$\begin{aligned} \hat{T}f(x) &= f(x + a) \\ \hat{T}^* f(x) &= f(x - a)\,. \end{aligned}$$

Now since expanding $f(x + a)$ about $f(x)$ in a series expansion yields

$$f(x + a) = f(x) + a\frac{\mathrm{d}f}{\mathrm{d}x} + \frac{a^2}{2!}\frac{\mathrm{d}^2 f}{\mathrm{d}x^2} + \cdots\,,$$

we conclude that the translation operator may be written as

$$\begin{aligned} \hat{T} &= 1 + a\frac{\mathrm{d}}{\mathrm{d}x} + \frac{a^2}{2!}\frac{\mathrm{d}^2}{\mathrm{d}x^2} + \cdots \\ &= \exp\left[a\frac{\mathrm{d}}{\mathrm{d}x}\right] \\ &= \mathrm{e}^{\mathrm{i}a\hat{p}_x/\hbar}\,. \end{aligned}$$

Since $\psi(x)$ is an eigenfunction of $\hat{\mathcal{H}}$, we have $\hat{\mathcal{H}}\psi(x) = E\psi(x)$, so that

$$\begin{aligned} \hat{T}\hat{\mathcal{H}}\psi(x) &= \hat{T}E\psi(x) \\ &= E\hat{T}\psi(x) \\ &= E\psi(x + a) \end{aligned}$$

and we also have

$$\begin{aligned} \hat{\mathcal{H}}\hat{T}\psi(x) &= \hat{\mathcal{H}}\psi(x + a) \\ &= \hat{\mathcal{H}}(x + a)\psi(x + a) \\ &= E\psi(x + a) \end{aligned}$$

so since $\hat{T}$ commutes with $\hat{\mathcal{H}}$, we know from Problem 1.29 that ψ is simultaneously an eigenfunction of $\hat{T}$ and $\hat{\mathcal{H}}$ and that $\hat{T}$ is invariant in time. We are interested in the **eigenvalues** of $\hat{T}$, so we observe that

$$\hat{T}\psi(x) = T_0\psi(x) \tag{8.8}$$

$$= \psi(x+a)\,, \tag{8.9}$$

so if we operate again with $\hat{T}$,

$$\begin{aligned}\hat{T}^2\psi(x) &= \hat{T}T_0\psi(x) = T_0\hat{T}\psi(x) = T_0^2\psi(x)\\ &= \hat{T}\psi(x+a) = \psi(x+2a)\,,\end{aligned}$$

and if we operate N times with $\hat{T}$, we obtain

$$\psi(x+Na) = T_0^N\psi(x)\,.$$

However, the periodic boundary condition of Equation (8.7) gives $\psi(x+Na) = \psi(x)$, so we must have $T_0^N = 1$, or

$$T_0 = 1^{1/N} = \mathrm{e}^{2\pi\mathrm{i}\ell/N}\,, \qquad \ell = 0, 1, \ldots, N-1\,. \tag{8.10}$$

Combining this result with Equations (8.8) and (8.9) leads to the conclusion that the wave function must be of the form

$$\psi(x) = \pi_\ell(x)\mathrm{e}^{2\pi\mathrm{i}\ell x/Na} \equiv \pi_k(x)\mathrm{e}^{\mathrm{i}kx}\,, \tag{8.11}$$

where we have defined $k = 2\pi\ell/Na$, and $\pi_k(x+a) = \pi_k(x)$. In other words, the total wave function is a product of an exactly periodic function, $\pi_k(x)$, and a phase shift given by the exponential term.

The actual potential function, $V(x)$, is in general too complicated to solve directly, but we will discover some of the features due to the periodicity by considering the Bloch potential, which is a periodic square well potential, sketched in Figure 8.3.

The potential to the left and right of the vertical axis are representative of the potential everywhere, so we will denote the wave function between $-b$ and 0 by ψ_ℓ and between 0 and c by ψ_r. Then from our previous examples in constant potential regions, we may write the solutions, assuming $E < V$, as

$$\begin{aligned}\psi_\ell(x) &= A\mathrm{e}^{\beta x} + B\mathrm{e}^{-\beta x}\,, \qquad -b \le x \le 0\,, \qquad \beta = \sqrt{2m(V-E)}/\hbar\,,\\ \psi_r(x) &= C\mathrm{e}^{\mathrm{i}\alpha x} + D\mathrm{e}^{-\mathrm{i}\alpha x}\,, \qquad 0 \le x \le c\,, \qquad \alpha = \sqrt{2mE}/\hbar\,.\end{aligned}$$

Matching the wave function and its first derivative at $x = 0$, we have

$$\begin{aligned}A + B &= C + D\,,\\ \beta A - \beta B &= \mathrm{i}\alpha C - \mathrm{i}\alpha D\,.\end{aligned}$$

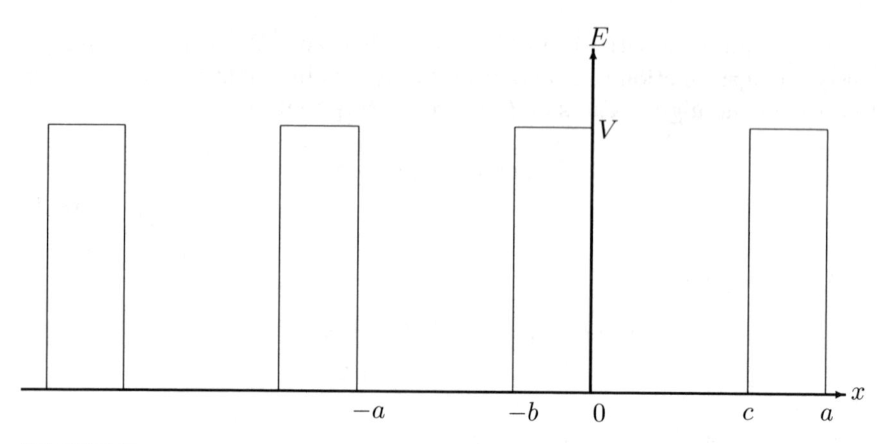

FIGURE 8.3
Sketch of Bloch potential. The barriers are of height V and width b while the wells are of width c, so that $a = b + c$, the lattice spacing.

and since $\pi_k(x) = \psi(x)\mathrm{e}^{-\mathrm{i}kx}$ is periodic, $\pi_k(c) = \pi_k(-b)$ and $\pi'_k(c) = \pi'_k(-b)$ so

$$\mathrm{e}^{\mathrm{i}kb}(\mathrm{e}^{-\beta b}A + \mathrm{e}^{\beta b}B) = \mathrm{e}^{-\mathrm{i}kc}(\mathrm{e}^{\mathrm{i}\alpha c}C + \mathrm{e}^{-\mathrm{i}\alpha c}D)\,,$$
$$\mathrm{e}^{\mathrm{i}kb}[(\beta - \mathrm{i}k)\mathrm{e}^{-\beta b}A - (\beta + \mathrm{i}k)\mathrm{e}^{\beta b}B] = \mathrm{i}\mathrm{e}^{-\mathrm{i}kc}[(\alpha - k)\mathrm{e}^{\mathrm{i}\alpha c}C - (\alpha + k)\mathrm{e}^{-\mathrm{i}\alpha c}D]\,.$$

The determinant of coefficients must vanish, so

$$\begin{vmatrix} 1 & 1 & -1 & -1 \\ \beta & -\beta & -\mathrm{i}\alpha & \mathrm{i}\alpha \\ \mathrm{e}^{\mathrm{i}kb}\mathrm{e}^{-\beta b} & \mathrm{e}^{\mathrm{i}kb}\mathrm{e}^{\beta b} & -\mathrm{e}^{-\mathrm{i}kc}\mathrm{e}^{\mathrm{i}\alpha c} & -\mathrm{e}^{-\mathrm{i}kc}\mathrm{e}^{-\mathrm{i}\alpha c} \\ (\beta - \mathrm{i}k)\mathrm{e}^{\mathrm{i}kb-\beta b} & -(\beta+\mathrm{i}k)\mathrm{e}^{\mathrm{i}kb+\beta b} & -\mathrm{i}(\alpha-k)\mathrm{e}^{-\mathrm{i}(kc-\alpha c)} & \mathrm{i}(\alpha+k)\mathrm{e}^{-\mathrm{i}(kc+\alpha c)} \end{vmatrix} = 0\,.$$

This may be reduced to

$$\begin{aligned} \cos ka &= \frac{\beta^2 - \alpha^2}{2\alpha\beta} \sinh \beta b \sin \alpha c + \cosh \beta b \cos \alpha c \\ &= F(E) \cos[\alpha c - \phi(E)]\,, \end{aligned} \tag{8.12}$$

where

$$F(E) = \left[1 + \frac{V^2}{4E(V - E)} \sinh^2 \beta b\right]^{1/2}, \qquad (> 1\,!)$$
$$\tan \phi(E) = \frac{V - 2E}{2\sqrt{E(V - E)}} \tanh \beta b\,.$$

A sketch of the right-hand side of Equation (8.12) is given in Figure 8.4, along with the limits imposed by the left-hand side where the shaded regions indicate that $|F(E)| < 1$ (allowed states) and the clear regions indicate that $|F(E)| > 1$ (forbidden states). The levels on the right in the figure correspond the levels of a single, isolated well of the same width and depth, so one can see how each single level is broadened into a band of energies from the effects of periodicity.

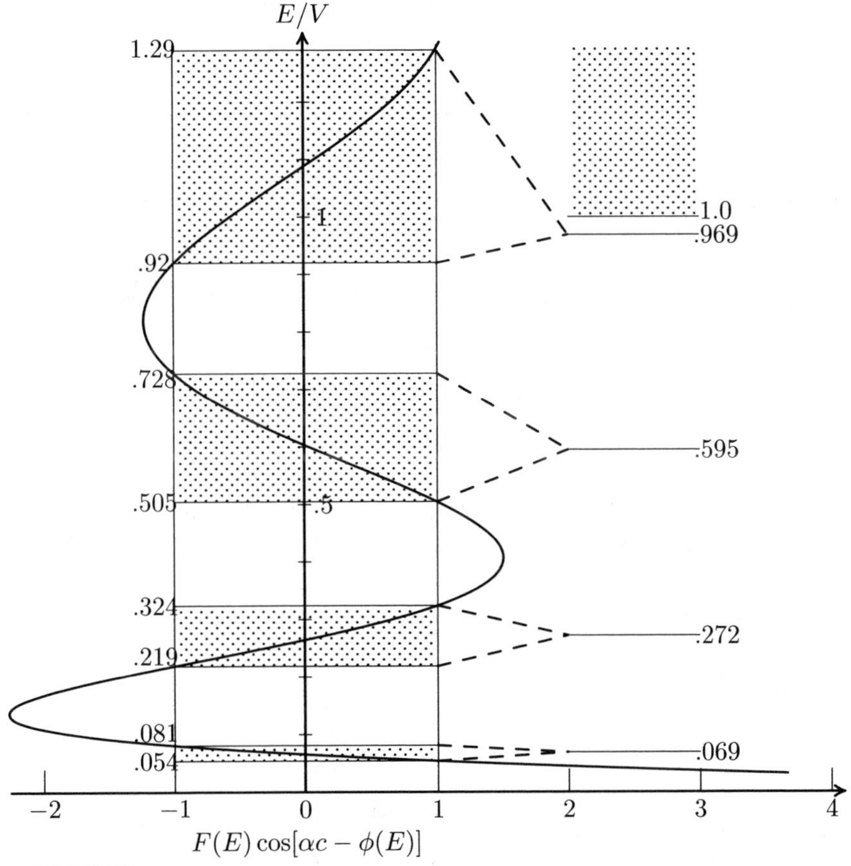

FIGURE 8.4
Plot of $F(E)\cos[\alpha c - \phi(E)]$ for a periodic well with well width c and barrier thickness $b = c/8$ and the corresponding energy levels for a single well with width c (see Section 3.1.4) with well depth of $V = 5h^2/4mc^2$. As $E \to 0$, $F(E)\cos[\alpha c - \phi(E)] \to 9.76$.

Problem 8.2 Show that $\hat{T}(x)$ is a linear, Hermitian operator if the operator $\hat{T}^*(x)$ is so defned that $\hat{T}^*(x)f(x) = f(x - a)$.

Problem 8.3 Show that $\hat{T}(x)$ commutes with $\hat{\mathcal{H}}$ if $\hat{\mathcal{H}}(x) = \hat{\mathcal{H}}(x + a)$.

Problem 8.4 Show that $\hat{T}(x)$ may be written formally as $\hat{T}(x) = \mathrm{e}^{a\mathrm{d}/\mathrm{d}x} = 1 + a\frac{\mathrm{d}}{\mathrm{d}x} + \cdots$ and may thus be regarded as the operator corresponding to the dynamical quantity $\exp(\mathrm{i}a\hat{p}_x/\hbar)$.

Problem 8.5 Show that the range of variation of ℓ can equally well be chosen to be symmetrical with respect to $\ell = 0$, that is,

$$\ell = 0, \pm 1, \pm 2, \ldots, \begin{cases} \pm\frac{N-1}{2} & N \text{ odd}, \\ \pm\frac{N}{2} & N \text{ even}. \end{cases}$$

A wave function having the form of Equation (8.11) is called a **Bloch function**.

Problem 8.6 Evaluate the average momentum of a particle whose wave function is a Bloch function (let ψ be normalized in a unit cell).

Problem 8.7 Evaluate the average kinetic energy of a particle that is in an eigenstate of the form of Equation (8.11).

8.2.1 δ-Function Barrier

In order to simplify the relation between E and k given in Equation (8.12), we consider a δ-function barrier, where we take the barrier between adjacent atoms to be an infinitely high but infinitesimally thin barrier, still with periodicity a. To this end, we define $U \equiv Vb = const.$ and let $b \to 0$. Hence, $c = a - b \to a$. Looking first at βb,

$$\beta b = \sqrt{2m(V-E)}b/\hbar = \sqrt{2m(U-Eb)}\sqrt{b}/\hbar \to \sqrt{2mU}\sqrt{b}/\hbar \to 0\,,$$

so we can let

$$\sinh\beta b \to \beta b\,, \qquad \text{and} \qquad \tanh\beta b \to \beta b\,,$$

so that

$$\begin{aligned} \tan\phi &= \frac{V-2E}{2\sqrt{E(V-E)}}\tanh\beta b \to \frac{U-2Eb}{2\sqrt{Eb(U-Eb)}}\sqrt{2m(U-Eb)}\sqrt{b}/\hbar \\ &\to \frac{\sqrt{m}U}{\hbar\sqrt{2E}} \equiv \sqrt{\frac{V_0}{E}}\,, \end{aligned}$$

with $V_0 \equiv mU^2/2\hbar^2$.

We next examine $F(E)$, which becomes

$$\begin{aligned} F(E) &= \sqrt{1 + \frac{V^2}{4E(V-E)}\sinh^2\beta b} \to \sqrt{1 + \frac{U^2}{4Eb(U-Eb)}\frac{2m(U-Eb)b}{\hbar^2}} \\ &\to \sqrt{1 + \frac{V_0}{E}}\,. \end{aligned}$$

Then using the trigonometric identities

$$\cos\phi = \frac{1}{\sqrt{1+\tan^2\phi}} = \frac{1}{\sqrt{1+V_0/E}}\,,$$

and

$$\sin\phi = \frac{\tan\phi}{\sqrt{1+\tan^2\phi}} = \frac{\sqrt{V_0/E}}{\sqrt{1+V_0/E}}\,,$$

we may write Equation (8.12) as

$$\begin{aligned}\cos ka &= F(E)(\cos\alpha a\cos\phi + \sin\alpha a\sin\phi)\\ &= \cos\alpha a + \sqrt{\frac{V_0}{E}}\sin\alpha a\,. \end{aligned} \tag{8.13}$$

This simpler relation can be reduced further in the limit as $E \to 0$, where

$$\alpha a = \sqrt{2mE}a/\hbar \to 0\,,$$

so $\cos\alpha a \to 1$ and $\sin\alpha a \to \alpha a$, so Equation (8.13) reduces to

$$\cos ka \to 1 + \sqrt{\frac{V_0}{E}}\frac{\sqrt{2mE}a}{\hbar} = 1 + \frac{mUa}{\hbar^2}\,.$$

It is thus clear that there are no solutions for real k as $E \to 0$ since the right-hand side exceeds unity. In general, since $F(E) > 1$ for all $E < V$ (finite barrier), or for all finite E (δ-function barrier), the right-hand side of Equation (8.12) and Equation (8.13) oscillates with amplitude greater than unity, and leaves only finite width bands where real values of k are possible.

Problem 8.8 Equation (8.12) also has a finite limit as $E \to 0$. Find this limiting value and show that for the values of Figure 8.4, the limiting value is 9.76.

8.2.2 Energy Bands and Electron Motion

If we assume that the amplitude of the oscillation is so much larger than unity that the function may be approximated by a straight line passing through the region between -1 and $+1$ in Figure 8.4, then in that range of ka, we may write the linear relationship as

$$\cos ka = A_n(E - E_n)\,,$$

or solving for E,

$$E = E_n + \frac{1}{A_n}\cos ka\,. \tag{8.14}$$

If we now plot E versus ka instead of versus $\cos ka$, then it will have the general appearance as shown in Figure 8.5, where the figure is left-right symmetric

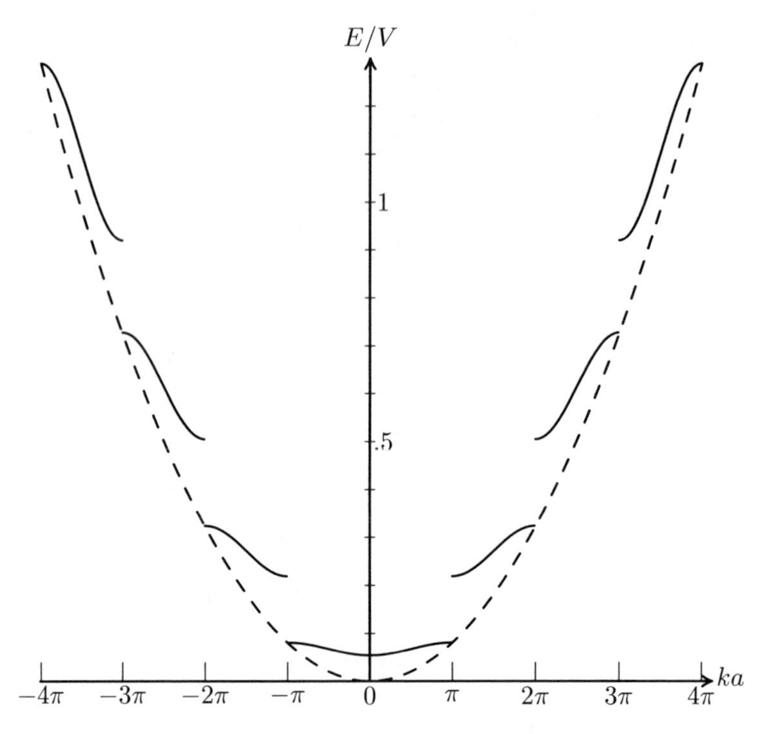

FIGURE 8.5
Plot of Equation (8.14) showing approximate dependence of $E(k)$ on ka (solid) for the data from Figure 8.4 and the free particle dependence, $E = \hbar^2k^2/2m$ (dashed).

because $\cos ka$ is symmetric, and the gaps occur between bands where there are no real values of k.

In addition to the explicit curves for each band, the general free particle relation, $E = \hbar^2k^2/2m$, is also sketched to show that in general, the kinetic energy of particles in these bands somewhat approximates the free particle energy. The deviations are very significant, however, since although the magnitude of E versus k is not so terribly different, the *dependence* of E on k is very different and has a dramatic effect on how particles move in response to external forces. To illustrate this, Figure 8.5 is usually "folded" so that each segment is moved by an integral multiple of 2π so that it lies between $-\pi$ and π. This is illustrated in Figure 8.6 where we again plot E versus ka, and this form of the plot is called a **Brillouin diagram**. From this figure, it is apparent that at the top and bottom of each band, near $k = 0$ (or a multiple of 2π for the higher bands), the curves may be approximated by a parabola.

We may evaluate the constant A_n in Equation (8.14) by differentiating both

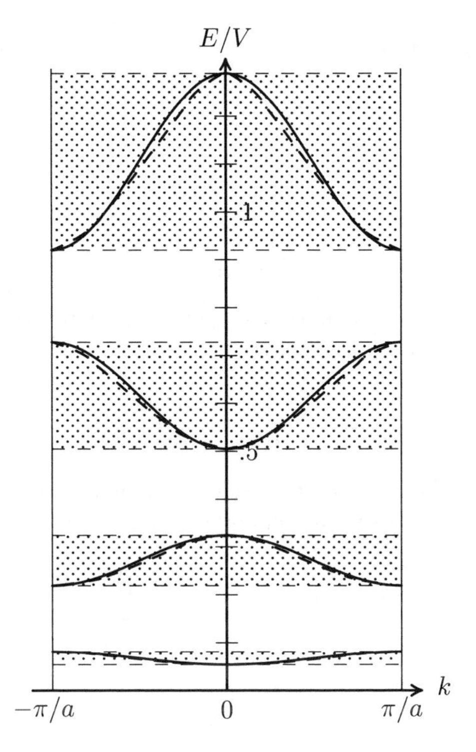

FIGURE 8.6
Plot of the reduced E vs. k curve from the approximate expression of Equation (8.14) (solid line) and from the exact expression of Equation (8.12) (dashed line).

sides and evaluating at $k = 0$ such that

$$\left.\frac{\mathrm{d}^2 E}{\mathrm{d}k^2}\right|_{k=0} = -\frac{a^2}{A_n},$$

so that near $k = 0$, we may approximate Equation (8.14) by expanding $\cos ka \simeq 1 - k^2a^2/2 + \cdots$ so that

$$E = E_n + \frac{1}{A_n} + \frac{k^2}{2}\left.\frac{\mathrm{d}^2 E}{\mathrm{d}k^2}\right|_{k=0} + \cdots . \tag{8.15}$$

This expression for $E(k)$ is almost identical to the expression for a free particle moving in a constant potential well,

$$E = V_0 + \frac{\hbar^2 k^2}{2m},$$

so we can represent particles near the bottom of each band in the periodic well as free particles if we identify $V_0 = E_n + 1/A_n$ and give the electrons an

effective mass m^*, where

$$\frac{k^2}{2}\frac{\mathrm{d}^2E}{\mathrm{d}k^2}\bigg|_{k=0} = \frac{\hbar^2k^2}{2m^*},$$

so that

$$m^* = \hbar^2 \bigg/ \frac{\mathrm{d}^2E}{\mathrm{d}k^2}\bigg|_{k=0}. \tag{8.16}$$

Another way of seeing that particles may move *as if* they have an effective mass is to consider that a wave packet has a group velocity given by

$$v_g = \frac{\mathrm{d}\omega}{\mathrm{d}k} = \frac{1}{\hbar}\frac{\mathrm{d}E}{\mathrm{d}k},$$

since $E = \hbar\omega$. Then in, for example, an electric field, where the electric force is represented by F, the force and energy are related by

$$\Delta E = F\Delta s = Fv_g\Delta t = \frac{F}{\hbar}\frac{\Delta E}{\Delta k}\Delta t,$$

which leads to

$$\frac{\mathrm{d}k}{\mathrm{d}t} = \frac{F}{\hbar}.$$

Then the acceleration is given by

$$a = \frac{\mathrm{d}v_g}{\mathrm{d}t} = \frac{\mathrm{d}}{\mathrm{d}t}\left(\frac{1}{\hbar}\frac{\mathrm{d}E}{\mathrm{d}k}\right) = \frac{1}{\hbar}\left(\frac{\mathrm{d}k}{\mathrm{d}t}\frac{\mathrm{d}^2E}{\mathrm{d}k^2}\right) = \frac{F}{\hbar^2}\frac{\mathrm{d}^2E}{\mathrm{d}k^2} = \frac{F}{m^*},$$

so again we have Equation (8.16) for the effective mass.

What this means is that electrons near the band edge will behave very much like free particles except that they will move either faster or slower than a free electron due to the interactions with the periodic potential well, characterized by their effective mass. This also is true near the *top* of each band, where the parabola is inverted. This means that near the top of each band, $m^* < 0$, and the electrons will move in the *opposite direction* from a free electron, or in the direction that a positively charged particle would move. Rather than talk about negative mass, we generally discuss these particles *as if* they have a positive charge and positive effective mass. We call these particles **holes**, and use the notation m_h^* for the hole effective mass and m_e^* for the electron effective mass. Except for these changes, we treat electrons as free particles, so that the periodic structure of the lattice affects only their effective mass.

8.2.3 Fermi Levels and Macroscopic Properties

If one examines the macroscopic properties of a solid material, in particular the conductivity which depends upon the electron motions, one finds materials grouped into three general classes: Conductors, Insulators, and Semiconductors. The differences between these types of behavior is determined by the Fermi level and the shape and separation of the energy bands. Quite simply, if the Fermi level is in the middle of a band somewhere, then the material is a conductor because the electrons near the Fermi level have available states nearby and can easily move under the influence of an external force. If the Fermi level lies near the bottom of a band, then it is characterized by electron conduction. If, on the other hand, it lies near the top of a band, then the material is characterized by hole conduction. From simple conductivity measurements, one cannot tell which is which. In order to tell, the Hall effect is generally employed.

8.2.3.1 Hall Effect

By a simple measurement of conductivity, one cannot determine whether the current is carried by electrons, holes, or a combination of both. With the addition of a magnetic field, however, we can determine the sign of the charge carriers (or their q/m ratio). We apply an external electric field $\mathcal{E}_L$ to a thin bar to induce a longitudinal current as in Figure 8.7a, where we assume the charge carriers are electrons, and in Figure 8.7b, where we assume the charge carriers to be holes. By adding a magnetic field transverse to the bar, the motions of the charged particles are deflected *in the same direction*, setting up a charge imbalance on the sides of the bar. This charge imbalance in turn sets up a transverse electric field that builds up until it prevents any further transverse flow so that the charges again travel down the bar undeflected. If the undeflected drift velocity is v_D, then the transverse electric field may be deduced by balancing the transverse forces, or

$$e\mathcal{E}_T = ev_D B \,.$$

The drift velocity is also related the the total current, such that

$$J = nev_D = ne\mathcal{E}_T/B \,,$$

where we have used the expression for v_D from the first expression. From this we can find the **Hall coefficient**,

$$R_H = \frac{\mathcal{E}_T}{JB} = \frac{1}{ne} \,,$$

which gives a value for the charge carrier density. The *direction* of $\mathcal{E}_T$ gives the *sign* of the charge carrier.

Problem 8.9 *Hall coefficient.*

(a) From the density of copper, $8.32 \cdot 10^3$ kg/m^3, and its atomic weight, 62.9 u, find the Hall coefficient for copper.

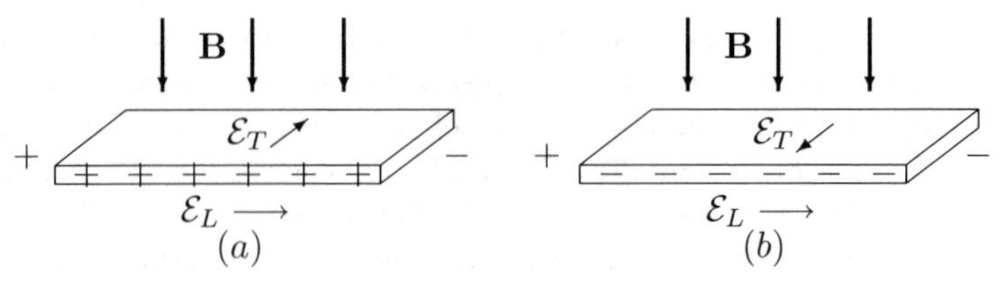

FIGURE 8.7
The Hall effect. (*a*) For negative carriers. (*b*) For positive carriers.

(b) For n-type silicon which is doped so that $n = 1000n_i$, find the Hall coefficient for this semiconductor.

(c) For these two substances above, consider a sample of each 1 mm thick and 1 cm wide with a 1.0 T vertical magnetic field as shown in Figure 8.7. Find the current necessary to generate 1 mV of Hall voltage for each sample.

8.3 Impurities in Semiconductors

8.3.1 Ionization of Impurities

Virtually all semiconductors are either Group IV materials, such as Germanium and Silicon, or combinations of Group III and Group V materials such as Indium Antimonide or Gallium Arsenide, since in each case, the average number of electrons in the outer shell, when shared with their four nearest neighbors, is eight, making a closed shell. For example, each Ge atom has four electrons in the outer shell, and it shares one electron with each of its four neighbors as they also share one of their outer electrons in a mutual bond so that there are four bonds with two shared electrons in each bond. The eight electrons associated with each atom makes a magic number, so the bonds are unusually strong, and are called covalent bonds.

Most impurities used for the doping of semiconductors are either Group V atoms, which have five electrons in the outer shell so that after using four for the covalent bonds there is one left over, or Group III atoms with three electrons in the outer shell that is one short for the full covalent bonding. This means that if those with one extra electron were ionized, the remaining ion is symmetric, and can be characterized rather well by a single positive charge, or effectively by an ionized hydrogen atom. For those lacking just one electron, if the atom were to capture one additional electron, it would be symmetric and effectively carry one unit of negative charge while the missing bond will travel around as a hole. While this latter picture is unlikely for an isolated atom, it is not unlikely in a semiconductor, since the energy associated with the attached electron is much less than in free space.

If we characterize either of these cases of **donors** (that *donate* an electron to the conduction band) or **acceptors** (that capture or *accept* an electron from the valence band, leaving behind a hole) as a hydrogen atom with one unit of either positive or negative charge, then the energy associated with the ionization is given by the ground state energy, since an electron is moved from (or into, for the hole) the ground state to a free particle, given by

$$|E| = \frac{m^* e^4}{32(\pi\epsilon\hbar)^2}, \tag{8.17}$$

where the effective mass and the dielectric constant $\epsilon = \kappa\epsilon_0$ are characteristic of the medium. If the radius of the orbit were of the order of the Bohr radius, it would be appropriate to use $\kappa = 1$, the value for free space, but if we assume that the radius is large enough for the electron to wander through many atoms, the dielectric constant of the medium is appropriate. As listed in Table 7.1, the relative dielectric constant for Germanium is 16 and the effective electron mass is $0.22m_e$, so the ionization energy for donors is

$$E_D = 13.6\left(\frac{m_e^*}{m_e}\right)\left(\frac{\epsilon_0}{\epsilon}\right)^2 = 0.0117 \text{ eV}, \tag{8.18}$$

and the radius is

$$r = a_0\left(\frac{\epsilon m_e}{\epsilon_0 m_e^*}\right) = 72.7a_0\,. \tag{8.19}$$

The ionization energy is so low that just thermal processes alone are likely to ionize virtually all impurities in Germanium. The radius is so large that thousands of ions are included in the volume that the electron traverses, so that the use of the relative dielectric constant is appropriate. For the ionization of acceptors, we use the effective hole mass and find $E_A = 0.0197$ eV. The corresponding estimates for Silicon are $E_D = 0.105$ eV, $E_A = 0.058$ eV, and the radius is ~ 10.9 Bohr radii, so the thermal ionization is not as likely, but still nearly all impurities are ionized. This is due to the fact that once ionized, the impurities are so rare that an electron must wander in the lattice for a great distance before it finds another impurity ion to recombine with. It is apparent from Table 7.1, that the results are similar for holes, but not identical, and for InSb and GaAs, the low electron effective masses lead to very low ionization energies.

There are some variations of the ionization energy due to the structure of the particular donor or acceptor atoms as illustrated in Table 8.1.

8.3.2 Impurity Levels in Semiconductors

In trying to estimate the densities of holes and electrons, which is important in determining the macroscopic properties of semiconductors, there are two levels

TABLE 8.1
Ionization energy levels for common dopants in Ge and Si.

Material	Type	Ionization Energy in eV	
		Germanium	Silicon
Boron	Acceptor	.0104	.045
Aluminum	Acceptor	.0102	.057
Gallium	Acceptor	.0108	.065
Indium	Acceptor	.0112	.16
Phosphorus	Donor	.0120	.044
Arsenic	Donor	.0127	.049
Antimony	Donor	.0096	.039

of estimates we can use. The first, approximate, method simply assumes that if the material is dominantly n-type, that is to say that the dominant impurity is a donor contributing electrons to the conduction band, all of the donors are assumed to be ionized, so that $n \approx n_D$ where n_D is the impurity density. The hole density in this case is obtained from Equation (7.40) so that $p \approx n_i^2/n_D$. In this case, the addition of electrons to the conduction band has *reduced* the hole concentration in the valence band. This can be seen either as due to the increase of recombination of hole-electron pairs where extra electrons "eat up" the holes or as due to the shift of the Fermi level. The opposite occurs when acceptor impurities predominate, where we approximate $p \approx n_A$, and then $n = n_i^2/n_A$ since now the excess holes lower the electron density due to recombination, producing a p-type semiconductor.

The second level of approximation is necessary when the dominance of one type of impurity is not so strong or the temperature is low. For this case, we must use charge neutrality along with $pn = n_i^2$ to discover the ionization level of the impurities and the hole and electron densities. Denoting the density of ionized donors as n_D^+ (since the ions are positively charged) and the density of ionized acceptors as n_A^-, charge neutrality gives

$$p + n_D^+ = n + n_A^- .$$

We know p and n in terms of ϵ_F from Equations (7.36) and (7.38). For the ionization levels of the donors and acceptors, we know that the probability of finding an ionized acceptor is equivalent to the probability of finding an electron at the acceptor energy, or

$$n_A^- = \frac{n_A}{e^{(\epsilon_A-\epsilon_F)/\kappa T} + 1} .$$

The density of ionized donors, on the other hand, is the probability of *not* finding an electron at the donor energy, so that

$$n_D^+ = n_D \left[1 - \frac{1}{e^{(\epsilon_D-\epsilon_F)/\kappa T} + 1}\right] = \frac{n_D}{e^{(\epsilon_F-\epsilon_D)/\kappa T} + 1} .$$

The picture is completed by noting that ϵ_D lies just below the conduction band edge by the amount of the ionization energy, while ϵ_A lies just *above* the valence band edge by the ionization energy since an electron must be raised from the valence band to leave a hole behind. A schematic representation of these levels is given in Figure 8.8. Charge neutrality is therefore given by

$$N_v e^{-\epsilon_F/\kappa T} + \frac{n_D}{e^{(\epsilon_F - \epsilon_D)/\kappa T} + 1} = N_c e^{-(\epsilon_g - \epsilon_F)/\kappa T} + \frac{n_A}{e^{(\epsilon_A - \epsilon_F)/\kappa T} + 1} \,. \tag{8.20}$$

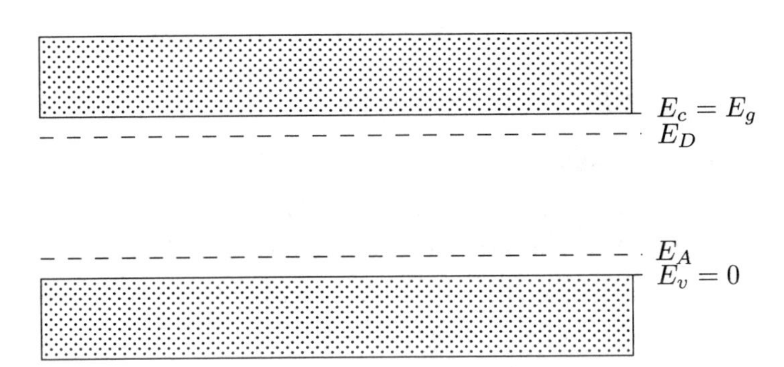

FIGURE 8.8
Donor and acceptor levels in impure semiconductors.

To illustrate how these impurities affect the Fermi level and the ionization levels in semiconductors, we consider a few examples.

Example 8.1
Effect of impurities on the Fermi level. Suppose a semiconductor's acceptor levels are 32% filled at 300 K. Where is the Fermi level?

Solution: The probability of an electron state being filled is given by the occupation index, so that

$$\frac{n_A^-}{n_A} = \frac{1}{e^{(\epsilon_A - \epsilon_F)/\kappa T} + 1} = 0.32 \,,$$

so that after inverting, we have

$$e^{(\epsilon_A - \epsilon_F)/\kappa T} + 1 = 3.125 \,,$$

and taking logarithms,

$$\epsilon_A - \epsilon_F = \kappa T \ln(3.125 - 1) = 0.0195 \text{ eV}.$$

Hence, the Fermi level is 0.0195 eV *below* ϵ_A. □

Example 8.2
Donors and Acceptors. In a case with both donors and acceptors, where $\epsilon_D - \epsilon_A = 0.12$ eV and the acceptor level is 97% ionized at 300 K, find the ionization level of the donors and locate the Fermi level.

Solution: From the given information, along with the example above, we have

$$\epsilon_A - \epsilon_F = \kappa T \ln(1/.97 - 1) = -.0899 \text{ eV}.$$

Then the donor ionization level is given from

$$\frac{n_D^+}{n_D} = \frac{1}{e^{(\epsilon_F - \epsilon_D)/\kappa T} + 1},$$

and the energy difference is obtained from

$$\epsilon_D - \epsilon_F = (\epsilon_D - \epsilon_A) + (\epsilon_A - \epsilon_F) = 0.12 - .0899 = 0.0301 \text{ eV}.$$

The Fermi level is thus 0.0899 eV *above* the acceptor level and 0.0301 eV *below* the donor level. The ionization level of the donors is therefore

$$\frac{n_D^+}{n_D} = \frac{1}{e^{-(.0301)11605/300} + 1} = 0.762,$$

so the donors are 76% ionized (where we used the equivalence 11,605 K$\leftrightarrow$1 eV). □

Problem 8.10 *Impurity levels in Germanium.* In Germanium, with an indium impurity that is 99% ionized at 300 K,

(a) Where is the Fermi level and what is the percent ionization of an Arsenic impurity?

(b) What is the donor density (assuming $n_D \gg n_A$), the electron density, and the hole density.

8.4 Drift, Diffusion, and Recombination

When either electrons or holes move under the influence of an electric field, the movement leads to **drift**, since the particles make frequent collisions with the lattice. Even though they accelerate uniformly between collisions, the plot of velocity versus time would yield a series of irregular sawteeth, as in Figure 8.9, with an average velocity which is the drift velocity. The slope of each tooth is proportional to the electric field, so the average velocity is also proportional, so that

$$v_D = \mu \mathcal{E}, \tag{8.21}$$

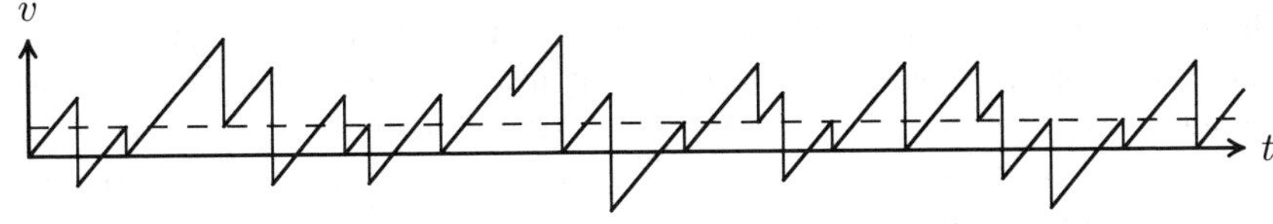

FIGURE 8.9
v vs. t for an electron in a uniform electric field with frequent collisions. $\langle v \rangle = v_D$ − − −.

where μ is the **mobility**. The current density is therefore given by

$$J = nev_D = ne\mu\mathcal{E} = \sigma\mathcal{E}\,,$$

so the conductivity is given by $\sigma = ne\mu$. In semiconductors, there is current due to both holes and electrons, so the more general relation is

$$\sigma = pe\mu_h + ne\mu_e\,. \tag{8.22}$$

In addition to drift motion, there can be an average flux of particles due to diffusion if the concentration is nonuniform. The diffusion process is described by

$$\Gamma = -D\frac{\mathrm{d}n}{\mathrm{d}x}\,,$$

where Γ is the particle flux density ($\langle nv \rangle$), D is the diffusion coefficient, and the flux of particles is opposite to the density gradient. For holes and electrons, the respective current densities are thus given by

$$J_e = eD_e\frac{\mathrm{d}n}{\mathrm{d}x}\,, \tag{8.23}$$

$$J_p = -eD_h\frac{\mathrm{d}p}{\mathrm{d}x}\,. \tag{8.24}$$

The total current is due to both drift and diffusion, but in semiconductors, one or the other is usually dominant, although it is not uncommon for drift to dominate for one species while diffusion dominates for the other.

Diffusion is not usually important unless the gradient is large, and this typically only occurs to minority carriers when a great many are injected at some point. An external source, such as light, may be able to produce a large number of hole-electron pairs, but if the material is n-type, which means $n \gg p$, the additional electrons may be only a small perturbation to the electron density, but the equal number of holes produced may increase the hole population greatly because the original level is so low. For extrinsic materials (where either $p \gg n$ or $n \gg p$), then, it may be generalized that minority current is dominated by diffusion, while majority current is dominated by drift. This latter generalization may be seen by noting that while the mobilities of electrons and holes are usually similar, the respective densities are generally not, so the majority carriers dominate the drift component.

When one considers the combined effect of drift and diffusion, then there are some relationships between the two that appear. For example, as holes diffuse away from a source, they create a charge imbalance that will produce an electric field which will hinder the diffusion by adding a drift component. For the total hole current, we have

$$J_p = -eD_h \frac{dp}{dx} + pe\mu_h \mathcal{E} .$$

In equilibrium, we require $J_p = 0$, so that we find

$$eD_h \frac{dp}{dx} = pe\mu_h \mathcal{E} . \tag{8.25}$$

From our analysis of statistics, we know that the hole density will be modified by a potential through a Boltzmann factor, so that Equation (7.38) will become

$$\begin{aligned} p(x) &= N_v \exp\left[\frac{-\epsilon_F - eV(x)}{\kappa T}\right] \\ &= p_0 \exp\left[\frac{-eV(x)}{\kappa T}\right] , \end{aligned}$$

where $V(x)$ is the self-consistent potential. The gradient of the density is then given by

$$\frac{dp}{dx} = -\frac{e}{\kappa T}\frac{dV}{dx}p = \frac{ep}{\kappa T}\mathcal{E} , \tag{8.26}$$

so that by comparing Equation (8.25) with Equation (8.26), one finds that

$$D_h = \mu_h \frac{\kappa T}{e} . \tag{8.27}$$

In the same way, it may be shown that

$$D_e = \mu_e \frac{\kappa T}{e} . \tag{8.28}$$

Together, these are called the **Einstein relations** and relate diffusion and mobility, which are both collisional processes, to one another.

If one considers what happens when equilibrium is disturbed by introducing a surplus of one species (typically the minority species) at one point, it is clear that at the same time they are diffusing, they are recombining with majority carriers as the system relaxes back toward equilibrium. The relaxation rate of holes back toward equilibrium is given by

$$\frac{dp}{dt} = -\frac{p - p_0}{\tau_p} ,$$

where p_0 is the equilibrium hole density, since the probability of recombination is proportional to the excess number of holes, and the proportionality constant is $1/\tau_p$, where τ_p is called the hole lifetime. The solution for the hole density versus time is simply given by

$$p - p_0 = (\Delta p)_0 e^{-t/\tau_p} ,$$

where $(\Delta p)_0$ is the excess density at $t = 0$.

We now combine the effects of both diffusion and recombination through the continuity equation,

$$\frac{\partial p}{\partial t} + \frac{\partial \Gamma_p}{\partial x} = -R ,$$

where $R = (p - p_0)/\tau_p$ is the recombination rate and Γ_p is the *particle* current density ($\Gamma_p = J_p/e$ where J_p is the electrical current density carried by the holes). This leads to the relation

$$\frac{\partial p}{\partial t} - D_h \frac{\partial^2 p}{\partial x^2} = -\frac{p - p_0}{\tau_p} ,$$

and in steady state, this reduces to

$$\frac{d^2 p}{dx^2} = \frac{p - p_0}{D_h \tau_p} ,$$

which has solutions, with a steady state imbalance $(\Delta p)_0$ at $x = 0$, of the form

$$p - p_0 = (\Delta p)_0 e^{-x/L_p} , \qquad L_p = \sqrt{D_h \tau_p} , \tag{8.29}$$

and L_p is called the **diffusion length**. This means, for example, that the excess density one diffusion length from a source of excess holes will be reduced to only 37% of the original excess.

Problem 8.11 *Diffusion lengths.* Determine the diffusion lengths for both holes and electrons for each of the materials in Table 7.1 at 300 K for $\tau = 10\ \mu$sec and for $\tau = 1000\ \mu$sec.

8.5 Semiconductor Devices

8.5.1 The *pn* Junction Diode

When semiconductor materials with different impurity levels are joined together, there is an instantaneous flow of holes and electrons due to a lack of equilibrium. As pictured in Figure 8.10a, when a p-type semiconductor is joined to an n-type semiconductor, electrons in the conduction band of the n-type material see vacant states in the p-type material, and immediately begin flowing to the left. Simultaneously, holes in the valence band on the p-side see empty hole states (or electrons in the valence band on the n-side see empty states in the valence band of the p-side) so holes flow to the right (electrons flow to the left). In both cases, there is a net flow of negative charge to the left, so an electric field is set up across the junction. This flow will continue until such time as the electric field (which points to the left, since the n-side is positively charged relative to the p-side) curtails any further flow.

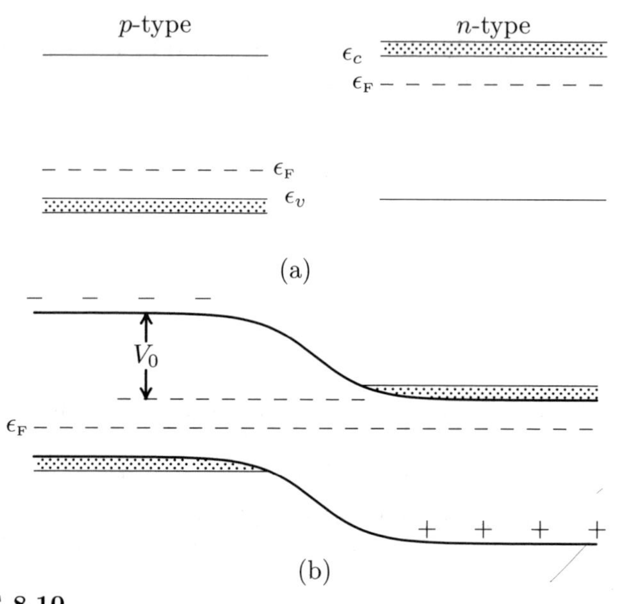

FIGURE 8.10
Energy levels in a *pn* junction. (a) Before joining. (b) After joining.

This will leave the junction with a built-in voltage as shown in Figure 8.10b. Actually, the negative voltage is plotted, so that Figure 8.10b gives an effective **potential energy** diagram for electrons, which must climb a barrier if they wish to move to the left. The equilibrium is established when the population densities for electrons in the conduction band differ by the Boltzmann factor, so that $n_{p0} = \mathrm{e}^{-eV_0/\kappa T} n_{n0}$, where n_{p0} is the electron density on the p-side (far from the junction) and n_{n0} is the electron density on the n-side, while V_0 is the equilibrium voltage.

When an external voltage V (called the **bias**) is applied to the p-side relative to the n-side, the equilibrium is upset and current will begin to flow. In Figure 8.11a, the effect of a negative bias is shown, where the barrier is effectively increased. This virtually shuts off all flow of the majority electrons on the n-side and the majority holes on the p-side. At the same time, it enhances the flow of the minority electrons on the p-side and the minority holes on the n-side. As fast as these minority carriers can diffuse to the edge of the layer, they immediately go to the opposite side as if falling over a waterfall. Because these minority carriers are so few in number, this current is very small, and since the voltage really only affects the majority carriers, this reverse current is constant and has the value I_0.

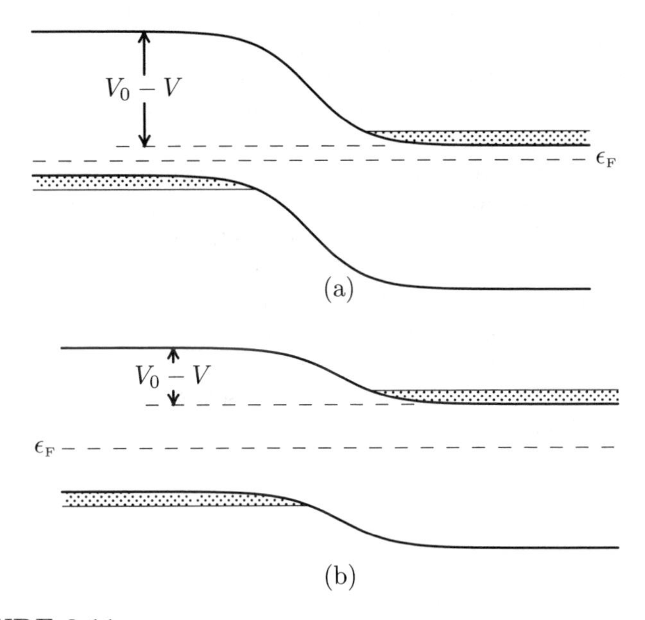

FIGURE 8.11
Effect of bias voltage V on *pn* junction. (a) Negative bias. (b) Positive bias.

In Figure 8.11b, the effect of a positive bias is shown, where the voltage barrier for the majority carriers is reduced. This enhances the flow of the majority carriers by the Boltzmann factor of $\exp(eV/\kappa T)$. The minority current in this case is virtually eliminated, but its magnitude is insignificant anyway. The forward current due to the enhanced flow of electrons to the left and holes to the right is therefore exponentially large compared to the reverse current, and varies as $I \sim A\exp(eV/\kappa T)$ but when the bias is zero, there must be zero current, so the relation between the voltage and current is given by

$$I = I_0(\mathrm{e}^{eV/\kappa T} - 1)\,. \tag{8.30}$$

Problem 8.12 *Junction diode.* If the reverse current for a junction diode is 1 μA, what bias voltage is required to produce 10 mA at room temperature (300 K)?

8.5.2 The *pnp* Transistor

The *pnp* transistor (and by simple analogies, the *npn* transistor) is a somewhat simple three-layer sandwich of what amounts to two *pn* junction diodes back-to-back, as shown in Figure 8.12. Generally the two ends, made of highly doped *p*-type material and called the **emitter** and **collector** are long, relatively speaking, and the thin *n*-type layer in between is called the **base**. The emitter junction is biased positive relative to the base so that hole current will flow into the base (and minority electron current back into the emitter, but this is usually ignored). Once inside the base, the holes, which are now minority carriers, diffuse across the base toward the collector side, which is strongly backward biased (*p*-side negative). This biasing sweeps any holes that make it across the base into the collector, so that there is a current from the emitter to the collector. The efficiency is primarily dependent on the thickness of the base (it should be short compared to a diffusion length so that few holes recombine) and the relative doping levels of the various regions. If there is too much recombination in the base, then more base current is required to produce the same amount of collector current. The idea is to produce as much collector current for as little base current as possible, and the ratio is called the gain. As the base-emitter junction is biased on more strongly, more base current and collector current is produced, but the ratio is relatively independent of bias voltage. Transistors are hence usually described in terms of current rather than in terms of voltage, since the current relations are effectively linear, whereas the voltage-current relations are generally exponential.

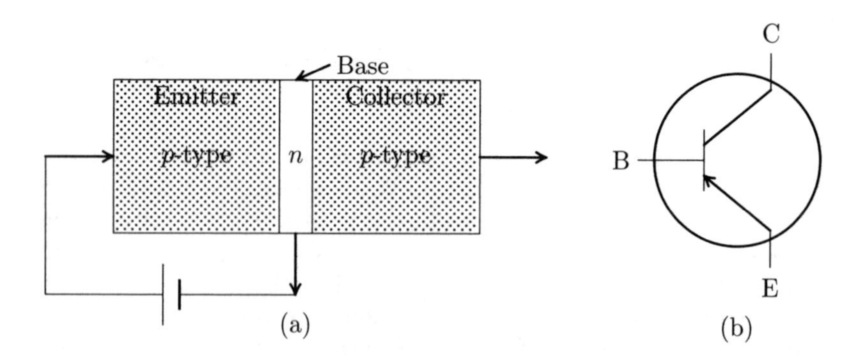

FIGURE 8.12
pnp Transistor. (a) Schematic of regions with battery showing forward bias. (b) Circuit Representation.

Another kind of transistor is the **field-effect transistor**, or FET. This transistor depends upon the fact that in a pn junction that is backward biased, a significant separation is formed between the regions where free holes and electrons reside, since majority carriers are held away from the layer because of the electric field, while minority carriers are drawn to the edge of the region and promptly swept over to the other side (forming a small reverse current). Because there are virtually no free carriers in this region, it is called the depletion region. The field effect transistor can be understood by considering it to be made of a rod of p-type material girdled by a small ring of n-type material as in Figure 8.13. One end is designated the **source** and the other end the **drain**, while the center band is called the **gate**. With no bias or modest bias on the gate relative to the source, the device is a simple conductor (although not a great one). When reverse bias is applied to the gate, however, the depletion layer begins to spread across the p-type rod, until it eventually closes off the current channel. Since there are no carriers in the depletion layer, the effect is to shrink the cross section of the conductor until there is finally no conductor left. The advantage of this type of device is that since it is reverse biased, the current in the gate is extremely low and the corresponding efficiency extremely high. Extensions to this kind of device include an insulated gate, so that although a voltage is applied which causes an electric field and its corresponding depletion layer, it draws *no* dc current! These are called IGFETs (Insulated Gate Field-Effect Transistors) or MOSFETs (Metal-Oxide-Silicon Field-Effect Transistors).

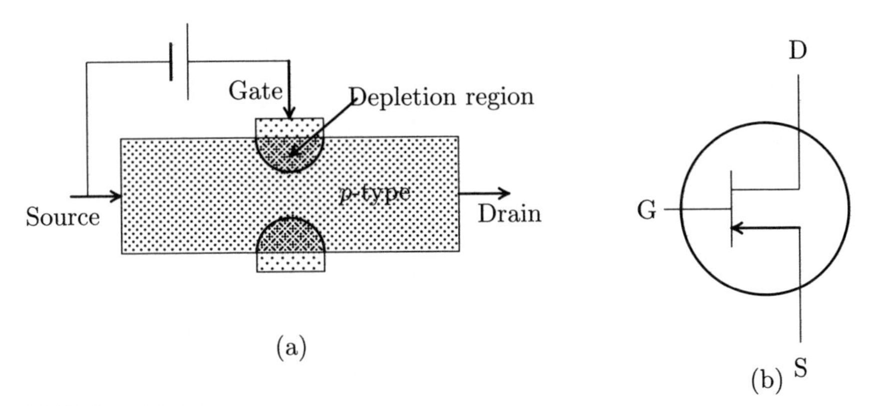

FIGURE 8.13
FET Transistor. (a) Schematic of regions with battery showing reverse bias. (b) Circuit Representation.

The MOSFET is pictured schematically in Figure 8.14 where in a p-type substrate, an n-type channel with a metal contact at each end allow current to flow between the source and the drain. The oxide coating insulates the metal gate from the channel, and by applying a voltage on the metal gate, the electric field produces a depletion region under the oxide that restricts the n-channel. The fact that the gate is insulated means there is no power to the gate, reducing the total power dissipation in the device. The gate is effectively a capacitor. Thousands or millions of such MOSFETs can be built on a single substrate leading to the massive chips we commonly use. For silicon-based chips, the fact that the oxide is silicon oxide is especially advantageous since the oxide layer adheres well and is an excellent insulator.

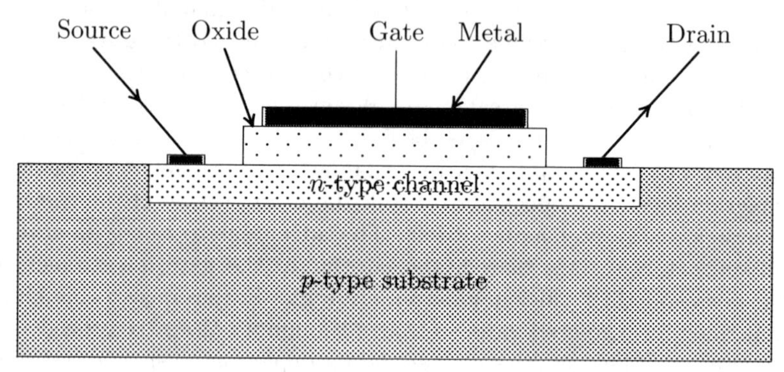

FIGURE 8.14
MOSFET Transistor.

9

Emission, Absorption, and Lasers

9.1 Emission and Absorption of Photons

In Section 5.4.2 where we treated time-dependent perturbation theory, we took the perturbation to be of the general form

$$\hat{\mathcal{H}}^{(1)}(\boldsymbol{r},t) = 2\hat{\mathcal{H}}^{(1)}(\boldsymbol{r})\cos\omega t \tag{9.1}$$

but for absorption and emission of photons, we want the perturbation to be an electromagnetic field of the form

$$\boldsymbol{E} = E_0 \cos\omega t\, \hat{e}_z \tag{9.2}$$

so that the perturbing potential is

$$V(\boldsymbol{r},t) = -\int E\,\mathrm{d}z = -qE_0 z\cos\omega t \tag{9.3}$$

and because of the factor of 2 in Equation (9.1), the matrix element is

$$\hat{\mathcal{H}}^{(1)}_{kj} = -\tfrac{1}{2}E_0\langle k|\wp_z|j\rangle \tag{9.4}$$

where $\wp_z \equiv qz$ is the z-component of the dipole moment. In the derivation of Equation (5.48) from Equation (5.47), we integrated over the energy $E = \hbar\omega$, whereas if we instead integrate over ω, the result is

$$R_{j\to k} = \frac{2\pi}{\hbar^2}|\hat{\mathcal{H}}^{(1)}_{kj}|^2\,, \tag{9.5}$$

where the δ-function in Equation (5.48) tells us to evaluate the expression at $\omega = \omega_{kj}$. The energy density in an electric field is given by

$$u = \tfrac{1}{2}\epsilon_0 E_0^2$$

and we let $u \to \rho(\omega)\mathrm{d}\omega$ so that

$$R_{j\to k} = \frac{\pi}{\epsilon_0\hbar^2}|\langle k|\wp_z|j\rangle|^2\rho(\omega)\,, \tag{9.6}$$

There is nothing special about the z-direction in evaluating the matrix element, so it is common to use the magnitude instead of the single component, where by symmetry,

$$|\wp|^2 = |\wp_x|^2 + |\wp_y|^2 + |\wp_z|^2 = 3|\wp_z|^2$$

so the final result is expressed as

$$R_{j \to k} = \frac{\pi}{3\epsilon_0 \hbar^2} |\langle k|\wp|j\rangle|^2 \rho(\omega) \, . \tag{9.7}$$

9.2 Spontaneous Emission

If we imagine a closed system in equilibrium, where there are atoms in both the lower state (ψ_k) and in the upper state (ψ_j), we can be assured that there is a balance between the rates of emission and absorption. If we define the spontaneous emission rate to be A* so that if there are N_j atoms in the upper state, there will be $N_j A$ transitions per second from state j to state k, as illustrated in Figure 5.2c. There are also transitions from state j to state k by stimulated emission as illustrated in Figure 5.2a. This rate is proportional to the energy density in the electromagnetic field of the wave and is given by $B_{jk}\rho(\omega)$, where B_{jk} is proportional to R_k from Equation (5.48) and $\rho(\omega)$ is the energy density in the electric field of the wave at frequency ω. The total number of transitions per second from state j to state k due to stimulated emission is therefore $N_j B_{jk}\rho(\omega)$. If there is an equilibrium, there must be an equal number of transitions from state k to state j by absorbing photons as illustrated in Figure 5.2b. The total number of transitions per second from state k to state j is then given by $N_k B_{kj}\rho(\omega)$ where N_k is the number of atoms in the lower state (ψ_k) and B_{kj} is the transition rate for the absorption process. The rate equation is therefore

$$\frac{\mathrm{d}N_j}{\mathrm{d}t} = -N_j A - N_j B_{jk}\rho(\omega) + N_k B_{kj}\rho(\omega) \, . \tag{9.8}$$

In equilibrium, there is a steady state, so $\mathrm{d}N_j/\mathrm{d}t = 0$, so that the energy density in the wave must be given by

$$\rho(\omega) = \frac{N_j A}{N_k B_{kj} - N_j B_{jk}} = \frac{A}{(N_k/N_j)B_{kj} - B_{jk}} \, . \tag{9.9}$$

Equilibrium also implies, however, that there is a temperature T and that the relative probability of finding an atom in one state compared to another

*The notation is due to Einstein who first treated this problem using his A and B coefficients for the spontaneous and stimulated emission rates.

is given by the Boltzmann factor of Equation (7.26), which depends on the energies of the two states such that

$$\frac{N_k}{N_j} = \mathrm{e}^{(E_j - E_k)/\kappa T}\,, \tag{9.10}$$

but with $E_j - E_k = \hbar\omega$, this becomes

$$\rho(\omega) = \frac{A}{\mathrm{e}^{\hbar\omega/\kappa T} B_{kj} - B_{jk}}\,. \tag{9.11}$$

We wish to relate this energy density expression to the black-body distribution since both are equilibrium expressions, but Equation (9.11) is in terms of the energy density of the electromagnetic field, while our black-body expression of Equation (1.3), which is the same as Equation (7.31) without the zero-point term, is given in terms of the *intensity*, $I(\lambda, T)$. In order to relate these two expressions, we revisit the expression for the density of states given by Equation (7.4) where $\mathrm{d}x\,\mathrm{d}y\,\mathrm{d}z = V$, $\mathrm{d}p_x\mathrm{d}p_y\mathrm{d}p_z = 4\pi p^2\,\mathrm{d}p$, and $p = \hbar k$. Then we have

$$N\,\mathrm{d}x\,\mathrm{d}y\,\mathrm{d}z\,\mathrm{d}p_x\mathrm{d}p_y\mathrm{d}p_z = \frac{4\pi V}{h^3}\hbar^3 k^2\,\mathrm{d}k = \frac{V}{2\pi^2}k^2\,\mathrm{d}k\,. \tag{9.12}$$

Then, since $k = \omega/c$, and since there are two independent photon states for each energy (spin 1 and spin -1), the density of photons is

$$\frac{N_\omega}{V} = \frac{\omega^2}{\pi^2 c^3}\frac{\mathrm{d}\omega}{\mathrm{e}^{\hbar\omega/\kappa T} - 1} \tag{9.13}$$

where we have included the Bose–Einstein factor since photons have integral spin, and the energy density is therefore

$$\rho(\omega)\mathrm{d}\omega = \hbar\omega\frac{N_\omega}{V} = \frac{\hbar\omega^3}{\pi^2 c^3}\frac{\mathrm{d}\omega}{\mathrm{e}^{\hbar\omega/\kappa T} - 1}\,. \tag{9.14}$$

Comparing this result with Equation (9.11), we first conclude that $B_{kj} = B_{jk}$, which means that the transition rates are the same for the upward and downward transitions. Solving for A, we find for the spontaneous emission rate,

$$A = \frac{\hbar\omega^3}{\pi^2 c^3}B_{kj}\,, \tag{9.15}$$

and since, from Equation (9.7)

$$B_{kj} = \frac{\pi|\langle k|\wp|j\rangle|^2}{3\epsilon_0\hbar^2}\,, \tag{9.16}$$

the spontaneous rate is given by

$$A = \frac{\omega^3|\langle k|\wp|j\rangle|^2}{3\pi\epsilon_0\hbar c^3}\,. \tag{9.17}$$

The lifetime in an excited state is the reciprocal of the spontaneous rate, although if there is more than one final state, the lifetime is the reciprocal of the *sum* of the various rates.

Example 9.1

Spontaneous decay for a harmonic oscillator. A particle with charge q and mass m moves in a harmonic oscillator potential with frequency ω_0.

(a) Find the rate of spontaneous emission from the nth excited state to the ground state.

Solution: The dipole maxrix element is

$$\langle k|\wp|j\rangle = q\langle n|x|0\rangle$$

and from Equation (2.93)

$$\hat{x}|0\rangle = \frac{1}{\sqrt{2}\alpha}(\hat{a}+\hat{a}^\dagger)|0\rangle = \frac{1}{\sqrt{2}\alpha}|1\rangle$$

so the only nonzero dipole element is for $n = 1$ in which case $\langle 1|x|0\rangle = 1/\sqrt{2}\alpha = \sqrt{\hbar/2m\omega_0}$. The transition rate is therefore

$$A = \frac{\omega_0^2 q^2}{6\pi\epsilon_0 mc^3}\ .$$

(b) Find the rate and lifetime of the spontaneous decay if the particle is an electron and the frequency is $\omega_0 = 3\times 10^{14}$ rad/s.

Solution: Using $m_e c^2/e = 511$ keV, we find for the rate (trans/sec)

$$A = \frac{9\times 10^{28} 1.602\times 10^{-19}}{6\pi\cdot 8.85\times 10^{-12} 3\times 10^8\cdot 5.11\times 10^5} = 5.6\times 10^5$$

and the lifetime is the reciprocal or $\tau = 1.77\,\mu$s. ▯

Problem 9.1 *Spontaneous decay rate.* Estimate the lifetime of an electron in the $n = 2$, $\ell = 1$ state of hydrogen. ($2p \to 1s$).

Problem 9.2 *Balmer-α transition.* Consider the hydrogen transition where $n = 3 \to n = 2$ (Balmer-α). The dipole selection rules require $\Delta\ell = \pm 1$ from Section 6.2.4 since the photon carries away one unit of angular momentum. Define $\boldsymbol{d} = e\boldsymbol{r}$. There are three transitions:

(a) $\ell = 2 \to \ell = 1$. Find all of the nonzero matrix elements for the dipole, $\langle 32m|\boldsymbol{d}|21m'\rangle$ (i.e., $\langle d_x\rangle$, $\langle d_y\rangle$, and $\langle d_z\rangle$). Then find the spontanteous lifetime for this transition.

(b) $\ell = 1 \rightarrow \ell = 0$. Find all of the nonzero matrix elements for the dipole, $\langle 31m|\boldsymbol{d}|20m'\rangle$. Then find the spontaneous lifetime for this transition.

(c) $\ell = 0 \rightarrow \ell = 1$. Find all of the nonzero matrix elements for the dipole, $\langle 300|\boldsymbol{d}|21m'\rangle$. Then find the spontaneous lifetime for this transition.

9.3 Stimulated Emission and Lasers

One of the most important applications of stimulated emission is its role in lasers and masers. Masers (Microwave Amplification by Stimulated Emission of Radiation) were discovered first, where the frequencies and energies were low enough that the cavities were microwave cavities, and the amplifying media were molecular systems where there are many energy levels relatively close to one another. Lasers (Light Amplification by Stimulated Emission of Radiation), on the other hand, are in the visible (or near visible) and require optical cavities because the wavelengths are too short for microwave cavities. The higher energy levels are typically atomic transitions rather than molecular transitions where the levels are more widely spaced.

We may note first that the simple system pictured in Figure 5.2, which indicates a system with two levels, can almost never result in laser action because the matrix element connecting the two levels is the same for absorption and stimulated emission, and without some outside source to preferentially fill the upper state, there will always be more absorption than emission because the Boltzmann factor always yields a higher equilibrium population in the lower state than in the upper state. What is needed for amplification is a *population inversion* which puts more electrons in the upper state than in the lower state. Then stimulated emission, with two photons out for each one in can overcome absorption and lead to gain. Since this can't happen with a two-level system, the next simplest system is a three-level system, depicted in Figure 9.1.

In thermal equilibrium, nearly all the electrons reside in level 1 since the Boltzmann factors give $N_2/N_1 \simeq \mathrm{e}^{(E_1-E_2)/\kappa T}$ and $N_3/N_1 \simeq \mathrm{e}^{(E_1-E_3)/\kappa T}$ where usually $\kappa T \ll \hbar\omega$ and the expressions are only approximate because the possible degeneracies at each level have been ignored. The first arrow in Figure 9.1, labeled W_{13}, corresponds to absorption, since an electron moves from state 1 to state 3. This rate is proportional to the energy density so we take $R_{13} = B_{13}\rho(\omega)$. The same wave energy density induces stimulated emission, so we have $R_{31} = B_{31}\rho(\omega)$. There is also spontaneous emission from level 3 back to level 1 where this rate is given by $S_{31} = A_{31}$. There is also spontaneous emission from state 3 to 2 and from state 2 to 1 where $S_{jk} = A_{jk}$. The rate equations then may be written as

$$\frac{\mathrm{d}N_3}{\mathrm{d}t} = R_{13}N_1 - (R_{31} + S_{31} + S_{32})N_3\,, \tag{9.18}$$

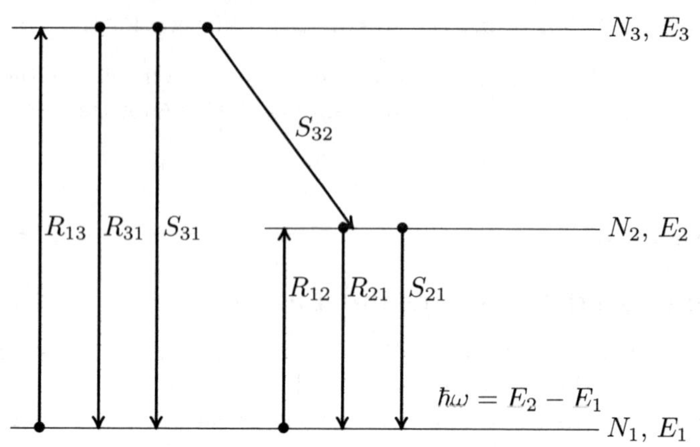

FIGURE 9.1
Three-level system for laser action.

$$\frac{\mathrm{d}N_2}{\mathrm{d}t} = S_{32}N_3 + R_{12}N_1 - (R_{21} + S_{21})N_2\,. \tag{9.19}$$

Since electrons are conserved, we also require

$$N = N_1 + N_2 + N_3 = N_0$$

be constant. In equilibrium, we have a steady-state situation, so the time derivatives vanish. The population densities are then the solutions of the system of equations

$$\begin{pmatrix} R_{13} & 0 & -(R_{31} + S_{31} + S_{32}) \\ R_{12} & -(R_{21} + S_{21}) & S_{32} \\ 1 & 1 & 1 \end{pmatrix} \begin{pmatrix} N_1 \\ N_2 \\ N_3 \end{pmatrix} = \begin{pmatrix} 0 \\ 0 \\ N_0 \end{pmatrix}. \tag{9.20}$$

We are interested only in the ratio N_2/N_1, which is given by

$$\frac{N_2}{N_1} = \frac{R_{13}S_{32} + R_{12}(R_{31} + S_{31} + S_{32})}{(R_{21} + S_{21})(R_{31} + S_{31} + S_{32})}. \tag{9.21}$$

Now the spontaneous emission rate, S_{31}, is nearly always much smaller than the stimulated emission rate, R_{31} (unless the energy density is extremely small), so we will neglect the spontaneous rate, S_{31}, and the material is normally chosen so that $S_{32} \gg R_{31}$ so that Equation (9.21) simplifies (with $R_{13} = R_{31}$ and $R_{12} = R_{21}$) to

$$\frac{N_2}{N_1} \approx \frac{R_{31} + R_{21}}{R_{21} + S_{21}}. \tag{9.22}$$

If $N_2/N_1 > 1$, we have population inversion, but it is often more convenient to indicate the percentage of population inversion given by the excess of N_2

over N_1 relative to the total population, which is

$$\frac{N_2 - N_1}{N_0} \approx \frac{R_{31} - S_{21}}{R_{31} + 2R_{21} + S_{21}}, \tag{9.23}$$

and this is usually expressed as a percentage. Population inversion will occur with $R_{31} > S_{21}$, but for strong population inversion, we want R_{31} large so that the inversion fraction will approach 1, which normally requires a strong exciting source.

9.3.1 Cavity Q and Power Balance

While population inversion can lead to gain, one must also consider loss, since any laser has some loss through its output. Other than the necessary output loss, however, it is desired to make the circulating power level inside the laser cavity high to promote the stimulated emission. It is therefore desirable to minimize any other loss. The most common measure of the intrinsic loss rate for a cavity (or resonant circuit) is the Q of the cavity. This quantity measures the amount of energy stored in the cavity relative to the energy lost per cycle. If one feeds energy into a cavity at a modest rate, but the loss rate is small in comparison, the energy will build until the fractional loss rate of the high accumulated energy is equal to the input rate. We define the Q by

$$Q = \frac{\omega \times \text{energy stored}}{\text{energy dissipated}}, \tag{9.24}$$

which is dimensionless because the dissipation is a rate of energy loss. For a laser, the power out may be written as

$$P_o = \text{power transmitted} = \frac{\text{energy transmitted}}{\text{energy dissipated}} \cdot \frac{\omega \int_V \rho_\omega \mathrm{d}V}{Q}$$

where ρ_ω is the radiation energy density per unit frequency interval. We may also define an effective Q_{ext} whose only loss is due to the output transmitted power which will be proportional to $T = 1 - R$ where T and R are the transmission and reflection coefficients of the laser mirrors. This is defined by

$$\frac{1}{Q_{\text{ext}}} = \frac{\text{power transmitted}}{\text{power stored}} = \frac{(1-R)\int_A c_\omega \rho_\omega \mathrm{d}A}{\omega \int_V \rho_\omega \mathrm{d}V}$$

where c_ω is the speed of light in the medium at angular frequency ω and the integration is over the surface area of the mirrors. Combining the losses due to dissipation and output, where the dissipated power is characterized by

$$\frac{1}{Q_o} = \frac{\text{power dissipated in cavity}}{\text{power stored}}$$

the total Q is obtained by combining the losses so that

$$\frac{1}{Q} = \frac{1}{Q_o} + \frac{1}{Q_{\text{ext}}}.$$

The primary loss which we have called dissipation is due to diffraction losses where a fraction of the beam is no longer perpendicular to the mirrors. Characterizing this loss by γ_D, we can quantify our expressions so that

$$\frac{1}{Q_{\text{ext}}} = \frac{\lambda(1-R)}{2\pi L\sqrt{\kappa}}$$

and

$$\frac{1}{Q_o} = \frac{\lambda\gamma_D}{2\pi L\sqrt{\kappa}}$$

where $\omega = 2\pi c/\lambda$ and $c_\omega = c/\sqrt{\kappa}$ and κ is the relative dielectric constant of the medium at frequency ω. The total Q is therefore

$$\frac{1}{Q} = \frac{\lambda(\gamma_D + 1 - R)}{2\pi L\sqrt{\kappa}} \tag{9.25}$$

and the total power output is

$$P_o = \frac{\omega\int_V \rho_\omega\,\mathrm{d}V}{Q}\cdot\frac{1/Q_{\text{ext}}}{1/Q} = \frac{\omega\int_V \rho_\omega\,\mathrm{d}V}{Q_{\text{ext}}}\,.$$

For the output power energy density, we know that the power is proportional to the product of the transition probability per unit time, the energy of the transition, and the number of stimulated emission photons minus the number of absorption photons, or

$$\rho_\omega = B_{21}\hbar\omega(N_2 - N_1)$$

so the power out is given by

$$P_{\text{out}} = \frac{\omega B_{21}\hbar\omega(N_2 - N_1)V}{Q_{\text{ext}}} = \frac{\hbar\omega^2 B_{21}(N_2 - N_1)(1-R)}{2\pi\sqrt{\kappa}}\left(\frac{V}{L}\right). \tag{9.26}$$

9.3.2 The He-Ne Laser

One of the first gaseous lasers was the Helium-Neon laser, which has the interesting feature that although the lasing medium is Neon, the excitation is accomplished by directly exciting the Helium (via an electrical discharge) which then excites the Neon via a collisional process due to the fact that each has an excited state at nearly the same energy so the collisional excitation is very efficient. The other interesting feature is that there are two different laser transitions, one in the visible and one in the infrared. The states, energy levels and laser transitions are illustrated in Figure 9.2. The collisional excitation is from the 2p state of He to the 3s state of Ne. Although the various energy levels in the figure are all shown as single levels, each of the Ne levels are multiplets, where the 3s level has 5 distinct states, the 2p and the 3p levels each have 10 distinct levels. The visible transition (0.6328 μ) is from $3s_2 \to 2p_4$ and the infrared transition (3.39 μ) is from $3s_2 \rightarrow 3p_4$. Thus both laser transitions have the same upper state and both are forbidden by simple dipole transitions, allowing the upper state to build the population inversion necessary for gain. In the schematic of Figure 9.3, the laser cavity involves two mirrors, one highly reflecting and one partially transmitting. The Helium states are excited by driving a gaseous discharge by a current through the electrodes. The ends of the discharge tube are shown at an angle which is the *Brewster angle* that is chosen to reflect the horizontally polarized light and transmit the vertically polarized light, resulting in a plane-polarized output since the cavity Q is high for the vertically polarized light and low for the horizontal polarization.

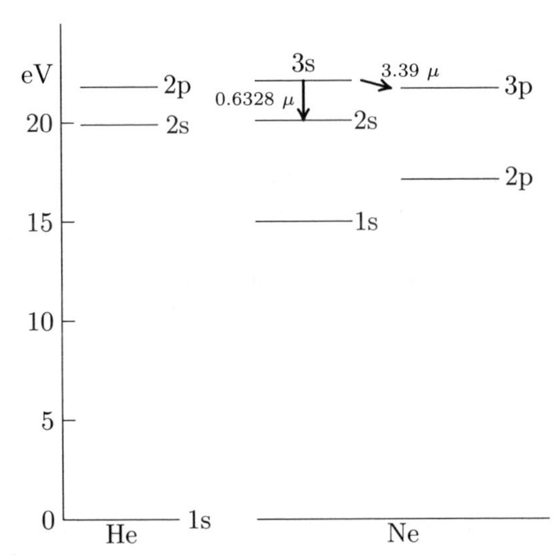

FIGURE 9.2
Energy levels and laser transitions for the He-Ne laser.

Problem 9.3 *Forbidden transitions in the He-Ne laser.*

(a) What dipole selection rule is violated in the He-Ne laser transitions?

(b) Why do forbidden transitions for the upper state promote population inversion?

(c) Both lower states have allowed transitions. Do these promote or hinder population inversion? Why?

9.3.3 The Ruby Laser

The ruby laser was the first optical laser discovered. It is both similar and yet very different from the He-Ne system. Whereas the He-Ne laser typically

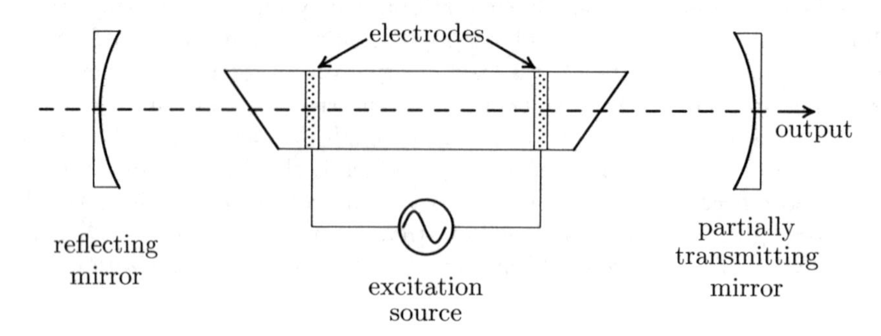

FIGURE 9.3
Schematic of the He-Ne laser.

operates as a steady-state laser, the ruby system is pulsed. The excitation is from a flash lamp that generates a very bright pulsed flash that fills the upper state of the Cr^{3+} which is present in the Al_2O_3 at about 0.05%. The ruby rod typically has parallel ends that may be made reflecting or it might reside in an optical cavity similar to the one in Figure 9.3, but the mirrors may be plane or concave and form a Fabry–Perot resonator whose length is given by

$$L = \tfrac{1}{2}n\lambda\,. \tag{9.27}$$

The energy level diagram for the Cr^{3+} is shown in Figure 9.4 where the upper levels represent a yellow (lower) and green (upper) band of states that absorb energy from a flash lamp. The absorption is inefficient with most of the energy going into heat, but with about 1 kJ to 4 kJ in the flash lamp, the laser pulse may contain several Joules. The typical pulse length may be 0.5 ms, but with any of several types of switches, such as a rapicly rotating mirror at one end, the length of the ouput pulse may be of the order of 10 ns so that while the energy is reduced (less than a Joule) the power is increased so that the instantaneous power may exceed 100 MW. Because of the effects of the Al_2O_3 lattice structure on the Cr^{3+}, the laser frequency may vary from 0.6933 μ to 0.6956 μ as the temperature is varied from -200 C° to 200 C°.

9.3.4 Semiconductor Lasers

Semiconductor lasers differ from the above solid-state and gaseous lasers in that they do not require an external pump or other source of excitation, since they are essentially two-level lasers where the population inversion is accomplished by the dc diode current. In order to see how population inversion

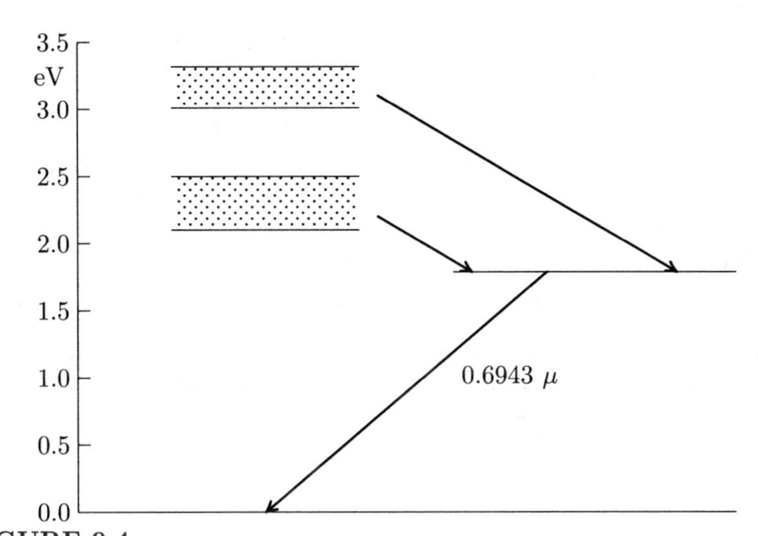

FIGURE 9.4
Energy level diagram for ruby.

occurs, we define the equilibrium electron and hole densities as

$$n_0(\mathcal{E}_i) = \frac{N_c}{1 + e^{(\mathcal{E}_i - E_F)/\kappa T}}$$

and

$$p_0(\mathcal{E}_j) = \frac{N_v}{1 + e^{(E_F - \mathcal{E}_j)/\kappa T}}$$

where $\mathcal{E}_i$ is in the conduction band and $\mathcal{E}_j$ is in the valence band. With forward bias on the diode, the electron density is increased so that $n > n_0$ and we may write

$$n(\mathcal{E}_i) = N_c P'_{FD}(\mathcal{E}_i) = \frac{N_c}{1 + e^{(\mathcal{E}_i - E_{Fe})/\kappa T}} \tag{9.28}$$

where E_{Fe} is the quasi-Fermi level due to the injection of electrons from the n-side to the p-side and P'_{FD} is the Fermi–Dirac distribution with the *quasi-Fermi level.* Since $n > n_0$, $E_{Fe} > E_F$. The corresponding expression for holes is

$$p(\mathcal{E}_j) = N_v P'_{FD}(\mathcal{E}_j) = \frac{N_v}{1 + e^{(E_{Fp} - \mathcal{E}_j)/\kappa T}} \tag{9.29}$$

where E_{Fp} is the quasi-Fermi level for holes. For this case, $E_{Fp} < E_F$ for $p > p_0$.

The gain and loss is due to stimulated emission of an electron in the conduction band from $\mathcal{E}_i$ to $\mathcal{E}_j$ in the valence band while absorption is due to an electron in the valence band making an upward transition to the conduction band. The absorption will be proportional to the number of electrons in the valence band able to make the upward transition times the probability that there is a vacancy in the final state in the conduction band, or

$$N_a = AW_v^c N_v(\mathcal{E}_j) N_c(\mathcal{E}_i) P'_{FD}(\mathcal{E}_j)[1 - P'_{FD}(\mathcal{E}_i)] \tag{9.30}$$

where W_v^c is the transition probability from the valence to the conduction band. The emission is proportional to the number of electrons in the conduction band able to make the downward transition times the probability that the the final state in the valence band is empty, or

$$N_e = AW_c^v N_c(\mathcal{E}_i) N_v(\mathcal{E}_j) P'_{FD}(\mathcal{E}_i)[1 - P'_{FD}(\mathcal{E}_j)]\,, \tag{9.31}$$

where W_c^v is the transition probability from the conduction to the valence band, and since these are reciprocal processes, we have $W_v^c = W_c^v$. In order for emission to dominate absorption, we require

$$N_e > N_a\,,$$

which requires

$$P'_{FD}(\mathcal{E}_i)[1 - P'_{FD}(\mathcal{E}_j)] > P'_{FD}(\mathcal{E}_j)[1 - P'_{FD}(\mathcal{E}_i)]$$

and canceling out the common factors, this gives the condition

$$P'_{FD}(\mathcal{E}_i) > P'_{FD}(\mathcal{E}_j)\,. \tag{9.32}$$

If we let $m_e \simeq m_h$, so that $N_c(\mathcal{E}_i) \simeq N_v(\mathcal{E}_j)$, then we need

$$e^{(\mathcal{E}_j - E_{Fp})/\kappa T} > e^{(\mathcal{E}_i - E_{Fe})/\kappa T}$$

and $\mathcal{E}_i - \mathcal{E}_j = \hbar\omega$, so that Equation (9.32) requires

$$E_{Fe} - E_{Fp} > h\nu = \hbar\omega\,. \tag{9.33}$$

This generally requires heavy doping so that the Fermi level resides near or preferably *inside* the valence band. The heavy injection of electrons into the p-region effectively guarantees that the hole density will be depleted sufficiently to create the population inversion.

The cavity for a semiconductor laser is made by polishing two faces of the semiconductor perpendicular to the plane of the junction. These faces are generally highly reflective and adequate for a high-Q cavity. The only losses are internal to the material. Because the junction layer is so narrow, the beam thickness is very small and leads to strong diffraction effects unless a cylindrical lens is used to make the output beam parallel.

Problem 9.4 *Population inversion in a semiconductor laser.*

(a) Fill in the steps leading to Equation (9.32).

(b) Show that Equation (9.32) leads to Equation (9.33).

10

Scattering Theory

An important application of quantum mechanics that has been used extensively in particle physics is the method of scattering. In principle, one fires a particle at an unknown potential and observes the scattering of the projectile, and from the analysis of the scattering data, one is able to determine what the potential was that it was scattered from. The problem of finding the scattering amplitudes for a given potential is called scattering, and the problem of obtaining the potential from the scattering data is called inverse scattering. In one dimension, the scattering data consists of measuring the transmission and reflection coefficients as a function of k, where the incident particle is represented by a plane wave of the form $\exp(\pm ikx)$. Again, in principle, one needs the transmission amplitude $a(k)$ and the reflection amplitude, $b(k)$, for all k (or all incident energies) in order to reconstruct the potential accurately. In one dimension, if one knows the $a(k)$ and the $b(k)$ for a sufficiently wide range of k, the potential can be reconstructed exactly. In three dimensions, which is the more typical case, the exact analysis is not available, but a series of approximations have been found that enable us to get a good idea of the potential the particle was scattered from. We will treat this more common problem first, beginning with a classical scattering problem, since the results are more useful and more easily obtained, and then treat the exact one-dimensional case which is more rigorous.

10.1 Scattering in Three Dimensions

10.1.1 Rutherford Scattering

The field of nuclear physics began with the discovery that there existed a nucleus inside the atom, which came about by the scattering of α-particles (from radioactive decay) by Rutherford. This scattering is simple Coulomb scattering, but the details of the classical, nonrelativistic analysis is instructive. We will subsequently solve the Coulomb scattering problem quantum mechanically, but we begin with the classical case. The dynamics of the encounter is shown in Figure 10.1 where an incident α-particle comes in from the left with velocity v_0 in such a direction that if the particles were uncharged, they would be a distance b apart at closest approach. The parameter b is called the **impact parameter**. The trajectory is curved due to the Coulomb repulsion, and the target nucleus (gold in the Rutherford experiments) recoils as shown. The objective is first to calculate the deflection angle θ as a function of the charges, masses, incident velocity and impact parameter b and then to calculate what fraction of incident particles have a deflection angle between θ and $\theta + \Delta\theta$ as the impact parameter is changed from b to $b + \Delta b$.

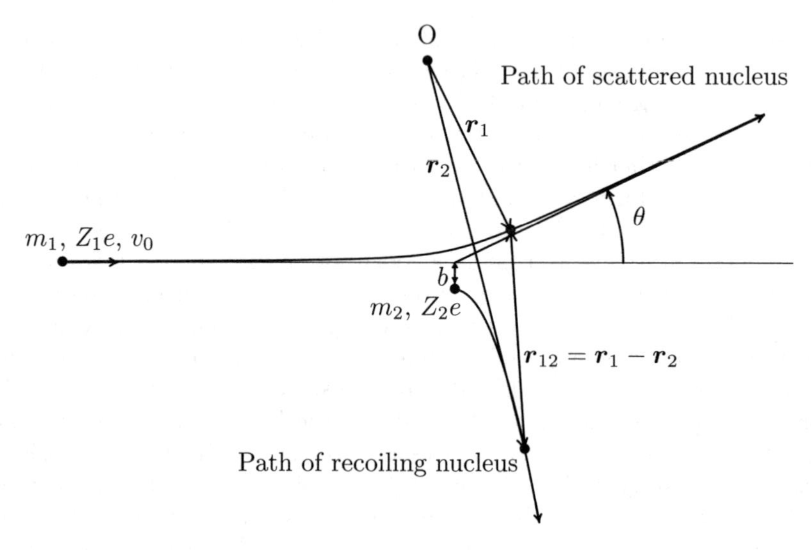

FIGURE 10.1
A Rutherford scattering encounter.

Labeling the location of the incident particle by $\boldsymbol{r}_1$ and the target particle's position by $\boldsymbol{r}_2$, the equations of motion are

$$m_1\ddot{\boldsymbol{r}}_1 = \frac{Z_1Z_2e^2(\boldsymbol{r}_1 - \boldsymbol{r}_2)}{4\pi\epsilon_0|\boldsymbol{r}_1 - \boldsymbol{r}_2|^3}, \tag{10.1}$$

$$m_2\ddot{\boldsymbol{r}}_2 = -\frac{Z_1Z_2e^2(\boldsymbol{r}_1 - \boldsymbol{r}_2)}{4\pi\epsilon_0|\boldsymbol{r}_1 - \boldsymbol{r}_2|^3}. \tag{10.2}$$

These two equations may be combined into the pair of equations

$$m_1\ddot{\boldsymbol{r}}_1 + m_2\ddot{\boldsymbol{r}}_2 = (m_1 + m_2)\ddot{\boldsymbol{r}}_{CM} = 0, \tag{10.3}$$

$$\ddot{\boldsymbol{r}}_2 - \ddot{\boldsymbol{r}}_1 = \ddot{\boldsymbol{r}}_{12} \tag{10.4}$$

$$= \frac{m_1 + m_2}{m_1m_2}\frac{Z_1Z_2e^2\boldsymbol{r}_{12}}{4\pi\epsilon_0 r_{12}^3}. \tag{10.5}$$

These two expressions show that the motion may be resolved into the motion of the center of mass which moves at a constant velocity and the motion of a single particle of mass $m_1 m_2/(m_1 + m_2)$ whose relative position is given by $\boldsymbol{r}_{12}$ as if it were moving about a *fixed* position at the center of mass. Letting the **reduced mass** be denoted $\mu = m_1 m_2/(m_1 + m_2)$, the equation of motion relative to the center of mass may be written from Equation (10.5) in the abbreviated form

$$\ddot{\boldsymbol{r}}_{12} = \frac{K}{r_{12}^3}\boldsymbol{r}_{12}\,, \tag{10.6}$$

where $K \equiv Z_1 Z_2 e^2/4\pi\epsilon_0\mu$.

It is convenient to switch to polar coordinates at this point to find the particle trajectory. With $\hat{\mathbf{e}}_r$ and $\hat{\mathbf{e}}_\theta$ as unit vectors, we have

$$\boldsymbol{r}_{12} = r\hat{\mathbf{e}}_r\,, \tag{10.7}$$

$$\dot{\boldsymbol{r}}_{12} = \dot{r}\hat{\mathbf{e}}_r + r\dot{\hat{\mathbf{e}}}_r = \dot{r}\hat{\mathbf{e}}_r + r\dot{\theta}\hat{\mathbf{e}}_\theta\,, \tag{10.8}$$

$$\ddot{\boldsymbol{r}}_{12} = \ddot{r}\hat{\mathbf{e}}_r + 2\dot{r}\dot{\theta}\hat{\mathbf{e}}_\theta + r\ddot{\theta}\hat{\mathbf{e}}_\theta - r\dot{\theta}^2\hat{\mathbf{e}}_r\,. \tag{10.9}$$

Using these relationships, Equation (10.6) may be written in component form as

$$\ddot{r} - r\dot{\theta}^2 = K/r^2\,, \tag{10.10}$$

$$r\ddot{\theta} + 2\dot{r}\dot{\theta} = 0\,. \tag{10.11}$$

If Equation (10.11) is multiplied by r, it may be integrated directly with the result

$$r^2\dot{\theta} = \text{const.} = -v_0 b\,, \tag{10.12}$$

which expresses the conservation of angular momentum.

In order to find the deflection angle, $\theta = \theta_c$, we need the dependence of r on θ. Taking θ as the independent variable and representing r by $u \equiv 1/r$ for convenience, we have

$$r = \frac{1}{u}\,, \tag{10.13}$$

$$\dot{r} = -\frac{\dot{u}}{u^2} = -\frac{\dot{\theta}}{u^2}\frac{\mathrm{d}u}{\mathrm{d}\theta} = +v_0 b\frac{\mathrm{d}u}{\mathrm{d}\theta}\,, \tag{10.14}$$

$$\ddot{r} = v_0 b\frac{\mathrm{d}^2 u}{\mathrm{d}\theta^2}\dot{\theta} = -v_0^2 b^2 u^2\frac{\mathrm{d}^2 u}{\mathrm{d}\theta^2}\,. \tag{10.15}$$

Therefore, Equation (10.10) becomes

$$-v_0^2 b^2 u^2\frac{\mathrm{d}^2 u}{\mathrm{d}\theta^2} - v_0^2 b^2 u^3 = K u^2\,,$$

which simplifies to

$$\frac{\mathrm{d}^2 u}{\mathrm{d}\theta^2} + u = -\frac{K}{v_0^2 b^2}\,, \tag{10.16}$$

which may be readily solved to obtain

$$u = \frac{1}{r} = A\cos(\theta + \delta) - \frac{K}{v_0^2 b^2}\,. \tag{10.17}$$

This equation is one form of the polar coordinate representation of a *conic section with the pole at one focus.* The eccentricity is given by e in the equation

$$r = \frac{ef}{1 - e\cos(\theta + \delta)}\,. \tag{10.18}$$

We may now find the deflection angle of the incident particle readily by noting that the angle θ_c corresponds to $r \to \infty$, which is found from Equation (10.17) by setting

$$A\cos(\theta_c + \delta) - \frac{K}{v_0^2 b^2} = 0\,.$$

One solution is $\theta = \pi$, which corresponds to the initial condition. The other solution must be $\pi - 2\delta$, since $\cos(\pi - \delta) = \cos(\pi + \delta)$ so that one obtains from Problem 10.4 the result

$$\tan\tfrac{1}{2}\theta_c = \cot\delta = \frac{K}{v_0^2 b}\,. \tag{10.19}$$

While this completes the problem in the center of mass (CM) system, it does not give the answer in the *lab* system, but for many problems, this answer may suffice if the ratio m_1/m_2 is sufficiently small that the CM may be considered stationary.

One may determine from these expressions that there is no limit to the possible deflection angle, since with $b \to 0$, the particles may backscatter. With a diffuse distribution of positive charge as the model for an atom, as proposed by J. J. Thomson, the maximum angle of deflection could never scatter with such large angles. In order to determine quantitatively what the scattering center looks like, however, it is necessary to find the *distribution of angular deflections.* While the incident particles may have constant incident velocity, they have a distribution of impact parameters, and we need to know the relative numbers at one angle with respect to any other angle, assuming the impact parameters are randomly distributed. We are thus looking for a distribution function $N(\theta_c)$, which is the number of particles scattered between angles θ_c and $\theta_c + \mathrm{d}\theta_c$ and relate this to the number of particles that have impact parameters between b and $b + \mathrm{d}b$. The annular region for this range of impact parameters has area $2\pi b\,\mathrm{d}b$. The total number of particles scattered from this annular region into angles between θ_c and $\theta_c + \mathrm{d}\theta_c$ is given by

$$\mathrm{d}N = N(\theta_c)\,\mathrm{d}\theta_c = N_0 nt\,2\pi b\,\mathrm{d}b\,, \tag{10.20}$$

$$N(\theta_c) = 2\pi N_0 nt\,b\frac{\mathrm{d}b}{\mathrm{d}\theta_c} \tag{10.21}$$

$$= \pi N_0 nt K^2 \frac{\cos\frac{1}{2}\theta_c}{v_0^4 \sin^3\frac{1}{2}\theta_c}\,, \tag{10.22}$$

where N_0 is the number of incident particles, t is the foil thickness, and n is the particle density of the foil, in particles per unit volume.

It is desired to express the number of particles scattered into a unit solid angle at a given angle θ_c. Using the relation relating the solid angle Ω with θ_c,

$$\Omega = 2\pi(1 - \cos\theta_c)\,,$$

so that

$$\mathrm{d}\Omega = 2\pi \sin\theta_c \,\mathrm{d}\theta_c\,,$$

we find

$$\frac{\mathrm{d}N}{\mathrm{d}\Omega} = \frac{N(\theta_c)}{2\pi\sin\theta_c} \tag{10.23}$$

$$= \frac{N_0 n t K^2}{4v_0^4 \sin^4 \frac{1}{2}\theta_c} \tag{10.24}$$

$$= \frac{N_0 n t Z_1^2 Z_2^2 e^4}{64\pi^2\epsilon_0^2 m_1^2 v_0^4 \sin^4 \frac{1}{2}\theta_c}\,. \tag{10.25}$$

This law is called the **Rutherford law of single scattering** and has the following properties:

1. Directly proportional to the thickness of the foil.
2. Directly proportional to the square of the atomic charge Z_2 of the scattering nuclei in the foil.
3. Inversely proportional to the square of the kinetic energy. $\frac{1}{2}m_1 v_0^2$ of the projectiles.
4. Inversely proportional to the *fourth power of the sine of the half angle of scattering angle* θ_c.

This law was carefully tested by Geiger and Marsden (1913) and verified the nuclear picture of the atom. The *size* of the nucleus could be observed by using energetic α-particles with very small impact parameters (large scattering angles) so that by observing at what angle this law fails as the incident particles begin to interact directly with the nucleus rather than only with its external Coulomb field.

This calculation is a purely classical result, and since we know the fundamental physics of the interaction is quantum mechanical, it was later repeated using quantum mechanics. Remarkably, the fundamental result is identical. The fact that both the Bohr atom model and the Rutherford scattering law both give the same results as quantum mechanics is unique to the Coulomb potential. For any other potential function, the classical results and the quantum mechanical results disagree both quantitatively and qualitatively.

Whereas Rutherford deliberately chose α-particles on a gold foil because gold can be beaten into foils so thin that the probability of multiple scattering is very small, in solid materials, multiple scattering is typical and leads to slowing of the incident particles. The rate of slowing after passing through a finite thickness can be used to estimate the original energy.

Problem 10.1 Show that $\dot{\hat{\mathbf{e}}}_r = \dot{\theta}\hat{\mathbf{e}}_\theta$ and that $\dot{\hat{\mathbf{e}}}_\theta = -\dot{\theta}\hat{\mathbf{e}}_r$.

Problem 10.2 Verify that the initial value of $r^2\dot{\theta}$ is $-v_0 b$.

Problem 10.3 Supply the missing steps leading to Equation (10.17).

Problem 10.4 Evaluate the constants A and δ in terms of the initial conditions $\theta_0 = \pi$, $r_0 = \infty$, $\dot{r}_0 = -v_0$, and $r_0^2\dot{\theta}_0 = -v_0 b$.

Ans.

$$A = -\frac{(1 + K^2/v_0^4 b^2)^{1/2}}{b}\,, \tag{10.26}$$

$$\tan\delta = \frac{v_0^2 b}{K}\,. \tag{10.27}$$

Problem 10.5 Use conservation of energy and angular momentum to evaluate the distance of closest approach of two charged particles. *Ans.:* $r_{\min} = (K/v_0^2)(1 + \csc\frac{1}{2}\theta_c)$.

Problem 10.6 When α-particles from RaC'(^{214}Po) (7.68 MeV) bombard aluminum foil, it is found that the Rutherford scattering law begins to break down at a lab scattering angle of about 60°. What is the approximate effective size of an aluminum nucleus in this experiment if an α-particle has a radius of about 2×10^{-15} m?

10.1.2 Quantum Scattering Theory in Three Dimensions

For this case, we again take a wave function to have the form $\psi(x) = A\mathrm{e}^{ikx}$ for the incident wave, but the *outgoing* wave in this case is a spherical wave, so we expect the solution to have the form

$$\psi(r,\theta) = A\left[\mathrm{e}^{ikx} + f(\theta)\frac{\mathrm{e}^{ikr}}{r}\right] \qquad \text{for large } r\,.$$

The $1/r$ factor is necessary to preserve the probability density since we require $|\psi|^2 \sim |A|^2/r^2$ for large r. Although there could be some dependence on the angle ϕ as well as θ, we will not investigate this most general case in detail.

Our task is to find $f(\theta)$ as it tells us the probability of scattering in the direction of θ. For the *incident* wave, the differential probability of passing through the cross sectional area $\mathrm{d}\sigma$ in time $\mathrm{d}t$ is

$$\mathrm{d}P = |\psi_{\text{incident}}|^2\,\mathrm{d}V = |A|^2(v\,\mathrm{d}t)\mathrm{d}\sigma\,,$$

and the corresponding differential probability of passing through the annular cross section given by the solid angle $\mathrm{d}\Omega$ is

$$\mathrm{d}P = |\psi_{\text{scattered}}|^2 \,\mathrm{d}V = \frac{|A|^2|f(\theta)|^2}{r^2}(v\mathrm{d}t)r^2\,\mathrm{d}\Omega\,,$$

so that the differential cross section $\mathrm{d}\sigma = |f|^2\,\mathrm{d}\Omega$ which we define as

$$D(\theta) = \frac{\mathrm{d}\sigma}{\mathrm{d}\Omega} = |f(\theta)|^2\,. \tag{10.28}$$

It is clear that the wave function must be a solution of the Schrödinger equation, so that in order to relate $f(\theta)$ to the potential that scattered the wave, we must investigate solutions of this general type. There are two general methods, one called the **partial wave analysis** and the other the **Born approximation**.

10.1.3 Partial Wave Analysis

If the scattering potential is spherically symmetric, we know from Chapter 3 that the solution is separable and of the form

$$\psi(r,\theta,\phi) = R(r)Y_\ell^m(\theta,\phi)\,.$$

Writing the radial function as we did before in terms of $u(r) = rR(r)$, we need to solve Equation (3.90)

$$u'' + \left[\frac{2\mu}{\hbar^2}(E-V) - \frac{\ell(\ell+1)}{r^2}\right]u = 0\,. \tag{10.29}$$

As $r \to \infty$, this becomes $u'' + k^2u = 0$ where $k^2 = 2\mu E/\hbar^2$ with solutions

$$u(r) = A\mathrm{e}^{\mathrm{i}kr} + B\mathrm{e}^{-\mathrm{i}kr}$$

so that

$$R \to \frac{A\mathrm{e}^{\mathrm{i}kr}}{r}$$

which is the form we expected for the asymptotic solution. This form is valid in the limit as $kr \gg 1$, which is called the **radiation zone**, but we need the solution close enough to the scattering center to see the effect of the potential. The exact solution of Equation (10.29) is

$$u(r) = Arj_\ell(kr) + Bry_\ell(kr)$$

where j_ℓ and y_ℓ are **spherical Bessel functions**, the first few of which are

$$j_0(z) = \frac{\sin z}{z}$$

$$j_1(z) = \frac{\sin z}{z^2} - \frac{\cos z}{z}$$
$$y_0(z) = -\frac{\cos z}{z}$$
$$y_1(z) = -\frac{\cos z}{z^2} - \frac{\sin z}{z}$$

so that none of these functions individually has the proper form as $r \to \infty$. A combination does have the proper form, however, since the combinations

$$h_n^{(1)}(z) = j_n(z) + \mathrm{i}y_n(z) \tag{10.30}$$
$$h_n^{(2)}(z) = j_n(z) - \mathrm{i}y_n(z) \tag{10.31}$$

that are called **spherical Hankel functions of the first and second kind**, of which the first two of the first kind are

$$h_0^{(1)}(z) = \frac{\mathrm{e}^{\mathrm{i}z}}{\mathrm{i}z} \tag{10.32}$$
$$h_1^{(1)}(z) = \left(\frac{1}{\mathrm{i}z^2} - \frac{1}{z}\right)\mathrm{e}^{\mathrm{i}z}, \tag{10.33}$$

which do have the proper form. Therefore, the radial solutions are of the form

$$R(r) = C_\ell h_\ell^{(1)}(kr). \tag{10.34}$$

The complete solution then takes the form

$$\psi(r,\theta,\phi) = A\left[\mathrm{e}^{\mathrm{i}kz} + \sum_{\ell,m} C_{\ell,m} h_\ell^{(1)}(kr) Y_\ell^m(\theta,\phi)\right]. \tag{10.35}$$

The asymptotic form of the Hankel functions takes the form

$$h_\ell^{(1)}(z) \approx (-\mathrm{i})^{\ell+1}\frac{\mathrm{e}^{\mathrm{i}z}}{z}$$

so our expression in the radiation zone is

$$\psi(r,\theta,\phi) = A\left[\mathrm{e}^{\mathrm{i}kz} + f(\theta,\phi)\frac{\mathrm{e}^{\mathrm{i}kr}}{r}\right] \tag{10.36}$$

where

$$f(\theta,\phi) = \frac{1}{k}\sum_{\ell,m}(-\mathrm{i})^{\ell+1} C_{\ell,m} Y_\ell^m(\theta,\phi). \tag{10.37}$$

When the potential is spherically symmetric, there is no ϕ-dependence, so this reduces to

$$f(\theta,\phi) = \frac{1}{k}\sum_{\ell=0}^{\infty}(-\mathrm{i})^{\ell+1} C_{\ell,0} Y_\ell^0(\theta,\phi). \tag{10.38}$$

and since

$$Y_\ell^0(\theta,\phi) = \sqrt{\frac{2\ell+1}{4\pi}}P_\ell(\cos\theta) \tag{10.39}$$

it is convenient to redefine

$$C_{\ell,0} = k\mathrm{i}^{\ell+1}\sqrt{4\pi(2\ell+1)}a_\ell \tag{10.40}$$

so that in terms of the a_ℓ, we have

$$\psi(r,\theta) = A\left[\mathrm{e}^{\mathrm{i}kz} + k\sum_{\ell=0}\mathrm{i}^{\ell+1}\sqrt{2\ell+1}a_\ell h_\ell^{(1)}(kr)P_\ell(\cos\theta)\right]. \tag{10.41}$$

In the extreme radiation zone, $h_\ell^{(1)}(kr) \to (-\mathrm{i})^{\ell+1}\mathrm{e}^{\mathrm{i}kr}/r$, so we may write Equation (10.37) as

$$f(\theta) = \sum_{\ell=0}^{\infty}(2\ell+1)a_\ell P_\ell(\cos\theta) \tag{10.42}$$

The differential cross section is, from Equation (10.28),

$$D(\theta) = |f(\theta)|^2 = \sum_{\ell'}\sum_{\ell}(2\ell'+1)(2\ell+1)a_{\ell'}^* a_\ell P_{\ell'}(\cos\theta)P_\ell(\cos\theta)$$

and using the orthogonality of the Legendre polynomials, we may integrate over θ [see Equation (B.34)] and ϕ to obtain the total cross section as

$$\sigma = 4\pi\sum_{\ell=0}^{\infty}(2\ell+1)|a_\ell|^2 . \tag{10.43}$$

The more general scattering function for potentials which are not spherically symmetric may be written in terms of the **partial wave amplitudes**, the $C_{\ell,m}$. It follows that the differential cross section for the general case is given by

$$D(\theta,\phi) = |f(\theta,\phi)|^2 = \frac{1}{k^2}\sum_{\ell,m}\sum_{\ell',m'}\mathrm{i}^{\ell-\ell'}C_{\ell,m}^* C_{\ell',m'}Y_\ell^{m*}Y_{\ell'}^{m'} \tag{10.44}$$

and the total cross section is

$$\sigma = \frac{1}{k^2}\sum_{\ell,m}\sum_{\ell',m'}\mathrm{i}^{\ell-\ell'}C_{\ell,m}^* C_{\ell',m'}\int Y_\ell^{m*}Y_{\ell'}^{m'}\,\mathrm{d}\Omega = \frac{1}{k^2}\sum_{\ell,m}|C_{\ell,m}|^2 \tag{10.45}$$

since the Y_ℓ^m form an orthonormal set.

To complete this section, we need to express $\psi(\boldsymbol{r})$ in a consistent set of coordinates, and $\exp(\mathrm{i}kz)$ is in cartesian coordinates instead of spherical coordinates. To finish this part of the problem, we use the expansion of a plane wave in spherical coordinates through the relation

$$\mathrm{e}^{\mathrm{i}kz} = \sum_{\ell=0}^{\infty}\mathrm{i}^\ell(2\ell+1)j_\ell(kr)P_\ell(\cos\theta) .$$

Since nearly all examples of $V(\boldsymbol{r})$ are independent of ϕ, we can simplify our expressions by using Equation (10.39) so that our general expression in the radiation zone can be represented as

$$\psi(r,\theta) = A\sum_{\ell=0}^{\infty}\left[\mathrm{i}^{\ell}(2\ell+1)j_{\ell}(kr) + \sqrt{\frac{2\ell+1}{4\pi}}C_{\ell}h_{\ell}^{(1)}(kr)\right]P_{\ell}(\cos\theta) \quad (10.46)$$

where we have let $C_{\ell,0} = C_{\ell}$.

10.1.3.1 Strategy for Finding the Partial Wave Amplitudes

To complete the problem, we need to find a way to match the solutions in the *exterior* zone to those in the *interior* zone where the potential is important. This will involve matching boundary conditions. As an example, we consider the problem of scattering from a hard sphere, where the potential is

$$V(r) = \begin{cases} \infty & r \le a \\ 0 & r > a\,. \end{cases}$$

Since the wave function must be continuous at $r = a$ and the wave function vanishes for $r < a$, we have the condition that

$$\psi(a,\theta) = 0$$

so that

$$\sum_{\ell=0}^{\infty}\left[\mathrm{i}^{\ell}(2\ell+1)j_{\ell}(ka) + \sqrt{\frac{2\ell+1}{4\pi}}C_{\ell}h_{\ell}^{(1)}(ka)\right]P_{\ell}(\cos\theta) = 0 \quad (10.47)$$

for all θ. Since the $P_{\ell}(\cos\theta)$ are orthogonal (or linearly independent), each coefficient in the sum must vanish individually, so that we can solve for C_{ℓ} as

$$C_{\ell} = -\mathrm{i}^{\ell}\sqrt{4\pi(2\ell+1)}\frac{j_{\ell}(ka)}{h_{\ell}^{(1)}(ka)}\,.$$

The total cross section is therefore

$$\sigma = \frac{4\pi}{k^2}\sum_{\ell=0}^{\infty}(2\ell+1)\left|\frac{j_{\ell}(ka)}{h_{\ell}^{(1)}(ka)}\right|^2. \quad (10.48)$$

Although this is exact, it is difficult to digest, so we will examine *low energy* scattering where $ka \ll 1$. This implies that $\lambda \gg a$ so this represents the long wavelength limit for scattering. To see where this leads, we examine the small argument behavior of both $j_{\ell}(z)$ and $h_{\ell}^{(1)}(z)$ where the leading terms are given by

$$\begin{aligned} j_{\ell}(z) &= \frac{z^{\ell}}{1\cdot 3\cdot 5\cdots(2\ell+1)}\left[1 - \frac{\frac{1}{2}z^2}{1!(2\ell+3)} + \cdots\right] \\ h_{\ell}(z) &\sim \mathrm{i}y_{\ell}(z) \\ &= -\mathrm{i}\frac{1\cdot 1\cdot 3\cdot 5\cdots(2\ell-1)}{z^{\ell+1}}\left[1 - \frac{\frac{1}{2}z^2}{1!(1-2\ell)} + \cdots\right] \end{aligned}$$

so the ratio is approximately

$$\frac{j_\ell(z)}{\mathrm{i}y_\ell(z)} \approx \frac{\mathrm{i}z^{2\ell+1}}{(2\ell+1)[1\cdot 1\cdot 3\cdot 5\cdots(2\ell-1)]^2}\,.$$

The cross section is therefore

$$\sigma \simeq \frac{4\pi}{k^2}\left\{(ka)^2 + \sum_{\ell=1}^{\infty}\frac{(ka)^{4\ell+2}}{2\ell+1}\left[\frac{2^{\ell-1}(\ell-1)!}{(2\ell-1)!}\right]^2\right\}. \tag{10.49}$$

With $z = ka \ll 1$, the higher order terms vary as $(ka)^{4\ell+2}$ so only the first term makes any real difference, and the sum collapses to

$$\sigma \approx 4\pi a^2\,. \tag{10.50}$$

This result is four times larger than the geometrical cross section of the hard sphere, and is due to the fact that we have scattered *waves* instead of particles, and diffraction effects account for the larger effective size.

If we examine the high-energy limit, the result of Equation (10.48) is still exact, but the sum over ℓ is limited, since the energy in angular momentum cannot exceed the total energy, so we may estimate

$$E \sim \frac{\hbar^2\ell_{\max}(\ell_{\max}+1)}{2\mu a^2} \sim \frac{\hbar^2k^2}{2\mu}$$

so $\ell_{\max} \sim ka \gg 1$. The sum therefore terminates at $\ell = \ell_{\max}$. Also for $ka \gg 1$, we need the asymptotic forms for $j_\ell(kr)$ and $h_\ell^{(1)}(kr)$ which are given by

$$j_\ell(kr) \approx \frac{\sin(kr-\ell\pi/2)}{kr}\,, \qquad kr \gg 1$$

$$y_\ell(kr) \approx -\frac{\cos(kr-\ell\pi/2)}{kr}\,, \qquad kr \gg 1$$

so

$$\frac{j_\ell(kr)}{h_\ell^{(1)}(kr)} \approx \mathrm{i}\sin(kr-\ell\pi/2)\mathrm{e}^{-\mathrm{i}(kr-\ell\pi/2)}$$

so the magnitude of the ratio is

$$\left|\frac{j_\ell(ka)}{h_\ell^{(1)}(ka)}\right|^2 \approx \sin^2(ka-\ell\pi/2)$$

and since there are many values of ℓ, we may approximate the average to be $\langle\sin^2(ka-\ell\pi/2)\rangle \sim \frac{1}{2}$, so Equation (10.48) becomes

$$\sigma \approx \frac{2\pi}{k^2}\sum_{\ell=0}^{\ell_{\max}}(2\ell+1)$$

and noting that [see Equation (3.88)]

$$\sum_{\ell=0}^{\ell_{\max}} (2\ell+1) = (\ell_{\max}+1)^2$$

and with $\ell_{\max} \sim ka \gg 1$, we find

$$\sigma \approx 2\pi a^2 \tag{10.51}$$

which is half the value for low energy scattering.

Problem 10.7 *Cross sections.* Consider the scattering from a hard sphere potential $V(r) \to \infty,\ r < a$, and $V(r) = 0,\ r > a$.

(a) Estimate the percentage correction to the cross section for a 100 keV proton on a hard sphere of radius 6 fm using the low energy limit with the $\ell = 1$ term included.

(b) Estimate the value of $\ell_{\max}$ and the correction to the cross section if we do not neglect 1 compared to $\ell_{\max}$ for a 700 MeV proton on a hard sphere of radius 6 fm using the high energy limit.

Example 10.1
Shallow well. If, instead of a hard sphere, we consider a shallow well such that the potential is given by

$$V(r) = \begin{cases} -V_0 & r \le a \\ 0 & r > a\,. \end{cases}$$

then the solution in the inner region that is regular at the origin is

$$R_\ell(r) = A_\ell j_\ell(k_0 r) \tag{10.52}$$

where $k_0^2 = 2\mu(E+V_0)/\hbar^2$. This model potential might be used to describe neutron-neutron scattering. In the low energy limit, we consider only the $\ell = 0$ case with solution

$$R_{\text{in}}(r) = A j_0(k_0 r) \tag{10.53}$$

in the interior region. The solution for $r > a$ with $\ell = 0$ is

$$R_{\text{out}}(r) = B\left[j_0(kr) + \frac{C_0}{\sqrt{4\pi}} h_0^{(1)}(kr)\right] \tag{10.54}$$

where $k^2 = 2\mu E/\hbar^2$. The matching at $r = a$ can be done either with the $R(r)$ or with $u(r) = rR(r)$ so that

$$\begin{aligned} u_{\text{in}}(r) &= A\sin k_0 r \\ u_{\text{out}}(r) &= B\left[\sin kr - \frac{\mathrm{i}C_0}{\sqrt{4\pi}}\mathrm{e}^{\mathrm{i}kr}\right] \\ u'_{\text{in}}(r) &= k_0 A\cos k_0 r \\ u'_{\text{out}}(r) &= kB\left[\cos kr + \frac{C_0}{\sqrt{4\pi}}\mathrm{e}^{\mathrm{i}kr}\right] \end{aligned}$$

so matching $u(r)$ and $u'(r)$ at $r = a$ we find

$$A \sin k_0 a = B \left[\sin ka - \frac{iC_0}{\sqrt{4\pi}} e^{ika} \right]$$

$$k_0 A \cos k_0 a = kB \left[\cos ka + \frac{C_0}{\sqrt{4\pi}} e^{ika} \right]$$

and taking the ratio, we eliminate A and B to obtain

$$\frac{\sqrt{4\pi} \cos ka + C_0 e^{ika}}{\sqrt{4\pi} \sin ka - iC_0 e^{ika}} = \frac{k_0}{k} \cot k_0 a \,.$$

Solving for C_0, we obtain

$$C_0 = \frac{\sqrt{4\pi} \sin ka (k_0 \cot k_0 a - k \cot ka) e^{-ika}}{k + ik_0 \cot k_0 a} \,. \tag{10.55}$$

The total cross section is then given by

$$\begin{aligned} \sigma &= \frac{1}{k^2} |C_0|^2 \\ &= 4\pi \frac{\sin^2 ka}{k^2} \frac{(k_0 \cot k_0 a - k \cot ka)^2}{k^2 + k_0^2 \cot^2 k_0 a} \\ &= 4\pi \frac{\sin^2 ka}{k^2} \frac{(k_0 - k \cot ka \tan k_0 a)^2}{k_0^2 + k^2 \tan^2 k_0 a} \,. \end{aligned} \tag{10.56}$$

In the limit as $ka \to 0$, $k \cot ka \to 1/a$ and $\sin^2 ka/k^2 \to a^2$ so the low energy approximation leads to

$$\sigma \approx 4\pi a^2 \frac{(k_0 a - \tan k_0 a)^2}{k_0^2 a^2 + k^2 a^2 \tan^2 k_0 a} \tag{10.57}$$

which for $k_0 \gg k$ leads to $\sigma \approx 4\pi a^2$. The result is not so simple, however, since $\tan k_0 a$ may pass through zero or approach infinity, so that either the numerator or denominator may vanish. When the denominator vanishes, there is a pole in the scattering cross section, and this corresponds to an eigenvalue where there is a bound state in the well for that particular value of k_0. When the numerator vanishes, the well is transparent for that particular energy. □

10.1.4 Born Approximation

For this method, we must first examine the character of **Green's functions** where we effectively invert a differential equation into an integral or an integral equation. If we take a general one-dimensional example for an inhomogeneous differential equation of the form

$$\hat{O} \psi(x) = f(x) \tag{10.58}$$

where $\hat{\mathcal{O}}$ is a differential operator, we write the solution as

$$\psi(x) = \int G(x,y) f(y)\,\mathrm{d}y \tag{10.59}$$

where $G(x,y)$ is the Green's function. If $f(x)$ is independent of ψ, then the solution has been reduced to the evaluation of an integral. If $f(x)$ is proportional to ψ so that $f(x) = g(x)\psi(x)$, then we have an integral equation of the form

$$\psi(x) = \int G(x,y) g(y) \psi(y)\,\mathrm{d}y\,. \tag{10.60}$$

The most common (but not the only) way of finding the Green's function is to use the δ-function method. For example, we write the Schrödinger equation as

$$-\frac{\hbar^2}{2\mu}\nabla^2\psi + V\psi = E\psi$$

or as

$$(\nabla^2 + k^2)\psi = Q$$

where $k^2 = 2\mu E/\hbar^2$ and $Q = (2\mu V/\hbar^2)\psi$. Then we assume the Green's function satisfies an equation of the same form except the right-hand side is replaced by a three-dimensional δ-function, so we assume that

$$(\nabla^2 + k^2) G(\boldsymbol{r}) = \delta^3(\boldsymbol{r}) \tag{10.61}$$

so that the solution of the original equation is

$$\psi(\boldsymbol{r}) = \int G(\boldsymbol{r} - \boldsymbol{r}_0) Q(\boldsymbol{r}_0)\,\mathrm{d}^3 r_0$$

since

$$\begin{aligned}(\nabla^2 + k^2)\psi &= \int (\nabla^2 + k^2) G(\boldsymbol{r} - \boldsymbol{r}_0) Q(\boldsymbol{r}_0)\,\mathrm{d}^3 r_0 \\ &= \int \delta^3(\boldsymbol{r} - \boldsymbol{r}_0) Q(\boldsymbol{r}_0)\,\mathrm{d}^3 r_0 \\ &= Q(\boldsymbol{r})\,.\end{aligned}$$

We will use Fourier transforms to solve for the Green's function, where in three dimensions we have

$$G(\boldsymbol{r}) = \frac{1}{(2\pi)^{3/2}} \int_{-\infty}^{\infty} \mathrm{e}^{\mathrm{i}\boldsymbol{s}\cdot\boldsymbol{r}} g(\boldsymbol{s})\,\mathrm{d}^3 s \tag{10.62}$$

and we then examine the left-hand side of Equation (10.61) to find

$$(\nabla^2 + k^2) G(\boldsymbol{r}) = \frac{1}{(2\pi)^{3/2}} \int_{-\infty}^{\infty} \mathrm{e}^{\mathrm{i}\boldsymbol{s}\cdot\boldsymbol{r}} (-s^2 + k^2) g(\boldsymbol{s})\,\mathrm{d}^3 s$$

and using the Fourier representation of the δ-function, the right-hand side of Equation (10.61) is

$$\delta^3(\boldsymbol{r}) = \frac{1}{(2\pi)^3}\int \mathrm{e}^{\mathrm{i}\boldsymbol{s}\cdot\boldsymbol{r}}\,\mathrm{d}^3s\,.$$

Equating these expressions, it is clear that

$$g(\boldsymbol{s}) = \frac{1}{(2\pi)^{3/2}}\frac{1}{k^2 - s^2}\,. \tag{10.63}$$

The Green's function is therefore

$$G(\boldsymbol{r}) = \frac{1}{(2\pi)^3}\int \frac{\mathrm{e}^{\mathrm{i}\boldsymbol{s}\cdot\boldsymbol{r}}}{k^2 - s^2}\,\mathrm{d}^3s\,. \tag{10.64}$$

In order to evaluate this integral over s, we will use spherical coordinates with the z-axis along the $\boldsymbol{r}$ direction so that $\boldsymbol{r}\cdot\boldsymbol{s} = rs\cos\theta$ and the integral over ϕ is trivial. After the integral over ϕ, we have

$$G(\boldsymbol{r}) = \frac{1}{(2\pi)^2}\int_0^\infty \mathrm{d}s\,s^2 \int_0^\pi \mathrm{d}\theta\,\sin\theta\,\frac{\mathrm{e}^{\mathrm{i}sr\cos\theta}}{k^2 - s^2}$$

but the integral over θ is

$$\begin{aligned}\int_0^\pi \sin\theta\,\mathrm{e}^{\mathrm{i}sr\cos\theta}\,\mathrm{d}\theta &= -\left.\frac{\mathrm{e}^{\mathrm{i}sr\cos\theta}}{\mathrm{i}sr}\right|_0^\pi \\ &= -\frac{\mathrm{e}^{-\mathrm{i}sr} - \mathrm{e}^{\mathrm{i}sr}}{\mathrm{i}sr} \\ &= \frac{2}{sr}\sin sr\,.\end{aligned}$$

We are left with the remaining integral

$$G(\boldsymbol{r}) = \frac{1}{(2\pi)^2}\frac{2}{r}\int_0^\infty \frac{s\sin sr}{k^2 - s^2}\,\mathrm{d}s\,.$$

This integral is most easily done using contour methods, and since the integrand is even in s, we may extend the range to $[-\infty,\infty]$ so that

$$G(\boldsymbol{r}) = \frac{1}{4\pi^2 r}\int_{-\infty}^\infty \frac{s\sin sr}{k^2 - s^2}\,\mathrm{d}s\,.$$

Now we break $\sin sr$ back into the exponential form so that we have two integrals of the form

$$G(\boldsymbol{r}) = \frac{\mathrm{i}}{8\pi^2 r}\left[\int_{-\infty}^\infty \frac{s\mathrm{e}^{\mathrm{i}sr}}{(s-k)(s+k)}\,\mathrm{d}s - \int_{-\infty}^\infty \frac{s\mathrm{e}^{-\mathrm{i}sr}}{(s-k)(s+k)}\,\mathrm{d}s\right]. \tag{10.65}$$

The first integral is

$$I_1 = \int_{-\infty}^\infty \frac{s\mathrm{e}^{\mathrm{i}sr}}{(s-k)(s+k)}\,\mathrm{d}s$$

and we may close the contour above because $\mathrm{Re}(\mathrm{i}sr) < 0$ for s in the upper half s-plane so that the exponential term vanishes as $|s| \to \infty$, and we pick the pole at $s = k$ (assuming k has a positive imaginary part for causality), so the result is

$$I_1 = +2\pi\mathrm{i}\frac{k\mathrm{e}^{\mathrm{i}kr}}{2k} = \pi\mathrm{i}\mathrm{e}^{\mathrm{i}kr}\,.$$

The second integral is almost the same, except that for this case we need to close below because $\mathrm{Re}(-\mathrm{i}sr) < 0$ for s in the lower half s-plane, and we pick up the pole at $s = -k$, so

$$I_2 = \int_{-\infty}^{\infty} \frac{s\mathrm{e}^{-\mathrm{i}sr}}{(s-k)(s+k)}\,\mathrm{d}s = -2\pi\mathrm{i}\frac{-k\mathrm{e}^{\mathrm{i}kr}}{-2k} = -\pi\mathrm{i}\mathrm{e}^{\mathrm{i}kr}$$

where the leading minus sign is due to the fact that the contour was clockwise for this case. Putting the pieces together, we have the result for the Green's function as

$$G(\boldsymbol{r}) = \frac{\mathrm{i}}{8\pi^2 r}(\pi\mathrm{i}\mathrm{e}^{\mathrm{i}kr} + \pi\mathrm{i}\mathrm{e}^{\mathrm{i}kr}) = -\frac{\mathrm{e}^{\mathrm{i}kr}}{4\pi r}\,. \tag{10.66}$$

We can then write the exact solution of the Schrödinger equation as

$$\psi(\boldsymbol{r}) = \psi_0(\boldsymbol{r}) + \frac{1}{4\pi}\int \frac{\mathrm{e}^{\mathrm{i}\boldsymbol{k}(\boldsymbol{r}-\boldsymbol{r}_0)}}{|\boldsymbol{r}-\boldsymbol{r}_0|}\frac{2\mu}{\hbar^2}V(\boldsymbol{r}_0)\psi(\boldsymbol{r}_0)\,\mathrm{d}^3r_0 \tag{10.67}$$

where ψ_0 is the free-particle wave function that satisfies

$$(\nabla^2 + k^2)\psi_0 = 0\,.$$

Equation (10.67) is an integral equation that is equivalent to the Schrödinger equation and is exact for any $V(\boldsymbol{r})$. It may not be any easier to solve, but it is equivalent.

10.1.4.1 First Born Approximation

To find a first-order solution to Equation (10.67), we will use the **first Born approximation**. The first step is to assume that $V(\boldsymbol{r})$ is *localized* so that the range of the integral is limited to the range of $\boldsymbol{r}_0$ where $r_0 \ll r$. We may use the law of cosines to establish

$$|\boldsymbol{r}-\boldsymbol{r}_0|^2 = r^2 + r_0^2 - 2\boldsymbol{r}\cdot\boldsymbol{r}_0 \simeq r^2(1 - 2\boldsymbol{r}\cdot\boldsymbol{r}_0/r^2)$$

so that $|\boldsymbol{r}-\boldsymbol{r}_0| \simeq r - \hat{\boldsymbol{e}}_r\cdot\boldsymbol{r}_0$ where $\hat{\boldsymbol{e}}_r$ is a unit vector in the $\boldsymbol{r}$-direction. Then we let $\boldsymbol{k} = \hat{\boldsymbol{e}}_r k$ so that the exponential factor becomes

$$\mathrm{e}^{\mathrm{i}\boldsymbol{k}\cdot(\boldsymbol{r}-\boldsymbol{r}_0)} \simeq \mathrm{e}^{\mathrm{i}kr}\mathrm{e}^{-\mathrm{i}\boldsymbol{k}\cdot\boldsymbol{r}_0}$$

and

$$\frac{\mathrm{e}^{\mathrm{i}\boldsymbol{k}\cdot(\boldsymbol{r}-\boldsymbol{r}_0)}}{|\boldsymbol{r}-\boldsymbol{r}_0|} \simeq \frac{\mathrm{e}^{\mathrm{i}kr}}{r}\mathrm{e}^{-\mathrm{i}\boldsymbol{k}\cdot\boldsymbol{r}_0}\,.$$

We now let $\psi_0 = A\mathrm{e}^{\mathrm{i}kz}$ represent the incoming plane wave so that

$$\psi(\boldsymbol{r}) \simeq A\mathrm{e}^{\mathrm{i}kz} - \frac{m}{2\pi\hbar^2}\frac{\mathrm{e}^{\mathrm{i}kr}}{r}\int \mathrm{e}^{-\mathrm{i}\boldsymbol{k}\cdot\boldsymbol{r}_0} V(\boldsymbol{r}_0)\psi(\boldsymbol{r}_0)\,\mathrm{d}^3 r_0\,. \tag{10.68}$$

The first Born approximation then assumes that the potential has little effect on the wave function, so that $\psi(\boldsymbol{r}_0) \simeq \psi_0(\boldsymbol{r}_0)$ *inside* the integral, or

$$\psi(\boldsymbol{r}_0) \simeq \mathrm{e}^{\mathrm{i}kz_0} \equiv \mathrm{e}^{\mathrm{i}\boldsymbol{k}'\cdot\boldsymbol{r}_0}$$

where $\boldsymbol{k}' = \hat{\boldsymbol{e}}_z k$. Then we have

$$f(\theta,\phi) = -\frac{m}{2\pi\hbar^2}\int \mathrm{e}^{\mathrm{i}(\boldsymbol{k}'-\boldsymbol{k})\cdot\boldsymbol{r}_0} V(\boldsymbol{r}_0)\,\mathrm{d}^3 r_0\,. \tag{10.69}$$

Example 10.2

Low energy. For low energy, we may assume that $kr_0 \ll 1$ so the exponent in Equation (10.69) may be neglected so that

$$f(\theta,\phi) = -\frac{m}{2\pi\hbar^2}\int V(\boldsymbol{r}_0)\,\mathrm{d}^3 r_0$$

and if we consider a central potential, $V(r)$, then

$$f = -\frac{2\mu}{\hbar^2}\int_0^\infty r^2 V(r)\,\mathrm{d}r\,.$$

For a simple potential of the form

$$V(r) = \begin{cases} V_0 & r \le a \\ 0 & r > a \end{cases}$$

$$f = -\frac{2\mu}{\hbar^2}\int_0^a r^2 V_0\,\mathrm{d}r = -\frac{2\mu V_0 a^3}{3\hbar^2}$$

and the differential cross section is

$$\frac{\mathrm{d}\sigma}{\mathrm{d}\Omega} = |f|^2 = \left(\frac{2\mu V_0 a^3}{3\hbar^2}\right)^2$$

and the total cross section is

$$\sigma = 4\pi\left(\frac{2\mu V_0 a^3}{3\hbar^2}\right)^2. \tag{10.70}$$

We note that this result does not reduce to the result from the partial wave analysis for a *hard* sphere as $V_0 \to \infty$, since it blows up. The reason for this difference is that we assumed here that the potential made little difference in the first Born approximation, while a hard sphere *cannot* be assumed to make little difference to the scattering. This example is sometimes called *soft sphere* scattering. ▯

Example 10.3
Spherical symmetry. Here we do *not* assume low energy, but we do assume spherical symmetry. Then it is useful to define

$$\boldsymbol{\kappa} \equiv \boldsymbol{k}' - \boldsymbol{k}$$

and let the polar axis for the integral over $\boldsymbol{r}_0$ be aligned with $\boldsymbol{\kappa}$ so that

$$(\boldsymbol{k}' - \boldsymbol{k}) \cdot \boldsymbol{r}_0 = \kappa r_0 \cos\theta_0 \,.$$

Then we have

$$f(\theta) \simeq -\frac{m}{2\pi\hbar^2} \int e^{i\kappa r_0 \cos\theta_0} V(r_0) r_0^2 \sin\theta_0 \, dr_0 \, d\theta_0 \, d\phi_0$$

and the integral over ϕ_0 is trivial and the integral over θ_0 we did in developing the Green's function, so that

$$f(\theta) \simeq -\frac{2\mu}{\hbar^2 \kappa} \int_0^\infty r V(r) \sin\kappa r \, dr \,. \tag{10.71}$$

The θ dependence is buried in κ since from Figure 10.2 we deduce that

$$\kappa = 2k \sin(\theta/2) \,. \tag{10.72}$$

▯

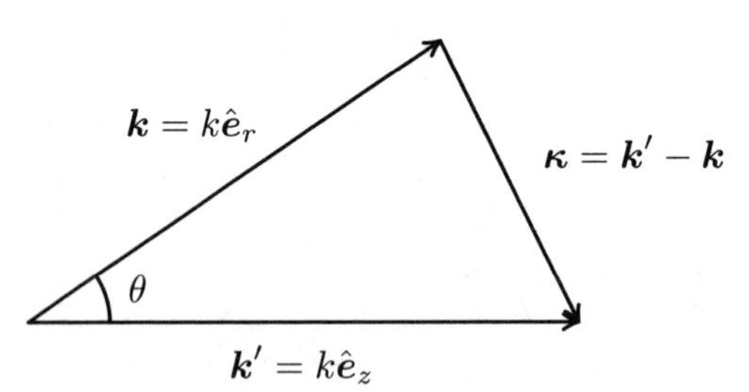

FIGURE 10.2
Wave vectors in the Born approximation where $\boldsymbol{k}'$ points in the *incident* direction and $\boldsymbol{k}$ points in the *scattered* direction

Problem 10.8 *Yukawa scattering.* Suppose the potential is of the form proposed by Yukawa to model the nuclear interaction potential

$$V(r) = \beta \frac{e^{-\mu_n r}}{r}$$

where β and μ_n are constants. Show that the Born approximation gives

$$f \simeq -\frac{2\mu\beta}{\hbar^2\kappa}\int_0^\infty e^{-\mu_n r}\sin\kappa r\,dr = -\frac{2\mu\beta}{\hbar^2(\mu_n^2+\kappa^2)}. \qquad (10.73)$$

Problem 10.9 Find the total cross section for Yukawa scattering in terms of E.

Problem 10.10 Find the differential and total cross sections in the first Born approximation for the potential $V(r) = V_0 e^{-r^2/a^2}$.

Problem 10.11 Find the differential cross section in the first Born approximation for neutron-neutron scattering in the case where the potential is approximated by $V(r) = V_0 \hat{\boldsymbol{S}}_1 \cdot \hat{\boldsymbol{S}}_2 e^{-r^2/a^2}$ where $\hat{\boldsymbol{S}}_1$ and $\hat{\boldsymbol{S}}_2$ are the spin vector operators for the two neutrons.

Example 10.4

Rutherford scattering. The Coulomb potential is equivalent to the Yukawa potential in the limit as $\mu_n \to 0$ if we let $\beta = q_2 q_2/4\pi\varepsilon_0$. From the result of Problem 10.8, the result in this limit is

$$f(\theta) \simeq -\frac{2\mu q_1 q_2}{4\pi\varepsilon_0\hbar^2\kappa^2}.$$

Using Equation (10.72) and $k^2 = 2\mu E/\hbar^2$, this becomes

$$f(\theta) \simeq -\frac{q_1 q_2}{16\pi\varepsilon_0 E\sin^2(\theta/2)}.$$

The differential cross section is therefore

$$\frac{d\sigma}{d\Omega} \simeq \left[\frac{q_1 q_2}{16\pi\varepsilon_0 E\sin^2(\theta/2)}\right]^2, \qquad (10.74)$$

which is the same as the classical result since $dN = N_0 nt\,d\sigma$. ☐

Problem 10.12 *Scattering problem.* Consider a square potential well of depth U_0 from $r = 0$ to $r = a$ so that

$$V(r) = \begin{cases} V(r) = \infty & r < 0 \\ V(r) = -U_0 & 0 \le r \le a \\ V(r) = 0 & r > a \end{cases}$$

with $k_0^2 = 2\mu(E + U_0)/\hbar^2$ and assume that $ka \ll 1$ so that only the $\ell = 0$ term is important.

(a) Find the appropriate radial solution $R(r)$ for $r \le a$ using the boundary condition at the origin.

(b) Match the interior and exterior solutions at $r = a$ and solve for C_0.

(c) Find the total cross section for this potential (exact).

(d) If the potential well of depth U_0 is changed to a potential hill, so that

$$V(r) = \begin{cases} V(r) = U_0 & r \leq a \\ V(r) = 0 & r > a \end{cases},$$

show that the total cross section expression is the same except that $k_0 \to \mathrm{i}\kappa$ where $\kappa^2 = 2\mu(U_0 - E)/\hbar^2$.

10.2 Scattering and Inverse Scattering in One Dimension*

In the following analysis, we will divide the range of k into two distinct ranges, consisting of the continuous spectrum, where there exists a solution of the Schrödinger equation for every value of k for $E \geq E_0$, and the discrete spectrum for $E < 0$ where there are N distinct eigenvalues E_n (or k_n).

10.2.1 Continuous Spectrum

We begin with the Schrödinger equation written as

$$\psi'' + \left(\frac{2mE}{\hbar^2} - \frac{2mV(x)}{\hbar^2}\right)\psi = 0$$

where we let $2mE/\hbar^2 \equiv \lambda$ and $2mV(x)/\hbar^2 \equiv u(x)$ so that it becomes

$$\psi'' + (\lambda - u)\psi = 0\,. \tag{10.75}$$

For the continuous spectrum, we let $\lambda = k^2 > 0$ and denote partial derivatives by subscripts so that we have

$$\psi_{xx} + (k^2 - u)\psi = 0\,. \tag{10.76}$$

We then *assume* a scattering solution in the form of a **Jost function**

$$\psi_+(x, k) = \mathrm{e}^{\mathrm{i}kx} + \int_x^\infty K(x, z)\mathrm{e}^{\mathrm{i}kz}\,\mathrm{d}z\,, \tag{10.77}$$

where the subscript + indicates that it satisfies a boundary condition as $x \to +\infty$. This $K(x, z)$ is nontrivial, but we still assume that it and its derivatives vanish as $|x| \to \infty$ and $K(x, z) = 0$ for $z < x$. In order to find the conditions on $K(x,z)$ so that ψ_+ will satisfy Equation (10.76), we first differentiate twice so that

$$\begin{aligned}
\psi_{+x} &= \mathrm{i}k\mathrm{e}^{\mathrm{i}kx} + \int_x^\infty K_x(x,z)\mathrm{e}^{\mathrm{i}kz}\,\mathrm{d}z - K(x,x)\mathrm{e}^{\mathrm{i}kx} \\
\psi_{+xx} &= -k^2\mathrm{e}^{\mathrm{i}kx} + \int_x^\infty K_{xx}(x,z)\mathrm{e}^{\mathrm{i}kz}\,\mathrm{d}z - K_x(x,x)\mathrm{e}^{\mathrm{i}kx} - \mathrm{i}kK(x,x)\mathrm{e}^{\mathrm{i}kx} \\
&\quad - \frac{\mathrm{d}K(x,x)}{\mathrm{d}x}\mathrm{e}^{\mathrm{i}kx} \\
&= \left(-k^2 - \frac{\mathrm{d}K(x,x)}{\mathrm{d}x} - K_x(x,x) - \mathrm{i}kK(x,x)\right)\mathrm{e}^{\mathrm{i}kx} \\
&\quad + \int_x^\infty K_{xx}(x,z)\mathrm{e}^{\mathrm{i}kz}\,\mathrm{d}z\,.
\end{aligned}$$

10.2.1.1 Notation

The partial derivatives of $K(x,z)$ and $K(x,x)$ are shorthand notation for

$$\begin{aligned}
K_x(x,z) &= \frac{\partial K(x,z)}{\partial x}, \\
K_z(x,z) &= \frac{\partial K(x,z)}{\partial z}, \\
K_x(x,x) &= \left.\frac{\partial K(x,z)}{\partial x}\right|_{z=x}, \\
K_z(x,x) &= \left.\frac{\partial K(x,z)}{\partial z}\right|_{z=x}, \\
\frac{\mathrm{d}K(x,x)}{\mathrm{d}x} &= \left.\frac{\partial K(x,z)}{\partial x}\right|_{z=x} + \left.\frac{\partial K(x,z)}{\partial z}\right|_{z=x}.
\end{aligned}$$

We then integrate Equation (10.77) by parts twice to obtain

$$\begin{aligned}
\psi_+ &= \mathrm{e}^{\mathrm{i}kx} + \frac{1}{\mathrm{i}k}K(x,z)\mathrm{e}^{\mathrm{i}kz}\Big|_x^\infty - \frac{1}{\mathrm{i}k}\int_x^\infty K_z(x,z)\mathrm{e}^{\mathrm{i}kz}\,\mathrm{d}z \\
&= \mathrm{e}^{\mathrm{i}kx} - \frac{1}{\mathrm{i}k}K(x,x)\mathrm{e}^{\mathrm{i}kx} - \frac{1}{(\mathrm{i}k)^2}K_z(x,z)\mathrm{e}^{\mathrm{i}kz}\Big|_x^\infty \\
&\quad + \frac{1}{(\mathrm{i}k)^2}\int_x^\infty K_{zz}(x,z)\mathrm{e}^{\mathrm{i}kz}\,\mathrm{d}z \\
&= \mathrm{e}^{\mathrm{i}kx}\left(1 + \frac{\mathrm{i}K(x,x)}{k} - \frac{K_z(x,x)}{k^2}\right) - \frac{1}{k^2}\int_x^\infty K_{zz}(x,z)\mathrm{e}^{\mathrm{i}kz}\,\mathrm{d}z\,.
\end{aligned}$$

Then Equation (10.76) becomes

$$\mathrm{e}^{\mathrm{i}kx}\left(-k^2 - \frac{\mathrm{d}K}{\mathrm{d}x} - K_x - \mathrm{i}kK + k^2 - K_z + \mathrm{i}kK - u\right)$$

$$+\int_x^\infty (K_{xx} - K_{zz} - uK)\mathrm{e}^{ikz}\,\mathrm{d}z = 0\,,$$

or

$$-\mathrm{e}^{ikx}\left(u + 2\frac{\mathrm{d}K}{\mathrm{d}x}\right) + \int_x^\infty (K_{xx} - K_{zz} - uK)\mathrm{e}^{ikz}\,\mathrm{d}z = 0\,,$$

since $K_x + K_z = \mathrm{d}K/\mathrm{d}x$. Hence if we have both

$$u = -2\frac{\mathrm{d}K}{\mathrm{d}x} = -2[K_x(x,x) + K_z(x,x)] \quad \text{and} \tag{10.78}$$

$$0 = K_{xx} - K_{zz} - uK(x,x)\,, \tag{10.79}$$

then we have a solution. These, along with the condition

$$K(x,z), K_z(x,z) \to 0 \quad \text{as} \quad z \to +\infty\,,$$

defines the problem for $K(x,z)$.

Whenever we are given $u(x)$ and wish to find $K(x,z)$, this is a **scattering problem**. When we are given $K(x,z)$ and wish to find $u(x)$, this is an **inverse scattering problem**.

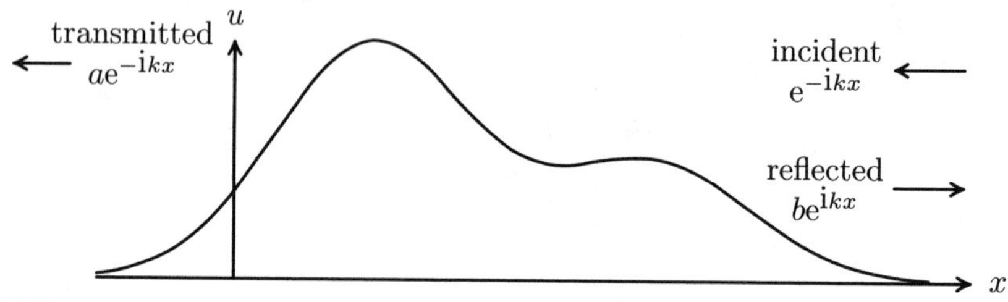

FIGURE 10.3
Scattering diagram for a wave incident from the right on a potential.

Our next step is to construct the function $\hat{\psi}(x,k)$ whose asymptotic behavior is

$$\hat{\psi}(x,k) \to \begin{cases} \mathrm{e}^{-ikx} + b(k)\mathrm{e}^{ikx} & \text{as } x \to \infty\,, \\ a(k)\mathrm{e}^{-ikx} & \text{as } x \to -\infty \end{cases} \tag{10.80}$$

which represents an incoming wave from $+\infty$ that is partially reflected and partially transmitted such that $b(k)$ is the amplitude of the reflected wave and $a(k)$ is the amplitude of the transmitted wave as illustrated in Figure 10.3. The transmission and reflection coefficients are related by a conservation law. We may find this law through the use of the Wronskian, which is defined as

$$W(\psi_1,\psi_2) \equiv \psi_1\psi_2' - \psi_2\psi_1'\,, \tag{10.81}$$

where ψ_1 and ψ_2 are both solutions of the same differential equation of the general form

$$\psi_{xx} + f(x)\psi = 0\,,$$

where in our case, $f(x) = \lambda - u(x)$. Taking the derivative of the Wronskian, we find

$$\begin{aligned} \frac{\mathrm{d}}{\mathrm{d}x}W &= \psi_1'\psi_2' + \psi_1\psi_2'' - \psi_2'\psi_1' - \psi_2\psi_1'' \\ &= -\psi_1 f(x)\psi_2 + \psi_2 f(x)\psi_1 \\ &= 0 \end{aligned} \tag{10.82}$$

so the Wronskian is constant. If we evaluate the Wronskian at $\pm\infty$ by using Equation (10.80), we find

$$W(\hat{\psi}, \hat{\psi}^*) = 2\mathrm{i}kaa^* = 2\mathrm{i}k(1 - bb^*)\,,$$

which is equivalent to

$$|a|^2 + |b|^2 = 1\,, \tag{10.83}$$

which is effectively a conservation of probability law stating that whatever is not reflected is transmitted.

The first step in discovering the properties of $\hat{\psi}(x,k)$ is to construct it from the function ψ_+ such that

$$\begin{aligned} \hat{\psi}(x,k) &\equiv \psi_+^* + b(k)\psi_+\,, \\ &= \mathrm{e}^{-\mathrm{i}kx} + \int_x^\infty K(x,z)\mathrm{e}^{-\mathrm{i}kz}\,\mathrm{d}z + b(k)\mathrm{e}^{\mathrm{i}kx} \\ &\quad + b(k)\int_x^\infty K(x,z)\mathrm{e}^{\mathrm{i}kz}\,\mathrm{d}z\,, \end{aligned} \tag{10.84}$$

where we have assumed $K^*(x,z) = K(x,z)$. With this choice, as $x \to \infty$

$$\hat{\psi}(x,k) \to \mathrm{e}^{-\mathrm{i}kx} + b(k)\mathrm{e}^{\mathrm{i}kx}\,,$$

so it has the proper form as $x \to +\infty$ and as $x \to -\infty$, we have

$$\hat{\psi}(x,k) \to \mathrm{e}^{-\mathrm{i}kx} + b(k)\mathrm{e}^{\mathrm{i}kx} + \int_{-\infty}^\infty K(x,z)\mathrm{e}^{-\mathrm{i}kz}\,\mathrm{d}z + b(k)\int_{-\infty}^\infty K(x,z)\mathrm{e}^{\mathrm{i}kz}\,\mathrm{d}z\,, \tag{10.85}$$

but since $K(x,z) = 0$ for all $z < x$, this last result is true for *any* x (not only in the limit). We rearrange the terms in Equation (10.85) so that

$$\int_{-\infty}^\infty K(x,z)\mathrm{e}^{-\mathrm{i}kz}\,\mathrm{d}z = \hat{\psi}(x,k) - \mathrm{e}^{-\mathrm{i}kx} - b(k)\mathrm{e}^{\mathrm{i}kx} - b(k)\int_{-\infty}^\infty K(x,z)\mathrm{e}^{\mathrm{i}kz}\,\mathrm{d}z\,. \tag{10.86}$$

Taking the inverse Fourier transform of Equation (10.86), we obtain

$$\begin{aligned} K(x,z) = \frac{1}{2\pi}\int_{-\infty}^\infty \Big[&\hat{\psi}(x,k) - \mathrm{e}^{-\mathrm{i}kx} - b(k)\mathrm{e}^{\mathrm{i}kx} \\ &- b(k)\int_{-\infty}^\infty K(x,y)\mathrm{e}^{\mathrm{i}ky}\,\mathrm{d}y\Big]\mathrm{e}^{\mathrm{i}kz}\,\mathrm{d}k\,. \end{aligned} \tag{10.87}$$

Note that we changed the dummy variable z in Equation (10.86) to y and used the variable z in the inverse transform.

Now we introduce the inverse Fourier transform of $b(k)$ through the function $F(x)$, where

$$F(x) \equiv \frac{1}{2\pi}\int_{-\infty}^{\infty} b(k)\mathrm{e}^{\mathrm{i}kx}\,\mathrm{d}k\,,$$

so that Equation (10.87) becomes

$$\begin{aligned} K(x,z) &= \frac{1}{2\pi}\int_{-\infty}^{\infty}\left[\hat{\psi}(x,k) - \mathrm{e}^{-\mathrm{i}kx}\right]\mathrm{e}^{\mathrm{i}kz}\,\mathrm{d}k - F(x+z) \\ &\quad - \int_{-\infty}^{\infty} K(x,y)F(y+z)\,\mathrm{d}y\,. \end{aligned} \tag{10.88}$$

The remaining integral we will evaluate by doing a contour integral, closing the path in the upper half k-plane at radius R as $R \to \infty$. For the continuous spectrum, $\hat{\psi}$ has no poles in the upper half plane, so there are no residues for this case. Now, in general, as $|k| \to \infty$, $|a(k)| \to 1$. We can understand this result, since for any bounded $u(x)$, as $|k| \to \infty$, $E \to \infty$ and the energy is so large that the potential has no effect on the particle. This is also apparent from the result of Equation (2.27) where $|F/A|(E) = |a(k)|$. This assures us that $b(k) \to 0$, since we found $|b|^2 = 1 - |a|^2$ from Equation (10.83). Then from Equation (10.84), we have that

$$\mathrm{e}^{\mathrm{i}kx}\hat{\psi} \to 1$$

as $|k| \to \infty$, so the integrand vanishes along the semicircular path. Since there are no residues, and the semicircular path has no contribution, the integral along the original path vanishes, so we have

$$K(x,z) + F(x+z) + \int_{-\infty}^{\infty} K(x,y)F(y+z)\,\mathrm{d}y = 0\,, \tag{10.89}$$

which is the **Marchenko Equation** or the **Gel'fand–Levitan Equation**. This is an inhomogeneous integral equation of the first kind for the quantity $K(x,z)$, and indicates that a knowledge of $b(k)$ from the scattering data gives us $F(x)$, and the solution of the linear integral equation gives us $K(x,z)$, which along with Equation (10.78) gives us $u(x)$.

10.2.2 Discrete Spectrum

Before the problem is complete, we must consider the discrete spectrum, where $\hat{\psi}$ may have poles in the upper half of the k-plane. In this case, we must reexamine the first integral in Equation (10.88) and evaluate

$$\int_{-\infty}^{\infty}(\hat{\psi} - \mathrm{e}^{-\mathrm{i}kx})\mathrm{e}^{\mathrm{i}kz}\,\mathrm{d}k = 2\pi\mathrm{i}\sum_{n=1}^{N} R_n\,, \tag{10.90}$$

where R_n is the residue at the nth pole, or equivalently, the residue of $\hat{\psi}\mathrm{e}^{\mathrm{i}kz}$ since $\mathrm{e}^{\mathrm{i}k(z-x)}$ has no pole. Letting $\lambda \equiv -\kappa^2$ when $\lambda < 0$ for the discrete spectrum, the poles occur where $k = \mathrm{i}\kappa_n$.

We first examine the other Jost function

$$\psi_-(x,k) \equiv \mathrm{e}^{-\mathrm{i}kx} + \int_{-\infty}^{x} L(x,z)\mathrm{e}^{-\mathrm{i}kz}\,\mathrm{d}z\,, \tag{10.91}$$

for some L (which is similar to K except that $L(x,z) = 0$ for $z > x$) so that as $x \to -\infty$, $\psi_- \to \mathrm{e}^{-\mathrm{i}kx}$. With this definition, it follows that

$$\hat{\psi} = a(k)\psi_- = \psi_+^* + b(k)\psi_+\,,$$

so that

$$\psi_- = a^{-1}\psi_+^* + ba^{-1}\psi_+\,. \tag{10.92}$$

Now since

$$\begin{aligned}
\psi_+(x,\mathrm{i}\kappa_n) &\sim \mathrm{e}^{-\kappa_n x} \quad \text{as} \quad x \to +\infty\,,\\
\psi_-(x,\mathrm{i}\kappa_n) &\sim \mathrm{e}^{\kappa_n x} \quad \text{as} \quad x \to -\infty\,,
\end{aligned}$$

for the discrete eigenfunctions and since both ψ_+ and ψ_- are solutions of Equation (10.76) *everywhere*, they must each have the same exponential character as $x \to \pm\infty$, so they differ at most by a constant. Thus we may write

$$\psi_n(x) = c_n\psi_+(x,\mathrm{i}\kappa_n) = d_n\psi_-(x,\mathrm{i}\kappa_n)\,, \tag{10.93}$$

where c_n and d_n are normalization constants.

The Wronskian for the two Jost functions is

$$W(\psi_-,\psi_+) = \psi_-\psi_+' - \psi_+\psi_-'\,,$$

and we find again that

$$\begin{aligned}
\frac{\mathrm{d}}{\mathrm{d}x}W &= \psi_-'\psi_+' + \psi_-\psi_+'' - \psi_+'\psi_-' - \psi_+\psi_-''\\
&= \psi_-(u-\lambda)\psi_+ - \psi_+(u-\lambda)\psi_-\\
&= 0
\end{aligned}$$

where we have used the fact that both ψ_+ and ψ_- are solutions of the Schrödinger equation. It follows that the Wronskian is a constant (with respect to x) for any two solutions which have the same value of $\lambda = k^2 = -\kappa_n^2$.

Using Equation (10.92), we may write the Wronskian as

$$\begin{aligned}
W(\psi_-,\psi_+) &= (a^{-1}\psi_+^* + ba^{-1}\psi_+)\psi_+' - \psi_+(a^{-1}\psi_+^{*\prime} + ba^{-1}\psi_+')\\
&= a^{-1}(\psi_+^*\psi_+' - \psi_+\psi_+^{*\prime})\\
&= a^{-1}W(\psi_+^*,\psi_+)\,.
\end{aligned}$$

Then, since as $x \to \infty$,

$$\psi_+ \sim \mathrm{e}^{\mathrm{i}kx}\,,\ \psi_+^* \sim \mathrm{e}^{-\mathrm{i}kx}$$
$$\psi'_+ \sim \mathrm{i}k\mathrm{e}^{\mathrm{i}kx}\,,\ \psi_+^* \sim -\mathrm{i}k\mathrm{e}^{-\mathrm{i}kx}\,,$$

we have the result that $W(\psi_+^*, \psi_+) = 2\mathrm{i}k$, and

$$W(\psi_-, \psi_+) = 2\mathrm{i}ka^{-1}\,. \tag{10.94}$$

Furthermore, since $a(k) = 2\mathrm{i}k/W(\psi_-, \psi_+)$, and since when $k \to \mathrm{i}\kappa_n$, ψ_+ is proportional to ψ_- from Equation (10.93), $W_n = 0$, so $a(k)$ has poles at each value of $k = \mathrm{i}\kappa_n$.

Now we differentiate the original Schrödinger equation, Equation (10.76), with respect to k, and obtain

$$\psi_{xxk} + 2k\psi + (k^2 - u)\psi_k = 0\,,$$

and multiply by ψ to find

$$\psi\psi_{xxk} + 2k\psi^2 + (k^2 - u)\psi\psi_k = 0\,. \tag{10.95}$$

Then multiply Equation (10.76) by ψ_k and subtract from Equation (10.95) to get

$$\psi\psi_{xxk} - \psi_k\psi_{xx} + 2k\psi^2 = 0\,,$$

but

$$\frac{\mathrm{d}}{\mathrm{d}x}W(\psi_k, \psi) = \frac{\mathrm{d}}{\mathrm{d}x}(\psi_k\psi_x - \psi_{kx}\psi) = \psi_k\psi_{xx} - \psi_{kxx}\psi\,,$$

so

$$\frac{\mathrm{d}}{\mathrm{d}x}W(\psi_k, \psi) = 2k\psi^2\,.$$

Integrating this, we have

$$W(\psi_k, \psi) = 2k\int^x \psi^2\,\mathrm{d}x\,. \tag{10.96}$$

If we evaluate Equation (10.96) at $x = \pm\infty$ with $k = \mathrm{i}\kappa_n$ and $\psi = \psi_+$, then

$$W(\psi_k, \psi)|_{-\infty}^{\infty} = 2\mathrm{i}\kappa_n \int_{-\infty}^{\infty} \psi_+^2\,\mathrm{d}x = \frac{2\mathrm{i}\kappa_n}{c_n^2}\,, \tag{10.97}$$

where we have defined the discrete eigenfunctions to be $\psi_n(x) = c_n\psi_+(x, \mathrm{i}\kappa_n)$ and $\int_{-\infty}^{\infty} \psi_n^2\,\mathrm{d}x = 1$. We now differentiate Equation (10.94) with respect to k to get

$$W(\psi_{-k}, \psi_+) + W(\psi_-, \psi_{+k}) = \frac{2\mathrm{i}}{a} - \frac{2\mathrm{i}ka'}{a^2}\,. \tag{10.98}$$

Now for $k = \mathrm{i}\kappa_n$, $a^{-1} \to 0$ since there are poles in a at these values of k, and we have for these same discrete values that $\psi_+ = (d_n/c_n)\psi_-$, and

$\psi_- = (c_n/d_n)\psi_+$. Then using the notation that $\lim_{x\to\pm\infty} W = W^{\pm}$ and W_n for W evaluated at $k = \mathrm{i}\kappa_n$, we may evaluate Equation (10.98) as $x \to -\infty$ to find

$$W_n^-(\psi_{-k}, \tfrac{d_n}{c_n}\psi_-) + W_n^-(\tfrac{c_n}{d_n}\psi_+, \psi_{+k}) = \left(\frac{2\kappa_n a'}{a^2}\right)_n . \tag{10.99}$$

Now if we write Equation (10.97) for ψ_+ as

$$W_n^+(\psi_{+k}, \psi_+) - W_n^-(\psi_{+k}, \psi_+) = \frac{2\mathrm{i}\kappa_n}{c_n^2}, \tag{10.100}$$

and multiply Equation (10.99) by d_n/c_n, so that it becomes

$$\frac{d_n^2}{c_n^2} W_n^-(\psi_{-k}, \psi_-) - W_n^-(\psi_{+k}, \psi_+) = \frac{2\kappa_n d_n}{c_n}\left(\frac{a'}{a^2}\right)_n ,$$

and subtract this from Equation (10.100), we find

$$W_n^+(\psi_{+k}, \psi_+) - \frac{d_n^2}{c_n^2} W_n^-(\psi_{-k}, \psi_-) = \frac{2\mathrm{i}\kappa_n}{c_n^2} - \frac{2\kappa_n d_n}{c_n}\left(\frac{a'}{a^2}\right)_n , \tag{10.101}$$

but since $\psi_+ \to \mathrm{e}^{\mathrm{i}kx}$ as $x \to \infty$ and $\psi_- \to \mathrm{e}^{-\mathrm{i}kx}$ as $x \to -\infty$, both Wronskians vanish, so

$$\left(\frac{a'}{a^2}\right)_n = \frac{2\mathrm{i}\kappa_n}{c_n^2}\frac{c_n}{2\kappa_n d_n} = \frac{\mathrm{i}}{c_n d_n}. \tag{10.102}$$

10.2.2.1 Evaluating the Residues

We may now integrate Equation (10.102) to establish that $a(\mathrm{i}\kappa_n)$ has only simple poles. We first write Equation (10.102) as

$$\frac{\mathrm{d}a}{\mathrm{d}k} = \frac{\mathrm{i}a^2}{c_n d_n},$$

so that

$$\frac{\mathrm{d}a}{a^2} = \frac{\mathrm{i}\mathrm{d}k}{c_n d_n},$$

which integrates to

$$-\frac{1}{a} = \frac{\mathrm{i}k}{c_n d_n} + A, \tag{10.103}$$

where A is a constant of integration. But we know that $1/a = 0$ for $k = \mathrm{i}\kappa_n$, so we have

$$0 = \frac{-\kappa_n}{c_n d_n} + A,$$

so that $A = \kappa_n/c_n d_n$ and solving for a, we find

$$a = \frac{\mathrm{i}c_n d_n}{k - \mathrm{i}\kappa_n}. \tag{10.104}$$

Clearly then, the residue of a at $k = \mathrm{i}\kappa_n$ is $\mathrm{i}c_n d_n$.

Now returning to the original problem, we write

$$\hat{\psi}\mathrm{e}^{\mathrm{i}kz} = a(k)\psi_-(x,k)\mathrm{e}^{\mathrm{i}kz} = a(k)\frac{c_n}{d_n}\psi_+\mathrm{e}^{\mathrm{i}kz}\,,$$

and only $a(k)$ has any poles. The residue at $k = \mathrm{i}\kappa_n$ is

$$R_n = \frac{c_n}{d_n}\psi_+(x, \mathrm{i}\kappa_n)\mathrm{e}^{-\kappa_n z} \times [\text{residue of } a(k)]\,, \tag{10.105}$$

and the residues of $a(k)$ are $\mathrm{i}c_n d_n$, so

$$R_n = \mathrm{i}c_n^2\psi_+(x, \mathrm{i}\kappa_n)\mathrm{e}^{-\kappa_n z}\,, \tag{10.106}$$

so we don't need d_n. With this result, we have

$$2\pi\mathrm{i}\sum_{n=1}^{N} R_n = -2\pi\sum_{n=1}^{N} c_n^2\psi_+(x, \mathrm{i}\kappa_n)\mathrm{e}^{-\kappa_n z}\,,$$

so that using this in Equation (10.90) leads to

$$\frac{1}{2\pi}\int_{-\infty}^{\infty}(\hat{\psi} - \mathrm{e}^{-\mathrm{i}kx})\mathrm{e}^{\mathrm{i}kz}\,\mathrm{d}k = -\sum_{n=1}^{N} c_n^2\mathrm{e}^{-\kappa_n z}\left[\mathrm{e}^{-\kappa_n x} + \int_x^{\infty} K(x,y)\mathrm{e}^{-\kappa_n y}\,\mathrm{d}y\right]. \tag{10.107}$$

Using this in Equation (10.88) leads again to the Marchenko equation if

$$F(X) = \sum_{n=1}^{N} c_n^2\mathrm{e}^{-\kappa_n X} + \frac{1}{2\pi}\int_{-\infty}^{\infty} b(k)\mathrm{e}^{\mathrm{i}kX}\,\mathrm{d}k\,. \tag{10.108}$$

If there is no discrete spectrum, then $c_n = 0$ and we return to the previous result. We conclude by noting that the sum over the discrete spectrum and the integral over the continuous spectrum of scattering data is sufficient to define the scattering potential through the solution of the Marchenko equation.

10.2.3 The Solution of the Marchenko Equation

We may write the Marchenko equation either as Equation (10.89) or as

$$K(x,z) + F(x+z) + \int_x^{\infty} K(x,y)F(y+z)\,\mathrm{d}y = 0\,, \tag{10.109}$$

since we have taken $K(x,z) = 0$ for $x < z$. Either way, we have a Fredholm integral equation, for which many theorems are available. One direct method is iteration, where we begin by taking

$$K_1(x,z) = \begin{cases} -F(x+z)\,, & z > x\,, \\ 0 & z < x\,, \end{cases}$$

and then the successive iterations are of the form

$$K_2(x,z) = -F(x+z) - \int_x^\infty K_1(x,y)F(y+z)\,\mathrm{d}y\,,$$

$$\vdots \quad = \quad \vdots \quad - \quad \vdots$$

$$K_{n+1}(x,z) = -F(x+z) - \int_x^\infty K_n(x,y)F(y+z)\,\mathrm{d}y\,.$$

For a bounded $F(X)$, this converges so that $K_n(x,z) \to K(x,z)$ as $n \to \infty$. This is called the **Neumann series**. Occasionally this leads to an analytic solution, but usually this is useful only for numerical techniques, but always represents a formal solution.

An important special case where the Neumann series does lead to a tractable solution is the case where the kernel, $F(x+z)$, is **separable**, so that

$$F(x+z) = \sum_{n=1}^{N} X_n(x)Z_n(z)\,, \tag{10.110}$$

where N is finite (if N is not finite, the method may still be correct, but just not useful). The Marchenko equation then becomes

$$K(x,z) + \sum_{n=1}^{N} X_n(x)Z_n(z) + \sum_{n=1}^{N} Z_n(z)\int_x^\infty K(x,y)X_n(y)\,\mathrm{d}y = 0\,, \tag{10.111}$$

and since the integrals are each some function of x multiplied by $Z_n(z)$, we conclude that the solution must take the form

$$K(x,z) = \sum_{n=1}^{N} L_n(x)Z_n(z)\,. \tag{10.112}$$

Putting this back into the Marchenko equation leads to a system of N *algebraic* equations, each of the form

$$L_n(x) + X_n(x) + \sum_{n=1}^{N} L_m(x)\int_x^\infty Z_m(y)X_n(y)\,\mathrm{d}y = 0\,, \tag{10.113}$$

where the solutions are $L_n(x)$ for $n = 1, 2, \ldots, N$, and x is just a parameter in the problem.

Example 10.5

One pole in the reflection coefficient. For our first example, we choose the reflection coefficient to be

$$b(k) = -\frac{\beta}{\beta + \mathrm{i}k} = \frac{\mathrm{i}\beta}{k - \mathrm{i}\beta}\,, \qquad \beta > 0\,,$$

and $\psi(x) \sim \sqrt{\beta}\mathrm{e}^{-\beta x}$ as $x \to \infty$. From this, we immediately have that there is only one bound state with eigenvalue $\kappa_1 = \beta$, and that $c_1 = \sqrt{\beta}$. The definition of $F(x)$, from Equation (10.108), leads to

$$F(X) = \beta\mathrm{e}^{-\beta X} + \frac{\mathrm{i}\beta}{2\pi}\int_{-\infty}^{\infty}\frac{\mathrm{e}^{\mathrm{i}kX}}{k-\mathrm{i}\beta}\,\mathrm{d}k\,.$$

For $X > 0$, we evaluate the integral by closing the contour above, and there is no contribution from the semicircular arc since $\mathrm{i}kX \sim -[\mathrm{Im}(k)X] \to -\infty$, and the only residue is at $k = \mathrm{i}\beta$ which is in the upper half k-plane. Thus, we have

$$\int_{-\infty}^{\infty}\frac{\mathrm{e}^{\mathrm{i}kX}}{k-\mathrm{i}\beta}\,\mathrm{d}k = 2\pi\mathrm{i}\mathrm{e}^{-\beta X}\,.$$

For $X < 0$, the contour must be closed below, but there is no pole below, so the integral vanishes and we have

$$F(X) = \beta\mathrm{e}^{-\beta X}H(-X)\,,$$

where $H(x)$ is the Heaviside step function such that $H(x) = 1$ for $x > 0$ and $H(x) = 0$ for $x < 0$. Inserting this result into the Marchenko equation, we have $K(x,z) = 0$ for $x + z > 0$ since $F(x+z) = 0$ for this case, and there is no finite range for the integral where $F(y+z) \neq 0$. When $x + z < 0$, the Marchenko equation is

$$K(x,z) + \beta\mathrm{e}^{-\beta(x+z)} + \beta\int_{x}^{-z} K(x,y)\mathrm{e}^{-\beta(y+z)}\,\mathrm{d}y = 0\,. \tag{10.114}$$

We cannot use the fact that F is separable, since the integral has one limit depending on z, and the assumption that $K(x,z) = f(x)\mathrm{e}^{-\beta z}$ leads to $f(x)$ depending on z, which is contradictory. We can solve this equation by differentiating Equation (10.114) to obtain

$$K_z(x,z) - \beta^2\mathrm{e}^{-\beta(x+z)} - \beta K(x,-z) - \beta^2\int_{x}^{-z} K(x,y)\mathrm{e}^{-\beta(y+z)}\,\mathrm{d}y = 0\,,$$

and replacing the integral by its value from Equation (10.114), we find

$$K_z(x,z) - \beta^2\mathrm{e}^{-\beta(x+z)} - \beta K(x,-z) - \beta[-K(x,-z) - \beta^2\mathrm{e}^{-\beta(x+z)}] = 0\,,$$

which leads to $K_z(x,z) = 0$ so K is a constant. Letting $K(x,z) = A$, Equation (10.114) becomes

$$A + \beta\mathrm{e}^{-\beta(x+z)} + \beta A\int_{x}^{-z}\mathrm{e}^{-\beta(y+z)}\,\mathrm{d}y = (\beta + A)\mathrm{e}^{-\beta(x+z)} = 0\,.$$

Hence, $A = -\beta$. Combining these results, we have $K(x,z) = -\beta H(-x-z)$. Then we have $K(x,x) = -\beta H(-2x) = -\beta(-x)$. The scattering potential is then

$$u(x) = -2\frac{\mathrm{d}}{\mathrm{d}x}K(x,x) = 2\beta\frac{\mathrm{d}}{\mathrm{d}x}H(-x) = -2\beta\delta(x)\,,$$

since the derivative of the step function is a δ-function. □

Example 10.6
Zero reflection coefficient. For this case, we assume $b(k) = 0$ for all k, but assume there are two discrete eigenvalues, $\kappa_1 \neq \kappa_2$, so that

$$\psi_1 \sim c_1 e^{-\kappa_1 x}, \qquad \psi_2 \sim c_2 e^{-\kappa_2 x}.$$

Then we have, from Equation (10.108) that

$$F(X) = c_1^2 e^{-\kappa_1 X} + c_2^2 e^{-\kappa_2 X}.$$

Using this in the Marchenko equation leads to

$$K(x,z) + c_1^2 e^{-\kappa_1(x+z)} + c_2^2 e^{-\kappa_2(x+z)} + \int_x^\infty K(x,y)[c_1^2 e^{-\kappa_1(y+z)} + c_2^2 e^{-\kappa_2(y+z)}]\,dy = 0. \quad (10.115)$$

For this case, $F(x+z)$ is separable, so we may write

$$K(x,z) = L_1(x)e^{-\kappa_1 z} + L_2(x)e^{-\kappa_2 z},$$

and hence Equation (10.115) may be written as the two separate equations

$$L_1 + c_1^2 e^{-\kappa_1 x} + c_1^2 \left[L_1 \int_x^\infty e^{-2\kappa_1 y}\,dy + L_2 \int_x^\infty e^{-(\kappa_1+\kappa_2)y}\,dy \right] = 0,$$
$$L_2 + c_2^2 e^{-\kappa_2 x} + c_2^2 \left[L_1 \int_x^\infty e^{-(\kappa_1+\kappa_2)y}\,dy + L_2 \int_x^\infty e^{-2\kappa_2 y}\,dy \right] = 0.$$

Evaluating the integrals, each equation may be written in the form

$$L_n + c_n^2 e^{-\kappa_n x} + c_n^2 \sum_{m=1}^{2} \frac{L_m e^{-(\kappa_n+\kappa_m)x}}{\kappa_n + \kappa_m} = 0, \quad (10.116)$$

where $n = 1, 2$. Solving the two equations for two unknowns leads to

$$L_1 = -c_1^2 \frac{1 + f_2(\frac{\kappa_1-\kappa_2}{\kappa_1+\kappa_2})e^{-\kappa_1 x}}{1 + f_1 + f_2 + f_1 f_2(\frac{\kappa_1-\kappa_2}{\kappa_1+\kappa_2})^2},$$
$$L_2 = -c_2^2 \frac{1 + f_1(\frac{\kappa_2-\kappa_1}{\kappa_1+\kappa_2})e^{-\kappa_2 x}}{1 + f_1 + f_2 + f_1 f_2(\frac{\kappa_1-\kappa_2}{\kappa_1+\kappa_2})^2},$$

where $f_n = (c_n^2/2\kappa_n)e^{-2\kappa_n x}$. In principle, this solves the problem for $K(x,z)$ and then $u(x)$, but these forms are very cumbersome to manipulate.

The solution may also be written in terms of the matrix A whose elements are

$$A_{mn} = \delta_{mn} + \frac{c_m^2}{\kappa_m + \kappa_n} e^{-(\kappa_n+\kappa_m)x},$$

as

$$\mathsf{A}\boldsymbol{L} + \boldsymbol{B} = 0\,,$$

where $\boldsymbol{L}$ and $\boldsymbol{B}$ are column vectors whose elements are L_n and $B_n = c_n^2 \mathrm{e}^{-\kappa_n x}$. From this, the solution of the Marchenko equation is $K(x,x) = \tilde{\boldsymbol{E}} \cdot \boldsymbol{L}$ where $\tilde{\boldsymbol{E}}$ is the transpose of $\boldsymbol{E}$ whose elements are $E_n = \mathrm{e}^{-\kappa_n x}$. In terms of these elements, it should be noted that

$$\frac{\mathrm{d}}{\mathrm{d}x} A_{mn} = -c_m^2 \mathrm{e}^{-(\kappa_m+\kappa_n)x} = -B_m E_n\,.$$

Then it follows that (using the summation convention)

$$K = E_m L_m = -E_m A_{mn}^{-1} B_n = A_{mn}^{-1} \frac{\mathrm{d}}{\mathrm{d}x} A_{mn}\,.$$

This result may be expressed in terms of the determinant of A, denoted by $|A|$, as

$$K = \mathrm{tr}\left(\mathsf{A}^{-1}\frac{\mathrm{d}\mathsf{A}}{\mathrm{d}x}\right) = \frac{1}{|A|}\frac{\mathrm{d}|A|}{\mathrm{d}x} = \frac{\mathrm{d}}{\mathrm{d}x}\ln|A|\,.$$

The scattering potential is then

$$u(x,t) = -2\frac{\mathrm{d}^2}{\mathrm{d}x^2}\ln|A|\,. \tag{10.117}$$

The result can then be written as

$$\begin{aligned} |A| &= \left[1 + \frac{c_1^2}{2\kappa_1}\mathrm{e}^{-2\kappa_1 x}\right]\left[1 + \frac{c_2^2}{2\kappa_2}\mathrm{e}^{-2\kappa_2 x}\right] - \frac{c_1^2 c_2^2}{(\kappa_1+\kappa_2)^2}\mathrm{e}^{-2(\kappa_1+\kappa_2)x}\,,\\ &= 1 + \frac{c_1^2}{2\kappa_1}\mathrm{e}^{-2\kappa_1 x}\,, \quad \text{with} \quad c_2 = 0\,. \end{aligned}$$

For this latter case with only one eigenvalue, we find

$$\begin{aligned} K &= -\frac{c_1^2 \mathrm{e}^{-2\kappa_1 x}}{1 + \frac{c_1^2}{2\kappa_1}\mathrm{e}^{-2\kappa_1 x}}\\ u(x) &= -\frac{4\kappa_1 c_1^2 \mathrm{e}^{-2\kappa_1 x}}{[1 + \frac{c_1^2}{2\kappa_1}\mathrm{e}^{-2\kappa_1 x}]^2} \end{aligned}$$

or

$$u(x) = -2\kappa_1^2\,\mathrm{sech}^2[\kappa_1(x+x_0)]\,,$$

where $x_0 = \ln(\sqrt{2\kappa_1}/c_1)$. This solution is the one-soliton solution to the Korteweg–DeVries (KdV) equation,

$$u_t - 6uu_x + u_{xxx} = 0\,, \tag{10.118}$$

which is a nonlinear partial differential equation that is unique to the Schrödinger equation since it may be shown that if the potential evolves in time according to the KdV equation, *all of the eigenvalues remain invariant in time.* For the case with the two eigenvalues above, we have a two-soliton solution, and from the structure of the determinants, it is clear that we can generalize to n-soliton solutions. □

Problem 10.13 *Soliton solution of KdV equation.* If the solution of the KdV equation is written as

$$u(x) = -2\kappa_1^2 \operatorname{sech}^2[\kappa_1(x - vt)]\,,$$

show that this is a solution of Equation (10.118) and find the relation between v and κ_1 showing that the velocity is amplitude dependent.

11

Relativistic Quantum Mechanics and Particle Theory

11.1 Dirac Theory of the Electron*

11.1.1 The Klein–Gordon Equation

Up to this point in our study of quantum mechanics, we have ignored relativistic effects and taken spin as an extra observed property of particles with no specific model. In formulating an appropriate theory for the extension of quantum mechanics to include relativistic effects, it is logical to begin with the Hamiltonian based on our knowledge of relativity, where we have

$$\hat{\mathcal{H}} = c[(\hat{\boldsymbol{p}} - e\boldsymbol{A})^2 + m^2c^2]^{1/2} + e\phi\,, \tag{11.1}$$

where the first portion portion comes from $W^2 = p^2c^2 + m^2c^4$ using the canonical momentum $\hat{\boldsymbol{p}} = \gamma m\boldsymbol{v} + e\boldsymbol{A}$ when there is a magnetic field present represented by the vector potential $\boldsymbol{A}$, and then we have added the scalar electrostatic potential. This form for the Hamiltonian is very formidable, since it includes a square root with an operator inside ($\hat{\boldsymbol{p}} \to (\hbar/i)\nabla$). By rearranging and squaring, Equation (11.1) can be put into a more symmetric form that avoids this difficulty, so that

$$(\hat{\mathcal{H}} - e\phi)^2 - (\hat{\boldsymbol{p}} - e\boldsymbol{A})^2c^2 = m^2c^4\,. \tag{11.2}$$

One possible interpretation of this equation is that we should make the usual substitutions

$$\hat{\boldsymbol{p}} \to \frac{\hbar}{\mathrm{i}}\nabla \qquad \text{and} \qquad \hat{\mathcal{H}} \to -\frac{\hbar}{\mathrm{i}}\frac{\partial}{\partial t}\,,$$

which leads to

$$\left[\left(-\frac{\hbar}{\mathrm{i}}\frac{\partial}{\partial t} - e\phi\right)^2 - \left(\frac{\hbar}{\mathrm{i}}\nabla - e\boldsymbol{A}\right)^2 c^2\right]\psi = m^2c^4\psi\,. \tag{11.3}$$

This equation is called the **Klein–Gordon** equation, and represents one possible way to generalize the Schrödinger equation to include relativistic effects.

In fact, this equation was found by Schrödinger, but it was immediately discarded since it allows both positive and *negative* total energy. This is evident by considering a free particle whose wave function must satisfy

$$\nabla^2\psi - \frac{1}{c^2}\frac{\partial^2\psi}{\partial t^2} = \frac{m^2c^2}{\hbar^2}\psi\,. \tag{11.4}$$

Separating variables leads then to the relatively simple solution

$$\psi = A\exp[(\mathrm{i}/\hbar)(\boldsymbol{p}\cdot\boldsymbol{r} - Wt)]\,,$$

where the separation constants, p_x, p_y, p_z, and W must satisfy the relation

$$W^2 - p^2c^2 = m^2c^4\,.$$

It follows then that for every given value of momentum, the total energy is given by

$$W = \pm\sqrt{p^2c^2 + m^2c^4}\,,$$

and there is no obvious way to eliminate the negative value for the total energy.

11.1.2 The Dirac Equation

As a way to avoid this apparently untenable conclusion, Dirac sought a linear, Hermitian operator to represent Equation (11.1) directly, assuming that the square root term included a *perfect square*, so that

$$[(\hat{\boldsymbol{p}} - e\boldsymbol{A})^2 + m^2c^2]^{1/2} = \hat{\boldsymbol{\alpha}}\cdot(\hat{\boldsymbol{p}} - e\boldsymbol{A}) + \hat{\beta}mc\,. \tag{11.5}$$

It is rather obvious that if $\hat{\boldsymbol{\alpha}}$ and $\hat{\beta}$ are an ordinary vector and a scalar, this is impossible, since it requires the apparently contradictory relationships

$$\alpha_x^2 = \alpha_y^2 = \alpha_z^2 = \beta^2 = 1\,, \tag{11.6}$$

$$\begin{aligned}
\alpha_x\alpha_y + \alpha_y\alpha_x &= 0\,,\\
\alpha_x\alpha_z + \alpha_z\alpha_x &= 0\,,\\
\alpha_y\alpha_z + \alpha_z\alpha_y &= 0\,,\\
\alpha_x\beta + \beta\alpha_x &= 0\,,\\
\alpha_y\beta + \beta\alpha_y &= 0\,,\\
\alpha_z\beta + \beta\alpha_z &= 0\,.
\end{aligned} \tag{11.7}$$

On the other hand, if these are not ordinary numbers, but **anticommuting operators**, then it is entirely possible that this linearization of Equation (11.5) can be achieved. It is customary to represent these operators in terms of the 4×4 matrices:

$$\hat{\alpha}_x = \begin{pmatrix} 0 & 0 & 0 & 1 \\ 0 & 0 & 1 & 0 \\ 0 & 1 & 0 & 0 \\ 1 & 0 & 0 & 0 \end{pmatrix} = \begin{pmatrix} \hat{0} & \hat{\sigma}_x \\ \hat{\sigma}_x & \hat{0} \end{pmatrix}, \tag{11.8}$$

$$\hat{\alpha}_y = \begin{pmatrix} 0 & 0 & 0 & -\mathrm{i} \\ 0 & 0 & \mathrm{i} & 0 \\ 0 & -\mathrm{i} & 0 & 0 \\ \mathrm{i} & 0 & 0 & 0 \end{pmatrix} = \begin{pmatrix} \hat{0} & \hat{\sigma}_y \\ \hat{\sigma}_y & \hat{0} \end{pmatrix}, \tag{11.9}$$

$$\hat{\alpha}_z = \begin{pmatrix} 0 & 0 & 1 & 0 \\ 0 & 0 & 0 & -1 \\ 1 & 0 & 0 & 0 \\ 0 & -1 & 0 & 0 \end{pmatrix} = \begin{pmatrix} \hat{0} & \hat{\sigma}_z \\ \hat{\sigma}_z & \hat{0} \end{pmatrix}, \tag{11.10}$$

$$\hat{\beta} = \begin{pmatrix} 1 & 0 & 0 & 0 \\ 0 & 1 & 0 & 0 \\ 0 & 0 & -1 & 0 \\ 0 & 0 & 0 & -1 \end{pmatrix} = \begin{pmatrix} \hat{1} & \hat{0} \\ \hat{0} & -\hat{1} \end{pmatrix}, \tag{11.11}$$

where $\hat{\sigma}_x$, $\hat{\sigma}_y$, and $\hat{\sigma}_z$ are the Pauli spin matrices of Equations (4.14)–(4.16) and $\hat{1}$ and $\hat{0}$ are the unit and null 2×2 matrices.

Since this theory is based on 4×4 matrices, the wave function is taken to be a column vector with 4 components, so that

$$\psi = \begin{pmatrix} \psi_1 \\ \psi_2 \\ \psi_3 \\ \psi_4 \end{pmatrix}. \tag{11.12}$$

The adjoint of ψ is then a row matrix, so that

$$\psi^\dagger = (\psi_1^* \; \psi_2^* \; \psi_3^* \; \psi_4^*), \tag{11.13}$$

and the expected value of a general operator $\hat{F}$ is

$$\langle F \rangle = \int \psi^\dagger \hat{F} \psi \, \mathrm{d}V, \tag{11.14}$$

where ψ is assumed normalized such that

$$\int \psi^\dagger \psi \, \mathrm{d}V = \int (\psi_1^*\psi_1 + \psi_2^*\psi_2 + \psi_3^*\psi_3 + \psi_4^*\psi_4) \, \mathrm{d}V = 1.$$

Using these basic elements for the theory, the **Dirac equation** is given by

$$\hat{\mathcal{H}}\psi = \left[c\hat{\boldsymbol{\alpha}} \cdot \left(\frac{\hbar}{\mathrm{i}} \nabla - e\boldsymbol{A} \right) + \hat{\beta} mc^2 + e\phi \right] \psi = -\frac{\hbar}{\mathrm{i}} \frac{\partial \psi}{\partial t}. \tag{11.15}$$

This set of four linear, first-order partial differential equations for the four wave functions has been solved exactly for a number of problems, and the hydrogen atom in particular. This solution is beyond the scope of this course, but we can examine the free-particle solutions. The free-particle wave functions are of the form

$$\psi = \mathrm{e}^{\mathrm{i}(\boldsymbol{p}\cdot\boldsymbol{r}-Wt)/\hbar}\begin{pmatrix} C_1 \\ C_2 \\ C_3 \\ C_4 \end{pmatrix}, \tag{11.16}$$

where at least one of the four amplitudes must be nonzero. If we substitute this into Equation (11.15) and factor out the common exponential, the four equations may be written as

$$\begin{pmatrix} -W+mc^2 & 0 & cp_z & c(p_x-\mathrm{i}p_y) \\ 0 & -W+mc^2 & c(p_x+\mathrm{i}p_y) & -cp_z \\ cp_z & c(p_x-\mathrm{i}p_y) & -W-mc^2 & 0 \\ c(p_x+\mathrm{i}p_y) & -cp_z & 0 & -W-mc^2 \end{pmatrix}\begin{pmatrix} C_1 \\ C \\ C_3 \\ C_4 \end{pmatrix} = 0\,. \tag{11.17}$$

For a nontrivial solution, the determinant of coefficients must vanish, which leads to $(W^2 - m^2c^4 - p^2c^2)^2 = 0$, or

$$W = \pm(p^2c^2 + m^2c^4)^{1/2}\,, \tag{11.18}$$

which is the same result obtained from the Klein–Gordon equation. This implies that the *permissible energy values for a free particle range from* $+mc^2$ *to* $+\infty$ *and from* $-mc^2$ *to* $-\infty$. The first, positive, range is expected, since it indicates the total energy varies anywhere from the rest energy upward. The second result, however seems to defy logic, since it describes negative total energy.

Dirac interpreted these negative energy states as lying in a *sea of negative energy states*, where below $-mc^2$, there exists a vast, continuous sea of energy states occupied by particles that are analogous to holes so that they move as if they have a positive charge. Called positrons, these particles at rest lay mc^2 *below* zero energy, and in order to see one, one would have to supply at least that much energy to bring an electron in the negative sea up to a positive energy, where it would appear as an electron, leaving behind an empty energy state that would behave as an antielectron. This was an amazing prediction at the time, but electron-positron pairs were soon seen in cosmic ray experiments (Anderson, 1932).

One might say that if the motivation for the linearization of the Hamiltonian by the introduction of the 4×4 matrices was only to resolve the negative energy states, then it was a complete failure, since it did not eliminate these states. On the basis of the discovery of the positron alone, one would then have no real reason to prefer the Dirac equation over the Klein–Gordon equation. The actual solution of the Dirac equation for the hydrogen atom and for its insights into the characteristics of spin, however, make it an obvious choice over the Klein–Gordon equation.

Problem 11.1 Show that squaring Equation (11.5) and comparing the coefficients of like terms leads to Equations (11.6) and (11.7).

Problem 11.2 Show that the matrices of Equations (11.8) through (11.11) satisfy Equations (11.6) and (11.7), where in the latter, 1 is taken to mean the unit 4×4 matrix.

Problem 11.3 Expand Equation (11.15) by writing out the four separate equations. Arrange each equation to be of the form $\sum_{\ell=1}^{4} \hat{O}_{k\ell}\psi_k = 0$ by finding the $\hat{O}_{k\ell}$ for $k, \ell = 1, 2, 3, 4$.

Problem 11.4 Verify that the vanishing of the determinant of coefficients of Equation (11.17) leads to Equation (11.18).

Problem 11.5 Use the procedures of the Correspondence Principle in Section 1.6.3 to show that the operator that represents the x-component of velocity is $c\hat{\alpha}_x$.

11.1.3 Intrinsic Angular Momentum in the Dirac Theory

The most remarkable result of the Dirac theory is that spin is automatically included so that electrons are endowed with intrinsic spin without any ad hoc additions from empirical observations. In order to see how this is included, we first examine some familiar operators, namely the Hamiltonian and the angular momentum. We first want to investigate the commutator of these two operators that commuted in the non-relativistic theory. Assuming only a central electrostatic force is present, then the two operators are represented by

$$\hat{L}_z = \hat{x}\hat{p}_y - \hat{y}\hat{p}_x = \frac{\hbar}{\mathrm{i}}\left(x\frac{\partial}{\partial y} - y\frac{\partial}{\partial x}\right),$$

$$\hat{\mathcal{H}} = c\frac{\hbar}{\mathrm{i}}\left(\hat{\alpha}_x\frac{\partial}{\partial x} + \hat{\alpha}_y\frac{\partial}{\partial y} + \hat{\alpha}_z\frac{\partial}{\partial z} + i\hat{\beta}\frac{mc}{\hbar}\right) + e\phi(r).$$

It is readily apparent that $\hat{L}_z$ commutes with $\hat{\alpha}_z(\partial/\partial z)$, $i\hat{\beta}(mc/\hbar)$, and $e\phi(r)$. The terms that do not obviously commute are

$$\begin{aligned}\hat{L}_z\hat{\mathcal{H}} &= -c\hbar^2\left(x\frac{\partial}{\partial y} - y\frac{\partial}{\partial x}\right)\left(\hat{\alpha}_x\frac{\partial}{\partial x} + \hat{\alpha}_y\frac{\partial}{\partial y}\right) \\ &= -c\hbar^2\left(\hat{\alpha}_x x\frac{\partial^2}{\partial x\partial y} + \hat{\alpha}_y x\frac{\partial^2}{\partial y^2} - \hat{\alpha}_x y\frac{\partial^2}{\partial x^2} - \hat{\alpha}_y y\frac{\partial^2}{\partial x\partial y}\right),\end{aligned}$$

$$\hat{\mathcal{H}}\hat{L}_z = -c\hbar^2\left(\hat{\alpha}_x\frac{\partial}{\partial x} + \hat{\alpha}_y\frac{\partial}{\partial y}\right)\left(x\frac{\partial}{\partial y} - y\frac{\partial}{\partial x}\right)$$

$$= -c\hbar^2 \left(\hat{\alpha}_x \frac{\partial}{\partial y} + \hat{\alpha}_x x \frac{\partial^2}{\partial x \partial y} + \hat{\alpha}_y x \frac{\partial^2}{\partial y^2} - \hat{\alpha}_x y \frac{\partial^2}{\partial x^2} - \hat{\alpha}_y \frac{\partial}{\partial x} - \hat{\alpha}_y y \frac{\partial^2}{\partial x \partial y} \right),$$

so that the commutator is given by

$$\hat{L}_z \hat{\mathcal{H}} - \hat{\mathcal{H}} \hat{L}_z = -c\hbar^2 \left(\hat{\alpha}_y \frac{\partial}{\partial x} - \hat{\alpha}_x \frac{\partial}{\partial y} \right) \neq 0 \,. \tag{11.19}$$

This has the rather surprising implication that L_z, and by similar arguments, L_x, L_y, and $|\boldsymbol{L}|^2$ are *not constants of the motion.* However, if we now introduce the matrix operators

$$\hat{\sigma}'_x \equiv \begin{pmatrix} \hat{0} & \hat{1} \\ \hat{1} & \hat{0} \end{pmatrix} \hat{\alpha}_x = \begin{pmatrix} \hat{\sigma}_x & \hat{0} \\ \hat{0} & \hat{\sigma}_x \end{pmatrix}, \tag{11.20}$$

$$\hat{\sigma}'_y \equiv \begin{pmatrix} \hat{0} & \hat{1} \\ \hat{1} & \hat{0} \end{pmatrix} \hat{\alpha}_y = \begin{pmatrix} \hat{\sigma}_y & \hat{0} \\ \hat{0} & \hat{\sigma}_y \end{pmatrix}, \tag{11.21}$$

$$\hat{\sigma}'_z \equiv \begin{pmatrix} \hat{0} & \hat{1} \\ \hat{1} & \hat{0} \end{pmatrix} \hat{\alpha}_z = \begin{pmatrix} \hat{\sigma}_z & \hat{0} \\ \hat{0} & \hat{\sigma}_z \end{pmatrix}, \tag{11.22}$$

and examine the commutator of, for example, $\hat{\sigma}'_z$ with $\hat{\mathcal{H}}$, we find

$$\begin{aligned}
\hat{\sigma}'_z \hat{\mathcal{H}} &= \frac{c\hbar}{\mathrm{i}} \hat{\sigma}'_z \left(\hat{\alpha}_x \frac{\partial}{\partial x} + \hat{\alpha}_y \frac{\partial}{\partial y} + \hat{\alpha}_z \frac{\partial}{\partial z} \right) + \hat{\sigma}'_z V \\
&= c\hbar \left(\hat{\alpha}_y \frac{\partial}{\partial x} - \hat{\alpha}_x \frac{\partial}{\partial y} \right) + \frac{c\hbar}{\mathrm{i}} \hat{\sigma}'_z \hat{\alpha}_z \frac{\partial}{\partial z} + \hat{\sigma}'_z V \,, \\
\hat{\mathcal{H}} \hat{\sigma}'_z &= \frac{c\hbar}{\mathrm{i}} \left(\hat{\alpha}_x \frac{\partial}{\partial x} + \hat{\alpha}_y \frac{\partial}{\partial y} + \hat{\alpha}_z \frac{\partial}{\partial z} \right) \hat{\sigma}'_z + V \hat{\sigma}'_z \\
&= -c\hbar \left(\hat{\alpha}_y \frac{\partial}{\partial x} - \hat{\alpha}_x \frac{\partial}{\partial y} \right) + \frac{c\hbar}{\mathrm{i}} \hat{\alpha}_z \hat{\sigma}'_z \frac{\partial}{\partial z} + V \hat{\sigma}'_z \,,
\end{aligned}$$

so that, since $\hat{\sigma}'_z$ commutes both with $\hat{\alpha}_z$ (see Problem 10.7) and V,

$$\hat{\sigma}'_z \hat{\mathcal{H}} - \hat{\mathcal{H}} \hat{\sigma}'_z = 2c\hbar \left(\hat{\alpha}_y \frac{\partial}{\partial x} - \hat{\alpha}_x \frac{\partial}{\partial y} \right). \tag{11.23}$$

By comparing this result with Equation (11.19), we can see that if we combine these two results together, we find

$$\left(\hat{L}_z + \frac{\hbar}{2} \hat{\sigma}'_z \right) \hat{\mathcal{H}} - \hat{\mathcal{H}} \left(\hat{L}_z + \frac{\hbar}{2} \hat{\sigma}'_z \right) = 0 \,, \tag{11.24}$$

so that the *dynamical quantity that is represented by the operator*

$$\hat{J}_z = \hat{L}_z + \frac{\hbar}{2} \hat{\sigma}'_z \tag{11.25}$$

is a constant of the motion. By similar procedures, one can similarly show that corresponding operators for the x and y components, as well as $|J|^2$ are also constants of the motion. Thus the Dirac theory *automatically endows the electron with the property of spin in exactly the amount required by experiment.* Furthermore, if one solves the Dirac equation in a magnetic field, the possible energy states correspond to a particle having a spin of $\frac{1}{2}$ and a *magnetic moment of one Bohr magneton* so that the magnetic moment for the electron is *no longer anomalous.*

The Dirac theory therefore, using only the charge and mass of the electron, successfully accounts for all of the observed properties of the electron, including its antiparticle, the positron. This success is one of the great triumphs of theoretical physics.

Problem 11.6 Show that $\frac{1}{2}\hbar\hat{\sigma}'_x$, $\frac{1}{2}\hbar\hat{\sigma}'_y$, and $\frac{1}{2}\hbar\hat{\sigma}'_z$ satisfy the usual commutation relations for angular momentum (e.g., $\boldsymbol{L} \times \boldsymbol{L} = \mathrm{i}\hbar\boldsymbol{L}$).

Problem 11.7 Show that $\hat{\sigma}'_z$ commutes with $\hat{\alpha}_z$ and $\hat{\beta}$, and that

$$[\hat{\sigma}'_z, \hat{\alpha}_x] = 2\mathrm{i}\hat{\alpha}_y\,, \quad \text{and} \quad [\hat{\sigma}'_z, \hat{\alpha}_y] = -2\mathrm{i}\hat{\alpha}_x\,.$$

11.2 Quantum Electrodynamics (QED) and Electroweak Theory

11.2.1 The Formulation of Quantum Electrodynamics

While the Dirac equation gives an accurate description of the motion of an electron in the presence of an electromagnetic field, it does not directly describe the electromagnetic field from the moving electrons. This is not a trivial concern, since every accelerated electron radiates photons, and this is largely due the the electronic charge cloud. To complete the description of the interactions between electrons or between an electron and another charged particle, we need a further equation to describe the behavior of the electromagnetic field under the stimulus of given electronic motions. The two sets of equations should give a virtually complete description of the behavior of electrons and the radiation they either absorb or emit.

This other set of equations is, of course, the Maxwell equations, written in 4-vector form. In terms of the 4-potential, this is

$$\nabla^2 A_\mu - \frac{1}{c^2}\frac{\partial^2 A_\mu}{\partial t^2} = -\mu_0 J_\mu\,, \qquad \mu = 1, 2, 3, 4. \tag{11.26}$$

This expression depends on the source current density, J_μ, and is due to the motions of the particles themselves, so we need to describe this quantity more carefully. Classically, of course, this is simple since the current density $\boldsymbol{J} = \rho \boldsymbol{v}$, where ρ is the charge density and $\boldsymbol{v}$ is the velocity (or average velocity if many particles). From quantum mechanics, however, the charge of an electron must be "smeared out" just as the probability density, $\psi^* \psi$, is smeared out and the velocity distribution, $\psi^* \hat{\boldsymbol{v}} \psi$ is likewise spread out. We should therefore reformulate these quantities so that

$$\rho = e\psi^* \psi \,, \tag{11.27}$$

$$\boldsymbol{J} = e\psi^* \hat{\boldsymbol{v}} \psi \,. \tag{11.28}$$

Even this needs to be modified slightly to be consistent with Dirac theory, so that the 4-vector current density is given by

$$J_4 = \mathrm{i}c\rho = \mathrm{i}ce\psi^\dagger \psi = \mathrm{i}ce(\psi_1^* \psi_1 + \psi_2^* \psi_2 + \psi_3^* \psi_3 + \psi_4^* \psi_4) \,, \tag{11.29}$$

$$J_i = ec\psi^\dagger \hat{\alpha}_i \psi \,, \qquad i = 1, 2, 3, \tag{11.30}$$

where we have used $\hat{v}_i = c\hat{\alpha}_i$ (see Problem 11.5). With these expressions, we can write the full set of equations of quantum electrodynamics as

$$\left[c\hat{\boldsymbol{\alpha}} \cdot \left(\frac{\hbar}{\mathrm{i}} \nabla - e\boldsymbol{A} \right) + \hat{\beta} mc^2 + e\phi \right] \psi = -\frac{\hbar}{\mathrm{i}} \frac{\partial \psi}{\partial t} \,, \tag{11.31}$$

$$\nabla^2 \boldsymbol{A} - \frac{1}{c^2} \frac{\partial^2 \boldsymbol{A}}{\partial t^2} = -\mu_0 ec\psi^\dagger \hat{\boldsymbol{\alpha}} \psi \,, \tag{11.32}$$

$$\nabla^2 \phi - \frac{1}{c^2} \frac{\partial^2 \phi}{\partial t^2} = -\frac{e}{\epsilon_0} \psi^\dagger \psi \,. \tag{11.33}$$

Problem 11.8 Find the charge and current densities represented by the wave function

$$\psi = f(x, y, z) \begin{pmatrix} A \\ 0 \\ B \\ 0 \end{pmatrix} .$$

11.2.2 Feynman Diagrams

The direct solution of the equations of quantum electrodynamics is virtually impossible for any but the most trivial cases (like the one-electron atom), but we need to know something about interactions between particles in general, and between particles and photons. Feynman developed a method for getting answers to these questions, where instead of writing down a general solution, he was able to break up a process into separable steps, and then evaluate the probability amplitudes for each step, from the initial state to the final state, and get quantitative estimates for each possible scenario that might lead from the initial to the final state. These are represented by **Feynman diagrams**, which pictorially illustrate the interaction. Once drawn, there are prescriptive rules to evaluate the effects of any particular diagram, and if more than one diagram can couple a specific initial state to a particular final state, the diagrams are "summed" for the net interaction.

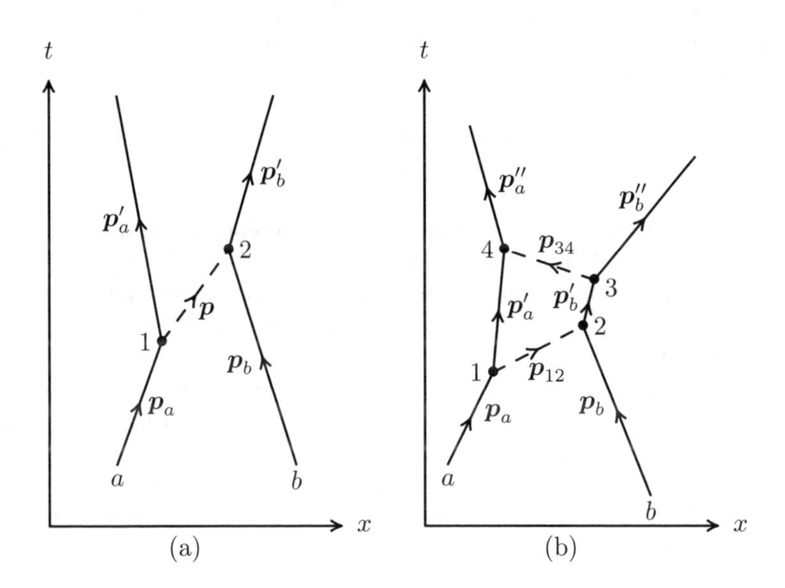

FIGURE 11.1
Feynman diagrams for electron–electron scattering. (a) One-photon exchange. (b) Two-photon exchange.

An example of a Feynman diagram is given in Figure 11.1 for two electrons interacting with each other through the Coulomb potential, which in QED is due to the exchange of one "virtual" photon in (a) and two virtual photons in (b), where "virtual" means a photon that is never directly observed since it is emitted and absorbed by the interacting particles, and the only evidence of its existence is the motion of the particles. The evaluation of the interaction is implemented through the evaluation of the perturbation-like integral at each vertex (bullet in the figure):

$$\langle f|V|i\rangle = \int \psi_f^\dagger \hat{V} \psi_i \, \mathrm{d}^N q \,. \tag{11.34}$$

The usual interpretation of an integral of this kind is that both ψ_f (final state wave function) and ψ_i (initial state wave function) are spread out in space and that $\hat{V}$ acts relatively smoothly over the trajectory of the electrons in Figure

11.1. Feynman proposed, however, that this could just as well be interpreted as a sequence of discontinuous events. In this picture,

1. ψ_i is the probability amplitude that electrons a and b *arrive* at space-time points 1 and 2 with momenta $\boldsymbol{p}_a$ and $\boldsymbol{p}_b$.

2. The operator $\hat{V}$ is the product of the probability amplitude (operators) for three independent events: (a) particle a *emits* a photon of momentum $\boldsymbol{p}$, (b) a photon of momentum $\boldsymbol{p}$ goes from 1 to 2, and (c) particle b *absorbs* a photon of momentum $\boldsymbol{p}$.

3. $\psi_f^\dagger$ is the probability amplitude that electrons a and b *depart* from space-time points 1 and 2 with momenta $\boldsymbol{p}'_a$ and $\boldsymbol{p}'_b$.

The operators are set up so that if momentum and energy are not conserved, the result vanishes. This is not guaranteed instantaneously, however, since after the virtual photon is emitted and before it is absorbed, momentum and energy are momentarily not conserved. There are limits to the violation, however, since the uncertainty principle sets bounds. This means that for short distances (Δx small) and short times (Δt small) between points 1 and 2, we can allow substantial violations of the conservation of momentum (Δp large) and energy (ΔE large) provided $(\Delta p)(\Delta x) \geq \frac{1}{2}\hbar$ and $(\Delta E)(\Delta t) \geq \frac{1}{2}\hbar$. For close encounters, then, the deviations may be large, while for distant encounters, the deviations must be small, since we can only allow low energy and momentum virtual photons for long range exchanges.

11.3 Quarks, Leptons, and the Standard Model

With the discovery of the top (t) quark, the search for elementary particles is nearly over according to the present understanding of our world. In Table 11.1, all of the elementary particles are shown along with their charge in units of e, the electronic charge, and are divided into leptons and quarks. The only remaining particle according to present theory is the **Higgs boson**, which is required by theory to account for the masses of the various particles. Theory also indicates that there are no more new families to be be found (there are three families: the electron and its neutrino along with the up and down quarks form the first family, etc.). The quark designations, d, u, s, c, b, and t are called *flavors*, and each quark comes in three *colors*, red, green and blue. Also each particle shown in Table 11.1 has a corresponding antiparticle, designated for the antiquarks by $\bar{d}$, $\bar{u}$, etc., the antineutrinos by $\bar{\nu}_e$, $\bar{\nu}_\mu$, etc., and the antielectron is the positron, denoted by e^+, since its charge is positive. Antiquarks are either antired, antigreen, or antiblue.

TABLE 11.1
Elementary particles in the standard model.

Leptons		Quarks					
		red		green		blue	
τ	-1	t	$+\frac{2}{3}$	t	$+\frac{2}{3}$	t	$+\frac{2}{3}$
ν_τ	0	b	$-\frac{1}{3}$	b	$-\frac{1}{3}$	b	$-\frac{1}{3}$
μ	-1	c	$+\frac{2}{3}$	c	$+\frac{2}{3}$	c	$+\frac{2}{3}$
ν_μ	0	s	$-\frac{1}{3}$	s	$-\frac{1}{3}$	s	$-\frac{1}{3}$
e	-1	u	$+\frac{2}{3}$	u	$+\frac{2}{3}$	u	$+\frac{2}{3}$
ν_e	0	d	$-\frac{1}{3}$	d	$-\frac{1}{3}$	d	$-\frac{1}{3}$

11.3.1 Hadrons

There are four fundamental forces: gravity, electomagnetism, the weak force and the strong force. While electromagnetism describes interactions between particles carrying *charge* (which occurs in integral multiples of the electronic charge and can be either plus or minus), the strong force describes interactions between particles carrying **color charge** (which comes in combinations of red, green, and blue, or their anticolors, antired, antigreen, or antiblue). All observed particles that are composed of quarks are called hadrons, and these come in two general types, as shown in Table 11.2. Those that are comprised of three quarks are called **baryons**, of which the proton and neutron are the simplest examples, and each of the quarks has a different color, so that the resulting color of the baryon is colorless (red+green+blue=white). The **mesons** are comprised of a quark-antiquark pair, and again are colorless, since the pair are either red-antired, green-antigreen, or blue-antiblue. Hence, we never *observe* any particles with color. The fact that we never observe single quarks or combinations other than the ones indicated above does not mean that quarks do not exist, since in scattering experiments with energetic electrons on protons, for example, it is possible to observe *three* scattering centers, not just one. Since baryons and mesons occur only in these configurations, one can also observe that this requires all observed particles to have integral charge although no quarks do. For example, the proton is made of the combination uud so the charge is $\frac{2}{3}+\frac{2}{3}-\frac{1}{3}=1$ unit of charge, while the neutron is made of the combination udd so the charge is $\frac{2}{3}-\frac{1}{3}-\frac{1}{3}=0$, so the neutron has no charge. In a similar way the mesons have integral charge, since the π^+ meson is made of the combination $u\bar{d}$, so has charge $\frac{2}{3}+\frac{1}{3}=1$, the π^0 is made of a linear combination $u\bar{u}$ and $d\bar{d}$, so has charge 0, while the π^- meson is made of the combination $d\bar{u}$, so has charge $-\frac{1}{3}-\frac{2}{3}=-1$. The combination $u\bar{s}$ is called the K^+ meson and has the property of strangeness. Each of the flavors has a corresponding *property*, so that the down quark carries "downness," although upness and downness are rarely mentioned, as the

TABLE 11.2
Strongly interacting particles — hadrons.

Hadrons		
	Baryons	Mesons
Constituents	3 Quarks	quark–Antiquark pair
Colors	1 red, 1 green, 1 blue	red–antired, etc.
Examples	proton (uud)	π^+ $(u\bar{d})$
	neutron (udd)	K^+ $(u\bar{s})$

flavor of both is sort of like vanilla, and too ordinary to mention. The strange particle carries "strangeness," the charmed quark carries "charm," the bottom quark carries "beauty," and, of course, the top quark therefore carries "truth."

11.3.2 Exchange Forces

Each force is mediated by an exchange of energy, momentum, etc., through an exchange particle. Whereas all of the particles in Table 11.1 are Fermions, all of the exchange particles are Bosons. A sketch of such an exchange is called a "Feynman diagram" showing space in one direction and time in the other direction as in Figure 11.1. In part (a), two electrons, a and b, are traveling upward (forward in time) in the figure and approaching one another. At point 1, particle a emits a photon with momentum $\boldsymbol{p}$ and recoils, changing its original momentum $\boldsymbol{p}_a$ to $\boldsymbol{p}'_a$, so that momentum is conserved. The photon is subsequently absorbed by particle b at point 2, so that its original momentum $\boldsymbol{p}_b$ is changed to $\boldsymbol{p}'_b$, again conserving momentum. However, *energy is not conserved* while the photon is propagating. This violation is brief, however, since the photon travels at the speed of light, unless the distance is great. In part (b) of the figure, a second photon is exchanged, and even higher order diagrams can be constructed, although each of lower probability.

Using Feynman diagrams with the uncertainty principle allows one to gain deeper insight into the *range* of various forces, which describes how the interaction forces vary with distance. Considering first the electromagnetic force, the force is transmitted by photons which travel at the speed of light. Hence, we may write $\Delta t = \Delta r/c$, so that the uncertainty relation becomes

$$\Delta E \sim \hbar c/\Delta r\,,$$

which means that the energy involved in the interaction is inversely proportional to the distance. The momentum of a photon is proportional to its energy, so a close encounter leads to large deflections, and for a distant encounter, (Δt) is large and (ΔE) small, leading to a small deflection.

Now one of the expansion parameters of quantum electrodynamics is the fine structure constant, α, which indicates that the term giving the probability of one photon being exchanged in the interaction between two electrons is of

TABLE 11.3
Exchange particles for each force.

Force	Exchange	Mass	Charge	Spin	Color
Electromagnetic	photon	no	no	1	no
Weak	$W^{\pm}$	yes	yes	0	no
	Z^0	yes	no	0	no
Strong	Gluon	no	yes	0	Color–anticolor pair (e.g. green–antired)
Gravity	graviton	no	no	2	no

order α, while the term describing a two-photon interaction is of order α^2, etc. If we then multiply the energy associated with a one-photon interaction by the probability of observing this interaction, we find

$$\Delta E = e\Delta V \sim \hbar\alpha c/\Delta r = \frac{e^2}{4\pi\epsilon_0 \Delta r},$$

which is Coulomb's Law for the potential energy. The corresponding force, which is proportional to the rate of change of the energy, is proportional to $1/r^2$. In most beginning textbooks, the Coulomb force is related to the electric field, which pictures forces as being due to a *field* being set up everywhere by one particle and the other particle responds to that field. This is a so-called "action at a distance" representation, and provides a useful way to describe the interactions, but this "field" approach is *not* representative of our fundamental understanding of the nature of physical forces. In classical cases, the two cases have identical results, but for very close encounters and high energy exchanges, the classical field approach ultimately fails.

Taking these ideas one step further, if one were now to assume that the interaction were to be mediated by a particle of finite mass, then $\Delta E \sim m_0c^2 + m_0 v\Delta v$ and $\Delta t \sim \Delta r/v$, so that the uncertainty relation takes the form

$$(m_0c^2 + m_0 v\Delta v)\Delta r/v \sim \hbar\,,$$

which gives the range of the force as

$$\Delta r \sim \left(\frac{\hbar}{m_0 c}\right)\frac{v}{c} < \frac{\hbar}{m_0 c}\,.$$

This means the range varies inversely as the mass of the mediating particle and is less than the Compton wavelength of the particle. The weak force is mediated by massive particles, either the $W^{\pm}$ or the Z^0, so the force has a very short range. For the photon, with $m_0 = 0$, the range for the Coulomb force is infinite, and similarly for the Gravitational force, which is mediated by a massless graviton.

Problem 11.9 Estimate the range of the weak force if the masses of the $W^{\pm}$ and the Z^0 are 80.6 GeV and 91.2 GeV, respectively.

11.3.3 Important Theories

- **Quantum Electrodynamics** — Electromagnetic force only (Feynman and Schwinger).
- **Electroweak Theory** — Weak plus Electromagnetic (Weinberg and Salam).
- **Quantum Chromodynamics** — Strong force (Georgi and Glashow).
- **Grand Unification Theories** (GUTs) — All Forces (do not exist yet).
- **Standard Model** — Present best model of everything except gravity.

A

Matrix Operations

Every operator can be represented as a matrix, but some require an infinite dimensional matrix. We consider here only finite dimensional matrices and will show how both the eigenvalues and eigenvectors may be found. The general method will be illustrated by a simple 2×2 matrix, and then a 3×3 example with a specific application to angular momentum will be analyzed.

A.1 General Properties

A square matrix (nonsquare matrices exist, but are not considered here), represented by A may be characterized by its elements, a_{ij}, where i denotes the row and j the column. This is illustrated for an $N \times N$ matrix by

$$\mathsf{A} = \overset{\downarrow}{i} \begin{pmatrix} a_{11} & a_{12} & \overset{\rightarrow j}{a_{13}} & \cdots & a_{1N} \\ a_{21} & a_{22} & a_{23} & \cdots & a_{2N} \\ a_{31} & a_{32} & a_{33} & \cdots & a_{2N} \\ \vdots & \vdots & \vdots & \ddots & \vdots \\ a_{N1} & a_{N2} & a_{N3} & \cdots & a_{NN} \end{pmatrix}$$

Some common designations of matrices are the transpose matrix, denoted $\tilde{\mathsf{A}}$, and the inverse matrix, denoted A^{-1}. The transpose of a matrix is obtained by interchanging the rows and columns so that in terms of the **elements** of a matrix, $\tilde{a}_{ij} = a_{ji}$. The inverse of a matrix (if it exists) has the property $\mathsf{AA}^{-1} = \mathsf{I}$ where I is the identity matrix whose elements are $I_{ij} = \delta_{ij}$ (diagonal elements only which are all unity). No inverse exists if the determinant of the matrix vanishes.

They are classified into several types:

1. **Symmetric Matrix**: A symmetric matrix has the property $a_{ij} = a_{ji}$, so that interchanging rows and columns leaves the matrix unchanged.

2. **Antisymmetric matrix**: An antisymmetric matrix has the property that $a_{ij} = -a_{ji}$ so that $a_{jj} = 0$ (no diagonal terms) and interchanging rows and columns changes the sign of the matrix.

3. **Adjoint Matrix**: The adjoint of A is the complex conjugate of the transposed matrix. This is denoted by $\mathsf{A}^\dagger = \tilde{\mathsf{A}}^*$.

4. **Hermitian Matrix**: A Hermitian matrix is self-adjoint, so that the transpose conjugate of a Hermitian matrix is the original matrix, or $\mathsf{A}^\dagger = \mathsf{A}$.

An important rule about the transpose operation relates to the transpose of the product of two matrices, where

$$\widetilde{\mathsf{AB}} = \tilde{\mathsf{B}}\tilde{\mathsf{A}} \,.$$

This is simply proved by noting that in terms of the elements,

$$\mathsf{AB} \equiv \mathsf{C}$$

is equivalent to

$$\sum_j a_{ij} b_{jk} = c_{ik} \,,$$

and the transpose is

$$\widetilde{\mathsf{AB}} \equiv \tilde{\mathsf{C}} \,,$$

where $\tilde{c}_{ik} = c_{ki}$. We also have $\tilde{a}_{ij} = a_{ji}$ and $\tilde{b}_{jk} = b_{kj}$ so that from the rules of matrix multiplication, we write the product as

$$\sum_j b_{kj} a_{ji} = c_{ki} \,,$$

which is equivalent to $\tilde{\mathsf{C}} = \tilde{\mathsf{B}}\tilde{\mathsf{A}}$.

In order to see that this definition of Hermitian corresponds to our previous definition of Hermitian, we note that the definition of Hermitian from Equation (1.32) in matrix form may be written

$$\psi^* \mathsf{F} \chi = (\mathsf{F}\psi)^* \chi \,.$$

In terms of components, the left-hand side may be written

$$\psi^* \mathsf{F} \chi = \sum_{i,j} \psi_i^* F_{ij} \chi_j \,,$$

and the right-hand side may be written

$$\begin{aligned}(\mathsf{F}\psi)^* \chi &= \sum_{i,j} (F_{ji}\psi_i)^* \chi_j \\ &= \sum_{i,j} F_{ji}^* \psi_i^* \chi_j \,.\end{aligned}$$

But we note that $F_{ji}^* = \tilde{F}_{ij}^* \to \mathsf{F}^\dagger$ (self-adjoint), then our definition of Hermitian means that a Hermitian matrix is equal to its transpose conjugate.

A.2 Eigenvalues and Eigenvectors

A general 2×2 Hermitian matrix can be represented by

$$\mathsf{A} = \begin{pmatrix} a_{11} & a_{12} \\ a_{12}^* & a_{22} \end{pmatrix},$$

where a_{11} and a_{22} are real, and the corresponding eigenvalue equation is $\mathsf{A}\psi = \lambda\psi$ where λ is a scalar eigenvalue and ψ is a two-component eigenvector. The eigenvalue equation can be written as

$$\begin{pmatrix} a_{11} & a_{12} \\ a_{12}^* & a_{22} \end{pmatrix} \begin{pmatrix} \phi_1 \\ \phi_2 \end{pmatrix} = \lambda \begin{pmatrix} \phi_1 \\ \phi_2 \end{pmatrix}. \tag{A.1}$$

Moving the term on the right to the other side, Equation (A.1) may be written as

$$\begin{pmatrix} a_{11} - \lambda & a_{12} \\ a_{12}^* & a_{22} - \lambda \end{pmatrix} \begin{pmatrix} \phi_1 \\ \phi_2 \end{pmatrix} = 0\,. \tag{A.2}$$

For this equation to have a nontrivial solution (i.e., at least one of the $\phi_j \neq 0$), the determinant of coefficients must vanish. This leads to

$$\begin{vmatrix} a_{11} - \lambda & a_{12} \\ a_{12}^* & a_{22} - \lambda \end{vmatrix} = (a_{11} - \lambda)(a_{22} - \lambda) - |a_{12}|^2 = 0. \tag{A.3}$$

The solution of the quadratic equation for λ leads to the two eigenvalues

$$\lambda_{1,2} = \frac{a_{11} + a_{22} \pm \sqrt{(a_{11} - a_{22})^2 + 4|a_{12}|^2}}{2}\,.$$

In order to find the eigenvectors, we use the first of the two equations implied by Equation (A.1) with λ_1 to find

$$a_{11}\phi_1 + a_{12}\phi_2 = \lambda_1\phi_1\,.$$

This equation indicates that ϕ_2 is related to ϕ_1 through the relation

$$\phi_2 = \alpha_1\phi_1, \qquad \alpha_1 = (\lambda_1 - a_{11})/a_{12}\,,$$

and from the normalization condition, we have

$$|\psi_1|^2 = |\phi_1|^2 + |\phi_2|^2 = (1 + |\alpha_1|^2)|\phi_1|^2 = 1\,,$$

so that

$$\phi_1 = 1/\sqrt{1 + |\alpha_1|^2}\,.$$

The eigenvector associated with eigenvalue λ_1 is therefore given by

$$\psi_1 = \frac{1}{\sqrt{1 + |\alpha_1|^2}} \begin{pmatrix} 1 \\ \alpha_1 \end{pmatrix}.$$

It may be shown with some additional algebra that

$$\psi_2 = \frac{1}{\sqrt{1+|\alpha_1|^2}} \begin{pmatrix} -\alpha_1^* \\ 1 \end{pmatrix} .$$

Example A.1

The rigid rotor. For a problem describing the rotation of a rigid body with different moments of inertia about each axis, the matrix is 3×3 and the eigenvalue equation is of the form

$$\begin{pmatrix} u & 0 & w \\ 0 & v & 0 \\ w & 0 & u \end{pmatrix} \begin{pmatrix} \phi_1 \\ \phi_2 \\ \phi_3 \end{pmatrix} = \lambda \begin{pmatrix} \phi_1 \\ \phi_2 \\ \phi_3 \end{pmatrix} , \tag{A.4}$$

where all components are real, so the matrix is symmetric. Setting the determinant of coefficients to zero, we find

$$\begin{vmatrix} u-\lambda & 0 & w \\ 0 & v-\lambda & 0 \\ w & 0 & u-\lambda \end{vmatrix} = (u-\lambda)^2(v-\lambda) - w^2(v-\lambda) = 0.$$

For this case, we find a common factor which must vanish for one of the eigenvalues, so we have $\lambda_1 = v$. The remaining two eigenvalues are found simply, given by $\lambda_{2,3} = u \pm w$.

We find the eigenfunctions for the first eigenvalue by examining the three equations from Equation (A.4),

$$\begin{aligned} u\phi_1 + w\phi_3 &= v\phi_1 , \\ v\phi_2 &= v\phi_2 , \\ w\phi_1 + u\phi_3 &= v\phi_3 . \end{aligned}$$

The first and third are inconsistent unless $(u-v)^2 = w^2$ or $\phi_1 = \phi_3 = 0$. The former is unlikely in general, leaving ϕ_2 undefined from the second equation, so we simply set $\phi_2 = 1$ to satisfy the normalization condition on ψ_1.

Using the second eigenvalue, we have

$$\begin{aligned} u\phi_1 + w\phi_3 &= (u+w)\phi_1 , \\ v\phi_2 &= (u+w)\phi_2 , \\ w\phi_1 + u\phi_3 &= (u+w)\phi_3 . \end{aligned}$$

The first and third of this set of equations give $\phi_1 = \phi_3$ and the second is inconsistent unless $v = u+w$, so we set $\phi_2 = 0$ and then normalization requires $\phi_1 = \phi_3 = 1/\sqrt{2}$.

For the third eigenvalue, we have

$$\begin{aligned} u\phi_1 + w\phi_3 &= (u-w)\phi_1 , \\ v\phi_2 &= (u-w)\phi_2 , \\ w\phi_1 + u\phi_3 &= (u-w)\phi_3 . \end{aligned}$$

The first and third of this set of equations give $\phi_1 = -\phi_3$ and the second is inconsistent unless $v = u - w$, so we again set $\phi_2 = 0$ and then normalization requires $\phi_1 = -\phi_3 = 1/\sqrt{2}$.

Summarizing these results, the eigenfunctions are

$$\psi_1 = \begin{pmatrix} 0 \\ 1 \\ 0 \end{pmatrix}, \qquad \psi_2 = \begin{pmatrix} \frac{1}{\sqrt{2}} \\ 0 \\ \frac{1}{\sqrt{2}} \end{pmatrix}, \qquad \psi_3 = \begin{pmatrix} \frac{1}{\sqrt{2}} \\ 0 \\ -\frac{1}{\sqrt{2}} \end{pmatrix}.$$

□

Example A.2

Degenerate eigenvalues. Consider a physical system whose Hamiltonian $\hat{\mathcal{H}}$ and initial state $|\psi_0\rangle$ are given by

$$\hat{\mathcal{H}} = \mathcal{E} \begin{pmatrix} 0 & \mathrm{i} & 0 \\ -\mathrm{i} & 0 & 0 \\ 0 & 0 & -1 \end{pmatrix} \qquad |\psi_0\rangle = \frac{1}{\sqrt{5}} \begin{pmatrix} 1-\mathrm{i} \\ 1-\mathrm{i} \\ 1 \end{pmatrix}$$

where $\mathcal{E}$ is a scalar constant with the dimensions of energy.

(a) What values will we obtain when measuring the energy and with what probabilities?

Solution: The eigenvalue equation may be written as

$$\begin{pmatrix} 0 & \mathrm{i}\mathcal{E} & 0 \\ -\mathrm{i}\mathcal{E} & 0 & 0 \\ 0 & 0 & -\mathcal{E} \end{pmatrix} \psi = \lambda \psi$$

so that the determinant of coefficients is

$$\begin{vmatrix} -\lambda & \mathrm{i}\mathcal{E} & 0 \\ -\mathrm{i}\mathcal{E} & -\lambda & 0 \\ 0 & 0 & -\mathcal{E} - \lambda \end{vmatrix} = 0$$

or expanding, we find

$$-\lambda^2(\mathcal{E} + \lambda) + \mathcal{E}^2(\mathcal{E} + \lambda) = 0$$

so the eigenvalues are $\lambda = -\mathcal{E}$ and $\lambda = \pm\mathcal{E}$, or one eigenvalue with $\lambda = \mathcal{E}$ and two degenerate eigenvalues with $\lambda = -\mathcal{E}$. For the eigenfunctions we first solve the equation with $\lambda = \mathcal{E}$ such that

$$\begin{pmatrix} 0 & \mathrm{i}\mathcal{E} & 0 \\ -\mathrm{i}\mathcal{E} & 0 & 0 \\ 0 & 0 & -\mathcal{E} \end{pmatrix} \begin{pmatrix} a_1 \\ a_2 \\ a_3 \end{pmatrix} = \mathcal{E} \begin{pmatrix} a_1 \\ a_2 \\ a_3 \end{pmatrix}$$

which leads to

$$\begin{aligned} \mathrm{i}\mathcal{E}a_2 &= \mathcal{E}a_1 \\ -\mathrm{i}\mathcal{E}a_1 &= \mathcal{E}a_2 \\ -\mathcal{E}a_3 &= \mathcal{E}a_3 \end{aligned}$$

which requires $a_3 = 0$ and $a_2 = -ia_1$ so that

$$|\psi_1\rangle = \frac{1}{\sqrt{2}} \begin{pmatrix} 1 \\ -\mathrm{i} \\ 0 \end{pmatrix}$$

For the degenerate cases, we have

$$\begin{pmatrix} 0 & \mathrm{i}\mathcal{E} & 0 \\ -\mathrm{i}\mathcal{E} & 0 & 0 \\ 0 & 0 & -\mathcal{E} \end{pmatrix} \begin{pmatrix} b_1 \\ b_2 \\ b_3 \end{pmatrix} = -\mathcal{E} \begin{pmatrix} b_1 \\ b_2 \\ b_3 \end{pmatrix}$$

which leads to

$$\begin{aligned} \mathrm{i}\mathcal{E}b_2 &= -\mathcal{E}b_1 \\ -\mathrm{i}\mathcal{E}b_1 &= -\mathcal{E}b_2 \\ -\mathcal{E}b_3 &= -\mathcal{E}b_3 \end{aligned}$$

One solution has $b_3 = 0$ and $b_2 = ib_1$. This solution may be written as

$$|\psi_2\rangle = \frac{1}{\sqrt{2}} \begin{pmatrix} 1 \\ \mathrm{i} \\ 0 \end{pmatrix}$$

or as

$$|\psi_2\rangle = \frac{1}{\sqrt{2}} \begin{pmatrix} -\mathrm{i} \\ 1 \\ 0 \end{pmatrix}$$

The other solution has $b_3 = 1$ and $b_1 = b_2 = 0$, or

$$|\psi_3\rangle = \begin{pmatrix} 0 \\ 0 \\ 1 \end{pmatrix}$$

The initial solution is a linear combination of these, or

$$|\psi_0\rangle = c_1|\psi_1\rangle + c_2|\psi_2\rangle + c_3|\psi_3\rangle$$

where

$$c_1 = \langle\psi_1|\psi_0\rangle = \frac{1}{\sqrt{10}}(1\ \mathrm{i}\ 0) \begin{pmatrix} 1-\mathrm{i} \\ 1-\mathrm{i} \\ 1 \end{pmatrix} = \sqrt{\frac{2}{5}}$$

$$c_2 = \langle\psi_2|\psi_0\rangle = \frac{1}{\sqrt{10}}(1\ \ -\mathrm{i}\ 0)\begin{pmatrix}1-\mathrm{i}\\1-\mathrm{i}\\1\end{pmatrix} = -\mathrm{i}\sqrt{\frac{2}{5}}$$

$$c_3 = \langle\psi_3|\psi_0\rangle = \frac{1}{\sqrt{5}}(0\ 0\ 1)\begin{pmatrix}1-\mathrm{i}\\1-\mathrm{i}\\1\end{pmatrix} = \sqrt{\frac{1}{5}}$$

so that

$$|\psi_0\rangle = \sqrt{\frac{2}{5}}|\psi_1\rangle - \mathrm{i}\sqrt{\frac{2}{5}}|\psi_2\rangle + \frac{1}{\sqrt{5}}|\psi_3\rangle$$

The probabilities of measuring the various values of the energy are given by

$$P_1(E_1) = |\langle\psi_1|\psi_0\rangle|^2 = \frac{2}{5}$$

$$P_2(E_2) = |\langle\psi_2|\psi_0\rangle|^2 + |\langle\psi_3|\psi_0\rangle|^2 = \frac{2}{5} + \frac{1}{5} = \frac{3}{5}$$

(b) Calculate $\langle\mathcal{H}\rangle$, the expectation value of the Hamiltonian.

Solution: The mean value of $\mathcal{H}$ can be calculated either by

$$\langle\mathcal{H}\rangle = P_1E_1 + P_2E_2 = \frac{2}{5}\mathcal{E} - \frac{3}{5}\mathcal{E} = -\frac{1}{5}\mathcal{E}$$

or by calculating $\langle\psi_0|\mathcal{H}|\psi_0\rangle$.

□

A.3 Diagonalization of Matrices

If we have a general $N \times N$ Hermitian matrix, one way of finding the eigenvalues and eigenvectors is through the use of a **unitary transformation**. A unitary transformation is one that changes one orthonormal basis set into another orthonormal basis set which might represent a rotation or translation of coordinates. We write the new set in terms of the old set by

$$\mathbf{\Psi}'_i = \sum_{j=1}^{N} u^*_{ij}\mathbf{\Psi}_j\,.$$

We require the new set to be orthonormal, so

$$\langle\mathbf{\Psi}'_i|\mathbf{\Psi}'_j\rangle = \delta_{ij}$$

or

$$\left\langle \sum_{k=1}^{N} u_{ik}^{*} \mathbf{\Psi}_k \middle| \sum_{\ell=1}^{N} u_{j\ell}^{*} \mathbf{\Psi}_\ell \right\rangle = \sum_{k,\ell} u_{ik} u_{j\ell}^{*} \langle \mathbf{\Psi}_k | \mathbf{\Psi}_\ell \rangle$$

$$= \sum_{k=1}^{N} u_{ik} u_{jk}^{*}$$

$$= \sum_{k=1}^{N} u_{ik} \tilde{u}_{kj}^{*}$$

$$= \delta_{ij}$$

which is equivalent to

$$\mathsf{U}\mathsf{U}^{\dagger} = \mathsf{I}.$$

This states that the adjoint of U is the inverse of U or $\mathsf{U}^{\dagger} = \mathsf{U}^{-1}$.

We now consider a general vector in the new space, $\mathbf{\Psi}_a'$ which is related to a corresponding vector in the space by

$$\mathbf{\Psi}_b' = \mathsf{Q}' \mathbf{\Psi}_a'$$

and further related to yet another vector by

$$\mathbf{\Psi}_c' = \mathsf{P}' \mathbf{\Psi}_b' = \mathsf{P}' \mathsf{Q}' \mathbf{\Psi}_a' = \mathsf{R}' \mathbf{\Psi}_a' .$$

In the original space, this is written as

$$\mathbf{\Psi}_b = \mathsf{Q} \mathbf{\Psi}_a$$

$$\mathbf{\Psi}_c = \mathsf{P} \mathbf{\Psi}_b = \mathsf{P}\mathsf{Q} \mathbf{\Psi}_a = \mathsf{R} \mathbf{\Psi}_a$$

where

$$\mathbf{\Psi}_a = \sum_{j=1}^{n} \alpha_j \mathbf{\Psi}_j = \sum_{i=1}^{N} \alpha_i' \mathbf{\Psi}_i' = \sum_{i,j} \alpha_i' u_{ij}^{*} \mathbf{\Psi}_j$$

Taking the scalar product with $\mathbf{\Psi}_k$, we find

$$\alpha_k = \sum_{i,j} \alpha_i' u_{ij}^{*} \delta_{jk} = \sum_{i=1}^{N} \tilde{u}_{ki}^{*} \alpha_i' \tag{A.5}$$

This is equivalent to

$$\boldsymbol{\alpha} = \mathsf{U}^{\dagger} \boldsymbol{\alpha}'$$

or

$$\mathbf{\Psi}_a = \mathsf{U}^{\dagger} \mathbf{\Psi}_a' .$$

The inverse transformation is obtained by multiplying Equation (A.5) by u_{jk} and summing over k to obtain

$$\sum_{k=1}^{N} u_{jk} \alpha_k = \sum_{i,k} u_{jk} \tilde{u}_{ki}^{*} \alpha_i'$$

or equivalently,

$$\mathsf{U}\boldsymbol{\alpha} = \boldsymbol{\alpha}'$$

since $\mathsf{U}\mathsf{U}^\dagger = \mathsf{I}$ or $\boldsymbol{\Psi}'_a = \mathsf{U}\boldsymbol{\Psi}_a$, and it follows that $\boldsymbol{\Psi}'_b = \mathsf{U}\boldsymbol{\Psi}_b$ and $\boldsymbol{\Psi}_b = \mathsf{U}^\dagger\boldsymbol{\Psi}'_b$. Then from $\boldsymbol{\Psi}_b = \mathsf{Q}\boldsymbol{\Psi}_a$, we have

$$\mathsf{U}^\dagger\boldsymbol{\Psi}'_b = \mathsf{Q}\mathsf{U}^\dagger\boldsymbol{\Psi}'_a$$

or multiplying by U on the left,

$$\boldsymbol{\Psi}'_b = \mathsf{U}\mathsf{Q}\mathsf{U}^\dagger\boldsymbol{\Psi}'_a$$

so that

$$\mathsf{Q}' = \mathsf{U}\mathsf{Q}\mathsf{U}^\dagger = \mathsf{U}\mathsf{Q}\mathsf{U}^{-1}\,.$$

The objective now is to choose Q to be diagonal so that $\mathsf{Q} = \mathsf{\Lambda} = \lambda\mathsf{I}$ where $\Lambda_{ij} = \lambda_i\delta_{ij}$. The equation to solve may then be written as

$$\mathsf{Q}' = \mathsf{U}\mathsf{\Lambda}\mathsf{U}^{-1}$$

or multiplying by U on the right,

$$\mathsf{Q}'\mathsf{U} = \mathsf{U}\mathsf{\Lambda}$$

Expanding this expression, it has the form

$$\begin{pmatrix} Q'_{11} & Q'_{12} & Q'_{13} & \cdots \\ Q'_{21} & Q'_{22} & Q'_{23} & \cdots \\ Q'_{31} & Q'_{32} & Q'_{33} & \cdots \\ \vdots & \vdots & \vdots & \ddots \end{pmatrix} \begin{pmatrix} u_{11} & u_{12} & u_{13} & \cdots \\ u_{21} & u_{22} & u_{23} & \cdots \\ u_{31} & u_{32} & u_{33} & \cdots \\ \vdots & \vdots & \vdots & \ddots \end{pmatrix}$$
$$= \begin{pmatrix} u_{11} & u_{12} & u_{13} & \cdots \\ u_{21} & u_{22} & u_{23} & \cdots \\ u_{31} & u_{32} & u_{33} & \cdots \\ \vdots & \vdots & \vdots & \ddots \end{pmatrix} \begin{pmatrix} \lambda_1 & 0 & 0 & \cdots \\ 0 & \lambda_2 & 0 & \cdots \\ 0 & 0 & \lambda_3 & \cdots \\ \vdots & \vdots & \vdots & \ddots \end{pmatrix}$$

From this representation, we may see the procedure in finding the appropriate elements u_{ij}. If we consider taking one column at a time so that each column of U is a separate vector,

$$\boldsymbol{u}_i = \begin{pmatrix} u_{1i} \\ u_{2i} \\ u_{3i} \\ \vdots \end{pmatrix}.$$

Then we need to solve the set of equations

$$\begin{aligned} \mathsf{Q}'\boldsymbol{u}_1 &= \lambda_1\boldsymbol{u}_1 \\ \mathsf{Q}'\boldsymbol{u}_2 &= \lambda_2\boldsymbol{u}_2 \\ \mathsf{Q}'\boldsymbol{u}_3 &= \lambda_3\boldsymbol{u}_3 \\ \vdots &= \vdots \end{aligned}$$

where each is of the form

$$(\mathsf{Q}' - \lambda_i \mathsf{I})\boldsymbol{u}_i = 0$$

so the procedure is

1. Find the eigenvalues of Q'.
2. Find the eigenvectors for each λ_i.
3. Construct U from the set of eigenvectors.

Example A.3
Diagonalization. Let

$$\mathsf{Q}' = \begin{pmatrix} 1 & 0 & -2 \\ 0 & 0 & 0 \\ -2 & 0 & 4 \end{pmatrix}$$

The eigenvalues are obtained from

$$\begin{vmatrix} 1-\lambda & 0 & -2 \\ 0 & -\lambda & 0 \\ -2 & 0 & 4-\lambda \end{vmatrix} = (1-\lambda)(-\lambda)(4-\lambda) - 2(-2\lambda) = 0$$

whose roots give $\lambda = 0,\ 0,\ 5$. Beginning with $\lambda_1 = 0$, we have

$$\begin{pmatrix} 1 & 0 & -2 \\ 0 & 0 & 0 \\ -2 & 0 & 4 \end{pmatrix} \begin{pmatrix} u_{11} \\ u_{21} \\ u_{31} \end{pmatrix} = 0$$

which leads to $u_{11} = 2u_{31}$ and $u_{21} = 0$. The normalized eigenvector is therefore

$$\boldsymbol{u}_1 = \frac{1}{\sqrt{5}} \begin{pmatrix} 2 \\ 0 \\ 1 \end{pmatrix}$$

We have the same equation for $\lambda_2 = 0$, but since the eigenvectors must be orthogonal, we let $u_{22} = 1$ and $u_{12} = u_{32} = 0$ so

$$\boldsymbol{u}_2 = \begin{pmatrix} 0 \\ 1 \\ 0 \end{pmatrix}$$

For $\lambda_3 = 5$, we must solve

$$\begin{pmatrix} -4 & 0 & -2 \\ 0 & -5 & 0 \\ -2 & 0 & -1 \end{pmatrix} \begin{pmatrix} u_{13} \\ u_{23} \\ u_{33} \end{pmatrix} = 0$$

or

$$
\begin{aligned}
-4u_{13} - 2u_{33} &= 0 \\
-5u_{23} &= 0 \\
-2u_{13} - u_{33} &= 0
\end{aligned}
$$

so $u_{23} = 0$ and $u_{33} = -2u_{13}$. The normalized eigenvector is therefore

$$
\boldsymbol{u}_3 = \frac{1}{\sqrt{5}} \begin{pmatrix} 1 \\ 0 \\ 2 \end{pmatrix}
$$

so the unitary matrix is

$$
\mathsf{U} = \begin{pmatrix} \frac{2}{\sqrt{5}} & 0 & \frac{2}{\sqrt{5}} \\ 0 & 1 & 0 \\ \frac{1}{\sqrt{5}} & 0 & -\frac{2}{\sqrt{5}} \end{pmatrix}
$$

For this case, $\mathsf{U}^\dagger = \mathsf{U} = \mathsf{U}^{-1}$, so $\mathsf{UU} = \mathsf{I}$. It is straightforward to verify that

$$
\mathsf{Q} = \begin{pmatrix} 0 & 0 & 0 \\ 0 & 0 & 0 \\ 0 & 0 & 5 \end{pmatrix} = \Lambda\,.
$$

□

Problem A.1 *Diagonalization of a Hermitian matrix.* Given the Hermitian matrix

$$
\Gamma = \begin{pmatrix} \gamma & 0 & \mathrm{i}\beta\gamma \\ 0 & 1 & 0 \\ -\mathrm{i}\beta\gamma & 0 & \gamma \end{pmatrix}
$$

(a) Find the eigenvalues of Γ.

(b) Find the unitary matrix U that will diagonalize the matrix Γ.

(c) Verify by matrix multiplication that $\mathsf{UU}^\dagger = \mathsf{I}$.

(d) Verify that $\mathsf{U}^{-1}\Gamma\mathsf{U}$ is diagonal.

B

*Generating Functions**

B.1 Hermite Polynomials

For every set of polynomials described by

$$P_n(x) = \sum_{j=0}^{n} a_{jn} x^j \,,$$

another set may be constructed where every polynomial is orthogonal to every other, where by orthogonal we mean

$$\int_a^b P_n(x) P_{n'}(x) w(x)\,\mathrm{d}x = \begin{cases} N_n & n = n' \\ 0 & n \neq n' \end{cases} \,,$$

where $w(x)$ is a weight function. For a large set of polynomials of this type that satisfy a variety of second order ordinary differential equations, there exist **Generating Functions** that are analytic functions related to a specific weighted sum of *all* of the individual polynomials. The usefulness of generating functions is that recursion formulas that relate each polynomial to others of the set, orthogonality relations that guarantee the independence of each polynomial, normalization constants, and other general relations may often be found for *all n* rather than treating each polynomial one at a time. The Hermite polynomials are merely one example of such a set. The generating function for the Hermite polynomials may be expressed as

$$G(\xi, s) = \mathrm{e}^{\xi^2 - (s-\xi)^2} = \sum_{n=0}^{\infty} \frac{H_n(\xi)}{n!} s^n \,. \tag{B.1}$$

We demonstrate this by establishing a series of recursion formulas that will lead us back to the original differential equation, Equation (2.35), with $\lambda_n = n + \frac{1}{2}$, or

$$H_n'' - 2\xi H_n' + 2nH_n = 0 \,. \tag{B.2}$$

This demonstration will therefore establish that the harmonic oscillator wave functions are related to the Hermite polynomials.

First we differentiate Equation (B.1) with respect to ξ to obtain

$$\frac{\partial G}{\partial \xi} = 2sG(\xi, s) \tag{B.3}$$

$$= 2\sum_{m=0}^{\infty} \frac{H_m(\xi)}{m!} s^{m+1} \tag{B.4}$$

$$= \sum_{m=0}^{\infty} \frac{H'_m(\xi)}{m!} s^m . \tag{B.5}$$

Now we note that since this is true for any ξ and any s, the coefficients of each power of s must vanish, so changing $m+1$ to n in Equation (B.4) and m to n in Equation (B.5), we find

$$\sum_n \frac{2H_{n-1}(\xi)}{(n-1)!} s^n = \sum_n \frac{H'_n(\xi)}{n!} s^n .$$

Multiplying the term on the left by n in both the numerator and denominator and comparing the two sides, the result is

$$H'_n(\xi) = 2nH_{n-1}(\xi) . \tag{B.6}$$

We next differentiate with respect to s, with the result

$$\frac{\partial G}{\partial s} = -2(s - \xi)G(\xi, s) \tag{B.7}$$

$$= 2\sum_{m=0}^{\infty} \frac{\xi H_m(\xi)}{m!} s^m - 2\sum_{m=0}^{\infty} \frac{H_m(\xi)}{m!} s^{m+1} \tag{B.8}$$

$$= \sum_{m=0}^{\infty} \frac{H_m(\xi)}{(m-1)!} s^{m-1} . \tag{B.9}$$

For this case, we must set $m = n$ in the first term of Equation (B.8), $m+1 = n$ in the second term of Equation (B.8) and $m - 1 = n$ in Equation (B.9). This leads to

$$2\xi H_n(\xi) - 2nH_{n-1}(\xi) = H_{n+1}(\xi) . \tag{B.10}$$

We may then combine Equation (B.6) and Equation (B.10) to obtain

$$H'_n(\xi) = 2\xi H_n(\xi) - H_{n+1}(\xi) . \tag{B.11}$$

To demonstrate that the functions defined by the generating function are in fact the Hermite polynomials, we first differentiate Equation (B.11) to obtain

$$H''_n = 2\xi H'_n + 2H_n - H'_{n+1} ,$$

and then index Equation (B.6) by one ($n \to n+1$) to get

$$H''_n = 2\xi H'_n + 2H_n - 2(n+1)H_n ,$$

which is equivalent to Equation (B.2) for the Hermite polynomials.

One advantage of the generating function is that we may use it to normalize the wave functions. We examine the integral

$$\int_{-\infty}^{\infty} \mathrm{e}^{-\xi^2} G(\xi,s)G(\xi,t)\,\mathrm{d}\xi = \sum_{m,n}\int_{-\infty}^{\infty} \mathrm{e}^{-\xi^2}\frac{H_m(\xi)}{m!}\frac{H_n(\xi)}{n!}s^m t^n\,\mathrm{d}\xi\,, \tag{B.12}$$

where the integrand of the left-hand side is

$$\begin{aligned}\mathrm{e}^{-\xi^2+\xi^2-(s-\xi)^2+\xi^2-(t-\xi)^2} &= \mathrm{e}^{-\xi^2+2(s+t)\xi-s^2-t^2}\\ &= \mathrm{e}^{2st-(\xi-s-t)^2}\,.\end{aligned}$$

The integral on the left is then simple, with the result

$$\sqrt{\pi}\mathrm{e}^{2st} = \sum_{m,n}\frac{s^m t^n}{m!n!}\int_{-\infty}^{\infty}\mathrm{e}^{-\xi^2}H_m(\xi)H_n(\xi)\,\mathrm{d}\xi\,.$$

The term on the left can be expanded in a power series so that we have

$$\sqrt{\pi}\sum_{n=0}^{\infty}\frac{2^n s^n t^n}{n!} = \sum_{m,n=0}^{\infty}\frac{s^m t^n}{m!n!}\int_{-\infty}^{\infty}\mathrm{e}^{-\xi^2}H_m(\xi)H_n(\xi)\,\mathrm{d}\xi\,. \tag{B.13}$$

We note that on the left side of Equation (B.13), s and t appear always to the same power, and for arbitrary s and t, it must be the same on the right, so we have at once the orthogonality relation

$$\int_{-\infty}^{\infty}\mathrm{e}^{-\xi^2}H_m(\xi)H_n(\xi)\,\mathrm{d}\xi = 0 \qquad m\neq n\,. \tag{B.14}$$

When $m = n$, we can identify term by term and obtain

$$\int_{-\infty}^{\infty}\mathrm{e}^{-\xi^2}[H_n(\xi)]^2\,\mathrm{d}\xi = \sqrt{\pi}2^n n!\,. \tag{B.15}$$

We therefore have

$$\int_{-\infty}^{\infty}|\psi_n(x)|^2\,\mathrm{d}x = \frac{N_n^2}{\alpha}\int_{-\infty}^{\infty}\mathrm{e}^{-\xi^2}[H_n(\xi)]^2\,\mathrm{d}\xi = \frac{N_n^2\sqrt{\pi}2^n n!}{\alpha} = 1\,, \tag{B.16}$$

which leads to Equation (2.45).

We thus have constructed an orthonormal set of eigenfunctions for the harmonic oscillator.

Problem B.1 *Potential Energy.* Evaluate $\langle V\rangle$ for a general eigenstate. [*Hint:* Use Equation (B.10) to eliminate ξ from the integral.]

Problem B.2 *Expectation values from recursion formulas.*

(a) Use the recursion relation Equation (B.10) to evaluate $\langle\psi_n|x|\psi_m\rangle$. (For a given n, find for which values of m there is a nonzero result, and find the result for those special cases.)

(b) Use the recursion relation Equation (B.10) to evaluate $\langle\psi_n|x^2|\psi_m\rangle$.

Problem B.3 *Expectation values from the generating function.*

(a) Use the generating function to evaluate $\langle\psi_n|x|\psi_m\rangle$. [For a given n, find for which values of m there is a nonzero result, and find the result for those special cases. The procedure is to multiply the integrands of Equation (B.12) by ξ and solve for the appropriate integrals on the right.]

(b) Use the generating function to evaluate $\langle\psi_n|x^2|\psi_m\rangle$. [For this case, multiply the integrands of Equation (B.12) by ξ^2.]

B.2 Generating Function for the Legendre Polynomials

We begin with the associated Legendre equation of Equation (3.54), which we write as

$$[(1-\mu^2)P']' + \left[\ell(\ell+1) - \frac{m^2}{1-\mu^2}\right]P = 0\,. \tag{B.17}$$

The solutions of Equation (B.17) with $m = 0$ are called the **Legendre polynomials** and the solutions with $m \neq 0$ are called the **associated Legendre functions**.

B.2.1 Legendre Polynomials

As we discovered in Section B.1, generating functions permit one to find recursion relations and normalization constants for a variety of functions which are solutions of second-order differential equations. The generating function for the Legendre polynomials is

$$G(t,\mu) = \sum_{k=0}^{\infty} P_k(\mu)t^k = \frac{1}{\sqrt{1-2\mu t+t^2}}\,. \tag{B.18}$$

This may be demonstrated by establishing several recursion formulas. We first examine

$$\frac{\partial G}{\partial t} = \sum_{k=0}^{\infty} P_k k t^{k-1} = \frac{\mu - t}{(1-2\mu t+t^2)^{3/2}} = \frac{\mu - t}{(1-2\mu t+t^2)}\sum_{k=0}^{\infty} P_k t^k$$

so that

$$\sum_{k=0}^{\infty}[(1-2\mu t+t^2)P_k k t^{k-1}-(\mu-t)P_k t^k]=\sum_{k=0}^{\infty}[kP_k t^{k-1}-\mu(2k+1)P_k t^k$$
$$+(k+1)P_k t^{k+1}]=0\,. \quad \text{(B.19)}$$

Changing indices so that $k-1=\ell$ in the first term, $k=\ell$ in the second, and $k+1=\ell$ in the third and factoring out t^ℓ, we find

$$\sum_{\ell}[(\ell+1)P_{\ell+1}-\mu(2\ell+1)P_\ell+\ell P_{\ell-1}]t^\ell=0\,,$$

which requires each coefficient to vanish since t is arbitrary, so that

$$(\ell+1)P_{\ell+1}-\mu(2\ell+1)P_\ell+\ell P_{\ell-1}=0\,. \quad \text{(B.20)}$$

Similarly, we find

$$\frac{\partial G}{\partial\mu}=\sum_{k=0}^{\infty}P_k' t^k=\frac{t}{(1-2\mu t+t^2)^{3/2}}=\frac{t}{(1-2\mu t+t^2)}\sum_{k=0}^{\infty}P_k t^k$$

so we may write the two sums as

$$\sum_{k=0}^{\infty}[(1-2\mu t+t^2)P_k' t^k-tP_k t^k]=\sum_{k=0}^{\infty}[P_k' t^k-(2\mu P_k'+P_k)t^{k+1}$$
$$+(k+1)P_k t^{k+2}]=0\,.$$

Again shift indices with $\ell=k$, $\ell=k+1$, $\ell=k+2$, respectively, and factor out t^ℓ so that

$$\sum_{\ell}[P_\ell'-2\mu P_{\ell-1}'+P_{\ell-2}'-P_{\ell-1}]t^\ell=0$$

which leads to the recursion formula

$$P_\ell'-2\mu P_{\ell-1}'+P_{\ell-2}'-P_{\ell-1}=0\,, \quad \text{(B.21)}$$

which may be written more conveniently by shifting the index from $\ell\to\ell+1$ so that

$$P_{\ell+1}'-2\mu P_\ell'+P_{\ell-1}'-P_\ell=0\,. \quad \text{(B.22)}$$

Now if we differentiate Equation (B.20) and then multiply Equation (B.22) by $\ell+1$ and subtract, we obtain

$$\mu P_\ell'-P_{\ell-1}'-\ell P_\ell=0\,. \quad \text{(B.23)}$$

We could also multiply Equation (B.22) by ℓ and subtract from the derivative of Equation (B.20) and obtain

$$P_{\ell+1}'-\mu P_\ell'-(\ell+1)P_\ell=0\,. \quad \text{(B.24)}$$

Now we shift the index in Equation (B.24) from $\ell \to \ell-1$ and then multiply Equation (B.23) by μ and subtract to obtain

$$(1-\mu^2)P'_\ell + \mu\ell P_\ell - \ell P_{\ell-1} = 0.$$

If we differentiate this, we find

$$[(1-\mu^2)P'_\ell]' + \ell(\mu P'_\ell + P_\ell - P'_{\ell-1}) = 0$$

but from Equation (B.23), $\mu P'_\ell - P'_{\ell-1} = \ell P_\ell$, so we have

$$[(1-\mu^2)P'_\ell]' + \ell(\ell+1)P_\ell = 0\,. \tag{B.25}$$

This proves that the functions represented by Equation (B.18) are Legendre polynomials since they satisfy the differential equation, and we may use this representation to define the arbitrary constant for each polynomial. The choice of constants from the generating function agrees with the expressions derived from the Rodrigues formula of Equation (3.55).

B.2.2 Relating the Legendre Equation with the Associated Legendre Equation

To solve this equation, we begin by relating the solutions of Equation (B.17) with $m \neq 0$ to those with $m = 0$. If we define $P(\mu)$ to be a solution of

$$(1-\mu^2)\frac{\mathrm{d}^2P}{\mathrm{d}\mu^2} - 2\mu\frac{\mathrm{d}P}{\mathrm{d}\mu} + \ell(\ell+1)P = 0\,, \tag{B.26}$$

which is the Legendre equation, then we wish to examine P^m where

$$P^m(\mu) = (1-\mu^2)^{m/2}\frac{\mathrm{d}^mP}{\mathrm{d}\mu^m}\,. \tag{B.27}$$

Since $m = 0, \pm1, \pm2, \ldots$, we should write $|m|$ everywhere, but instead shall understand that $m \geq 0$ in all that follows in this section in order to avoid the cumbersome notation. We first need to differentiate Equation (B.26) m times. Term by term, we find, denoting $D = \mathrm{d}/\mathrm{d}\mu$,

$$\begin{aligned}
D[(1-\mu^2)D^2P] &= (1-\mu^2)D^3P - 2\mu D^2P \\
D^2[(1-\mu^2)D^2P] &= (1-\mu^2)D^4P - 4\mu D^3P - 2D^2P \\
D^3[(1-\mu^2)D^2P] &= (1-\mu^2)D^5P - 6\mu D^4P - 6D^3P \\
D^4[(1-\mu^2)D^2P] &= (1-\mu^2)D^6P - 8\mu D^5P - 12D^4P \\
&\vdots \\
D^m[(1-\mu^2)D^2P] &= (1-\mu^2)[D^mP]'' - 2m\mu[D^mP]' - m(m-1)[D^mP]\,,
\end{aligned}$$

for the first term, and

$$\begin{aligned}
D(2\mu P') &= 2\mu D^2 P + 2DP \\
D^2(2\mu P') &= 2\mu D^3 P + 4D^2 P \\
D^3(2\mu P') &= 2\mu D^4 P + 6D^3 P \\
&\vdots \\
D^m(2\mu P') &= 2\mu[D^m P]' + 2m[D^m P]
\end{aligned}$$

for the second term. The remaining term is trivial, so the resulting equation is

$$(1-\mu^2)[D^m P]'' - 2(m+1)\mu[D^m P]' + [\ell(\ell+1) - m(m+1)][D^m P] = 0\,. \tag{B.28}$$

Then we let $P^m = (1-\mu^2)^{m/2} W$ in Equation (B.17). The appropriate derivatives are

$$\begin{aligned}
DP^m &= (1-\mu^2)^{m/2} W' - m\mu(1-\mu^2)^{m/2-1} W \\
D^2 P^m &= (1-\mu^2)^{m/2} W'' - 2m\mu(1-\mu^2)^{m/2-1} W' - m(1-\mu^2)^{m/2-1} W \\
&\quad + m(m-2)\mu^2(1-\mu^2)^{m/2-2} W\,,
\end{aligned}$$

so the first two terms of Equation (B.17) are

$$\begin{aligned}
-2\mu DP^m &= -2\mu(1-\mu^2)^{m/2} W' + 2m\mu^2(1-\mu^2)^{m/2-1} W, \\
(1-\mu^2) D^2 P^m &= (1-\mu^2)^{m/2+1} W'' - 2m\mu(1-\mu^2)^{m/2} W' \\
&\quad - m(1-\mu^2)^{m/2} W + m(m-2)\mu^2(1-\mu^2)^{m/2-1} W\,.
\end{aligned}$$

Using these in Equation (B.17) and dividing out the common factor $(1-\mu^2)^{m/2}$, we find

$$\begin{aligned}
(1-\mu^2)W'' - 2m\mu W' - mW + \frac{m(m-2)\mu^2}{1-\mu^2} W - 2\mu W' + \frac{2m\mu^2}{1-\mu^2} W \\
+ \left[\ell(\ell+1) - \frac{m^2}{1-\mu^2}\right] W = 0.
\end{aligned}$$

Collecting terms, this becomes

$$(1-\mu^2)W'' - 2(m+1)\mu W' + [\ell(\ell+1) - m(m+1)]W = 0\,, \tag{B.29}$$

which, comparing with Equation (B.28), means that $W = D^m P$, so that comparing with Equation (B.27), we can conclude that $P^m = (1-\mu^2)^{m/2} D^m P$ is a solution of Equation (B.17) if P is a solution of Equation (B.26).

B.2.3 Orthogonality

To prove that the functions are orthogonal, we do not need the generating function, since we will use only the differential equation for the Associated

Legendre Polynomials. We may write

$$P_{\ell'}^m \left\{ [(1-\mu^2) P_\ell^{m\prime}]' + [\ell(\ell+1) - \frac{m^2}{1-\mu^2}] P_\ell^m \right\}$$
$$- P_\ell^m \left\{ [(1-\mu^2) P_{\ell'}^{m\prime}]' + \left[\ell'(\ell'+1) - \frac{m^2}{1-\mu^2} \right] P_{\ell'}^m \right\} = 0$$

or

$$(1-\mu^2)[P_{\ell'}^m P_\ell^{m\prime\prime} - P_\ell^m P_{\ell'}^{m\prime\prime}] - 2\mu[P_{\ell'}^m P_\ell^{m\prime} - P_\ell^m P_{\ell'}^{m\prime}]$$
$$= [\ell'(\ell'+1) - \ell(\ell+1)] P_{\ell'}^m P_\ell^m$$

or equivalently

$$\frac{\mathrm{d}}{\mathrm{d}\mu}\left[(1-\mu^2) \left(P_{\ell'}^m \frac{\mathrm{d}}{\mathrm{d}\mu} P_\ell^m - P_\ell^m \frac{\mathrm{d}}{\mathrm{d}\mu} P_{\ell'}^m \right) \right] = [\ell'(\ell'+1) - \ell(\ell+1)] P_{\ell'}^m P_\ell^m .$$

Now if we integrate from -1 to 1,

$$\int_{-1}^1 \frac{\mathrm{d}}{\mathrm{d}\mu}\left[(1-\mu^2) \left(P_{\ell'}^m \frac{\mathrm{d}}{\mathrm{d}\mu} P_\ell^m - P_\ell^m \frac{\mathrm{d}}{\mathrm{d}\mu} P_{\ell'}^m \right) \right] \mathrm{d}\mu = 0$$

so we have

$$[\ell'(\ell'+1) - \ell(\ell+1)] \int_{-1}^1 P_{\ell'}^m P_\ell^m \, \mathrm{d}\mu = 0 , \tag{B.30}$$

so the polynomials with different ℓ are orthogonal. For different m, the $\Phi_m(\phi)$ are orthogonal.

B.2.4 Normalization of the P_ℓ

We begin by letting $\ell \to \ell - 1$ in Equation (B.20) and solve for P_ℓ

$$P_\ell = \frac{1}{\ell}[\mu(2\ell-1)P_{\ell-1} - (\ell-1)P_{\ell-2}]$$

so that we find

$$\int_{-1}^1 |P_\ell|^2 \, \mathrm{d}\mu = \frac{1}{\ell} \int_{-1}^1 [\mu(2\ell-1)P_\ell P_{\ell-1} - (\ell-1)P_\ell P_{\ell-2}] \, \mathrm{d}\mu$$
$$= \frac{2\ell-1}{\ell} \int_{-1}^1 \mu P_\ell P_{\ell-1} \, \mathrm{d}\mu \tag{B.31}$$

since the polynomials are orthogonal. Then using Equation (B.20) to solve for μP_ℓ, we find

$$\mu P_\ell = \frac{1}{2\ell+1}[(\ell+1)P_{\ell+1} + \ell P_{\ell-1}]$$

so the integral

$$\int_{-1}^{1} \mu P_\ell P_{\ell-1}\,\mathrm{d}\mu = \frac{1}{2\ell+1}\int_{-1}^{1}[(\ell+1)P_{\ell+1}P_{\ell-1} + \ell P_{\ell-1}^2]\,\mathrm{d}\mu$$
$$= \frac{\ell}{2\ell+1}\int_{-1}^{1} P_{\ell-1}^2\,\mathrm{d}\mu\,, \tag{B.32}$$

so combining Equation (B.31) and Equation (B.32), we find

$$\int_{-1}^{1} |P_\ell|^2\,\mathrm{d}\mu = \frac{2\ell-1}{2\ell+1}\int_{-1}^{1}|P_{\ell-1}|^2\,\mathrm{d}\mu\,. \tag{B.33}$$

This is a recursion formula which we may lower to $\ell = 0$ such that

$$\int_{-1}^{1} |P_\ell|^2\,\mathrm{d}\mu = \left(\frac{2\ell-1}{2\ell+1}\right)\left(\frac{2\ell-3}{2\ell-1}\right)\cdots\left(\frac{3}{5}\right)\left(\frac{1}{3}\right)\int_{-1}^{1}|P_0|^2\,\mathrm{d}\mu$$
$$= \frac{2}{2\ell+1}\,. \tag{B.34}$$

B.2.5 Normalization of the P_ℓ^m

We begin by noting that (for $m \geq 0$)

$$\frac{\mathrm{d}}{\mathrm{d}\mu}P_\ell^m = \frac{\mathrm{d}}{\mathrm{d}\mu}\left[(1-\mu^2)^{m/2}\frac{\mathrm{d}^m}{\mathrm{d}\mu^m}P_\ell\right]$$
$$= (1-\mu^2)^{m/2}\frac{\mathrm{d}^{m+1}}{\mathrm{d}\mu^{m+1}}P_\ell - \mu m(1-\mu^2)^{m/2-1}\frac{\mathrm{d}^m}{\mathrm{d}\mu^m}P_\ell\,,$$

so that

$$(1-\mu^2)^{1/2}\frac{\mathrm{d}}{\mathrm{d}\mu}P_\ell^m = (1-\mu^2)^{m+1/2}\frac{\mathrm{d}^{m+1}}{\mathrm{d}\mu^{m+1}}P_\ell - \frac{\mu m(1-\mu^2)^{m/2}}{(1-\mu^2)^{1/2}}\frac{\mathrm{d}^m}{\mathrm{d}\mu^m}P_\ell$$
$$= P_\ell^{m+1} - \frac{\mu m}{(1-\mu^2)^{1/2}}P_\ell^m\,. \tag{B.35}$$

We may then write

$$(P_\ell^{m+1})^2 = (1-\mu^2)\left(\frac{\mathrm{d}P_\ell^m}{\mathrm{d}\mu}\right)^2 + 2\mu m P_\ell^m\frac{\mathrm{d}P_\ell^m}{\mathrm{d}\mu} + \frac{\mu^2 m^2}{1-\mu^2}(P_\ell^m)^2\,. \tag{B.36}$$

Integrating the first term on the right of Equation (B.36), we have

$$\int_{-1}^{1}(1-\mu^2)\left(\frac{\mathrm{d}P_\ell^m}{\mathrm{d}\mu}\right)^2\mathrm{d}\mu = (1-\mu^2)P_\ell^m\frac{\mathrm{d}P_\ell^m}{\mathrm{d}\mu}\bigg|_{-1}^{1}$$
$$- \int_{-1}^{1}P_\ell^m\frac{\mathrm{d}}{\mathrm{d}\mu}\left[(1-\mu^2)\frac{\mathrm{d}P_\ell^m}{\mathrm{d}\mu}\right]\mathrm{d}\mu$$
$$= \int_{-1}^{1}\left[\ell(\ell+1) - \frac{m^2}{1-\mu^2}\right](P_\ell^m)^2\,\mathrm{d}\mu \tag{B.37}$$

where the first term on the right of Equation (B.37) vanishes because of the $(1-\mu^2)$ factor and we have used the differential equation in the second term. Then for a general function $f(x)$,

$$\int xf\frac{\mathrm{d}f}{\mathrm{d}x}\,\mathrm{d}x = xf^2 - \int\left(f^2 + xf\frac{\mathrm{d}f}{\mathrm{d}x}\right)\,\mathrm{d}x$$

so

$$2\int xf\frac{\mathrm{d}f}{\mathrm{d}x}\,\mathrm{d}x = xf^2 - \int f^2\,\mathrm{d}x$$

so we may write for the integral of the second term on the right of Equation (B.36)

$$\int_{-1}^{1} 2\mu m P_\ell^m \frac{\mathrm{d}P_\ell^m}{\mathrm{d}\mu}\,\mathrm{d}\mu = m\mu(P_\ell^m)^2|_{-1}^{1} - m\int_{-1}^{1}(P_\ell^m)^2\,\mathrm{d}\mu\,. \tag{B.38}$$

The first term on the right of Equation (B.38) vanishes for $m = 0$ because it is multiplied by m and for $m \neq 0$ because $P_\ell^m \propto (1-\mu^2)^{|m|/2}$ that vanishes at the end points.

We now assemble the components of the integral of Equation (B.36) to obtain

$$\begin{aligned}\int_{-1}^{1}(P_\ell^{m+1})^2\,\mathrm{d}\mu &= \int_{-1}^{1}\left[\ell(\ell+1) - \frac{m^2}{1-\mu^2} - m + \frac{\mu^2 m^2}{1-\mu^2}\right](P_\ell^m)^2\,\mathrm{d}\mu \\ &= \int_{-1}^{1}[\ell(\ell+1) - m(m+1)](P_\ell^m)^2\,\mathrm{d}\mu \\ &= (\ell-m)(\ell+m+1)\int_{-1}^{1}(P_\ell^m)^2\,\mathrm{d}\mu\,.\end{aligned} \tag{B.39}$$

This is in the form of a recursion formula, so lowering the index by one to start,

$$\begin{aligned}\int_{-1}^{1}(P_\ell^m)^2\,\mathrm{d}\mu &= (\ell-m+1)(\ell+m)\int_{-1}^{1}(P_\ell^{m-1})^2\,\mathrm{d}\mu \\ &= [(\ell-m+1)(\ell-m+2)\cdots(\ell-1)\ell \\ &\quad \times(\ell+m)(\ell+m-1)\cdots(\ell+1)]\int_{-1}^{1}(P_\ell)^2\,\mathrm{d}\mu \\ &= \frac{\ell!}{(\ell-m)!}\frac{(\ell+m)!}{\ell!}\frac{2}{2\ell+1} \\ &= \frac{(\ell+m)!}{(\ell-m)!}\frac{2}{2\ell+1}\,.\end{aligned} \tag{B.40}$$

With this result, we can finally write the normalized wave functions as

$$Y_\ell^m(\theta,\phi) = \sqrt{\frac{(2\ell+1)(\ell-m)!}{4\pi(\ell+m)!}}P_\ell^m(\cos\theta)\mathrm{e}^{\mathrm{i}m\phi}\,, \tag{B.41}$$

which are commonly referred to as spherical harmonics.

B.3 Laguerre Polynomials

The generating function for the associated Laguerre polynomials is

$$U_s(\rho, u) = \sum_{r=s}^{\infty} \frac{L_r^s(\rho) u^r}{r!} = \frac{(-1)^s \mathrm{e}^{-\rho u/(1-u)} u^s}{(1-u)^{s+1}} . \tag{B.42}$$

B.3.1 Recursion Formulas for the Laguerre Polynomials

We begin with $s = 0$ and differentiate $U_0 \equiv U(\rho, u)$ with respect to u.

$$\begin{aligned}
\frac{\partial U}{\partial u} = \sum_{r=0}^{\infty} \frac{L_r(\rho) u^{r-1}}{(r-1)!} &= \frac{\mathrm{e}^{-\rho u/(1-u)}}{1-u} \left[\frac{1}{1-u} - \frac{\rho}{1-u} - \frac{\rho u}{(1-u)^2} \right] \\
&= \frac{\mathrm{e}^{-\rho u/(1-u)}}{(1-u)^3} (1 - \rho - u) \\
&= \frac{(1-\rho-u)}{(1-u)^2} U
\end{aligned}$$

so that multiplying by $(1-u)^2$, this is equivalent to

$$(1 - 2u + u^2) \sum_{r=0}^{\infty} \frac{L_r(\rho) u^{r-1}}{(r-1)!} = (1 - \rho - u) \sum_{r=0}^{\infty} \frac{L_r(\rho) u^r}{r!}$$

so that collecting terms proportional to $u^k/k!$, we obtain

$$\sum_k [L_{k+1} - 2kL_k + k(k-1)L_{k-1} - L_k + \rho L_k + kL_{k-1}] \frac{u^k}{k!} = 0$$

or simplifying, since u is arbitrary,

$$L_{k+1} + (\rho - 1 - 2k) L_k + k^2 L_{k-1} = 0 . \tag{B.43}$$

Similarly, we differentiate with respect to ρ to find

$$\frac{\partial U}{\partial \rho} = \sum_{r=0}^{\infty} \frac{L_r'(\rho)}{r!} u^r = -\frac{u}{1-u} \sum_{r=0}^{\infty} \frac{L_r(\rho)}{r!} u^r ,$$

so we multiply by $1 - u$ and collect terms proportional to $u^k/k!$ to obtain

$$\sum_k (L_k' - kL_{k-1}' + kL_{k-1}) \frac{u^k}{k!} = 0$$

so that

$$L_k' - kL_{k-1}' + kL_{k-1} = 0 . \tag{B.44}$$

Now we index Equation (B.44) by one to obtain

$$L'_{k+1}(\rho) = (k+1)[L'_k(\rho) - L_k(\rho)],$$

and then differentiate

$$L''_{k+1}(\rho) = (k+1)[L''_k(\rho) - L'_k(\rho)].$$

We index one more time to obtain

$$L''_{k+2} = (k+2)(L''_{k+1} - L'_{k+1}),$$

and then use the expressions for L''_{k+1} and L'_{k+1} to obtain

$$L''_{k+2} = (k+1)(k+2)(L''_k - 2L'_k + L_k).$$

Now we index Equation (B.43) to obtain

$$L_{k+2} + (\rho - 3 - 2k)L_{k+1} + (k+1)^2 L_k = 0$$

and then differentiate twice to get

$$L''_{k+2} + (\rho - 3 - 2k)L''_{k+1} + 2L'_{k+1} + (k+1)^2 L''_k = 0.$$

Using the expressions for L''_{k+2}, L''_{k+1}, and L'_{k+1}, this becomes

$$(k+2)(k+1)(L''_k - 2L'_k + L_k) + (\rho - 3 - 2k)(k+1)(L''_k - L'_k) \\ +2(k+1)(L'_k - L_k) + (k+1)^2 L''_k = 0$$

and after factoring out the $(k+1)$ factor and gathering terms, the result is

$$\rho L''_k + (1-\rho)L'_k + kL_k = 0 \tag{B.45}$$

which is the differential equation for the Laguerre Polynomials.

By means of the generating function, it may be shown that

$$L^{2\ell+1}_{n+\ell}(\rho) = \sum_{k=0}^{n-\ell-1} (-1)^{k+1} \frac{[(n+\ell)!]^2}{(n-\ell-1-k)!(2\ell+1+k)!k!}\rho^k. \tag{B.46}$$

B.3.2 Normalization

Consider the generating functions

$$U_s(\rho, u) = \sum_{r=s}^{\infty} \frac{L^s_r(\rho)u^r}{r!} = \frac{(-1)^s e^{-\rho u/(1-u)} u^s}{(1-u)^{s+1}},$$

and

$$V_s(\rho, v) = \sum_{t=s}^{\infty} \frac{L^s_t(\rho)v^t}{t!} = \frac{(-1)^s e^{-\rho v/(1-v)} v^s}{(1-v)^{s+1}}.$$

Integrate the product

$$I = \int_0^\infty e^{-\rho}\rho^{s+1}U_s(\rho,u)V_s(\rho,v)\,d\rho$$

$$= \sum_{r,t=s}^{\infty}\frac{u^r v^t}{r!t!}\int_0^\infty e^{-\rho}\rho^{s+1}L_r^s(\rho)L_t^s(\rho)\,d\rho\,. \tag{B.47}$$

Using the forms for $U_s(\rho,u)$ and $V_s(\rho,v)$, the integral is

$$I = \frac{(uv)^s}{[(1-u)(1-v)]^{s+1}}\int_0^\infty \rho^{s+1}e^{-\rho[1+u/(1-u)+v/(1-v)]}\,d\rho\,.$$

Using the definition of the Gamma function,

$$\Gamma(z) = \int_0^\infty x^{z-1}e^{-x}\,dx\,,$$

and letting $\rho[1+u/(1-u)+v/(1-v)] = x = \alpha\rho$, so that $dx = \alpha\,d\rho$, and $\rho = x/\alpha$, then

$$\int_0^\infty \rho^{s+1}e^{-\alpha\rho}\,d\rho = \int_0^\infty \frac{x^{s+1}}{\alpha^{s+1}}e^{-x}\frac{dx}{\alpha} = \frac{1}{\alpha^{s+2}}\Gamma(s+2) = \frac{(s+1)!}{\alpha^{s+2}}\,.$$

Then $\alpha = 1+u/(1-u)+v/(1-v) = (1-uv)/(1-u)(1-v)$ so that

$$I = \frac{(uv)^s(s+1)![(1-u)(1-v)]^{s+2}}{[(1-u)(1-v)]^{s+1}(1-uv)^{s+2}} = \frac{(uv)^s(s+1)!(1-u)(1-v)}{(1-uv)^{s+2}}\,,$$

but using the binomial expansion

$$(1-uv)^{-(s+2)} = \sum_{k=0}^{\infty}\frac{(s+k+1)!}{(s+1)!k!}(uv)^k\,,$$

We obtain the final result

$$I = (1-u-v+uv)(s+1)!\sum_{k=0}^{\infty}\frac{(s+k+1)!}{(s+1)!k!}(uv)^{s+k}$$

$$= (1-u-v+uv)\sum_{k=0}^{\infty}\frac{(s+k+1)!}{k!}(uv)^{s+k}\,. \tag{B.48}$$

For the normalization constant, we want the case for $t = r$, so

$$I = \sum_{r=s}^{\infty}\frac{(uv)^r}{(r!)^2}\int_0^\infty e^{-\rho}\rho^{s+1}[L_r^s(\rho)]^2\,d\rho \tag{B.49}$$

so picking terms in both Equation (B.48) and Equation (B.49) proportional to $(uv)^r$, we find

$$\begin{aligned}\int_0^\infty e^{-\rho}\rho^{s+1}[L_r^s(\rho)]^2\,d\rho &= (r!)^2\left[\frac{(r+1)!}{(r-s)!}+\frac{r!}{(r-s-1)!}\right]\\ &= \frac{(r!)^3}{(r-s)!}(r+1+r-s)\\ &= \frac{(r!)^3(2r+1-s)}{(r-s)!}\,.\end{aligned}$$

Then using $r = n+\ell$ and $s = 2\ell+1$, this leads to

$$A_{n\ell}^2\int_0^\infty e^{-\rho}\rho^{2\ell+2}[L_{n+\ell}^{2\ell+1}(\rho)]^2\,d\rho = A_{n\ell}^2\frac{[(n+\ell)!]^3(2n)}{(n-\ell-1)!} = 1$$

so finally

$$A_{n\ell} = \left[\frac{(n-\ell-1)!}{2n[(n+\ell)!]^3}\right]^{1/2}\,. \tag{B.50}$$

In order to complete the normalization, we need to return to the original variables. With $\rho = \alpha_n r = (2Z/na_0')r$, using the variable r in the normalizing integral ($r^2\,dr \to \rho^2\,d\rho/\alpha_n^3$), the final normalization constant is

$$A_{n\ell} = \left[\left(\frac{2Z}{na_0'}\right)^3\frac{(n-\ell-1)!}{2n[(n+\ell)!]^3}\right]^{1/2}\,. \tag{B.51}$$

B.3.3 Mean Values of r^k

The mean values of r^k for many values of k and any value of n and ℓ may be obtained by means of a recursion formula. We begin with the radial equation for $u(\rho) = \rho R_{n\ell}(\rho)$,

$$u'' + \left[-\frac{1}{4}+\frac{n}{\rho}-\frac{\ell(\ell+1)}{\rho^2}\right]u = 0 \tag{B.52}$$

Multiplying by $\rho^{k-1}u$ and integrating, the first integral is

$$\begin{aligned}\int_0^\infty \rho^{k-1}u\frac{du'}{d\rho}\,d\rho &= \rho^{k-1}uu'\Big|_0^\infty - \int_0^\infty u'[u'\rho^{k-1}+(k-1)\rho^{k-2}u]\,d\rho\\ &= -\int_0^\infty \rho^{k-1}(u')^2\,d\rho - (k-1)\int_0^\infty \rho^{k-2}uu'\,d\rho\\ &= \frac{1}{4}\langle\rho^{k-1}\rangle - n\langle\rho^{k-2}\rangle + \ell(\ell+1)\langle\rho^{k-3}\rangle\end{aligned} \tag{B.53}$$

since

$$\langle\rho^k\rangle = A^2\int_0^\infty \rho^k[R_{n\ell}(\rho)]^2\rho^2\,d\rho = A^2\int_0^\infty \rho^k u^2\,d\rho$$

and we assume $k \geq 0$ so that the end point limits vanish. We define

$$I \equiv \int_0^\infty \rho^{k-1}(u')^2 \, \mathrm{d}\rho \,.$$

Integrating the second integral on the second line of Equation (B.53) by parts, we have

$$\int_0^\infty \rho^{k-2} u u' \, \mathrm{d}\rho = \rho^{k-2} u^2 |_0^\infty - \int_0^\infty u[\rho^{k-2} u' + (k-2)\rho^{k-3} u] \mathrm{d}\rho$$

but the first integral on the right is identical to the integral on the left, so

$$\int_0^\infty \rho^{k-2} u u' \, \mathrm{d}\rho = -\frac{k-2}{2} \langle \rho^{k-3} \rangle \tag{B.54}$$

so the integral I is given by

$$I = \frac{(k-1)(k-2)}{2} \langle \rho^{k-3} \rangle - \frac{1}{4} \langle \rho^{k-1} \rangle + n \langle \rho^{k-2} \rangle - \ell(\ell+1) \langle \rho^{k-3} \rangle \,. \tag{B.55}$$

Next we multiply Equation (B.52) by $\rho^k u'$ and again integrate so that the integral on the left is

$$\int_0^\infty \rho^k u' \frac{\mathrm{d}u'}{\mathrm{d}\rho} \mathrm{d}\rho = \rho^k (u')^2 \big|_0^\infty - \int_0^\infty u' [\rho^k u'' + k\rho^{k-1} u'] \, \mathrm{d}\rho$$

but the first integral on the right is again identical to the integral on the left, so

$$\int_0^\infty \rho^k u' u'' \mathrm{d}\rho = -\frac{1}{2} k \int_0^\infty \rho^{k-1} (u')^2] \, \mathrm{d}\rho = -\tfrac{1}{2} k I \,. \tag{B.56}$$

The integrals of the remaining terms in Equation (B.52) are each of the form of Equation (B.54), or

$$\int_0^\infty \rho^k u u' \mathrm{d}\rho = -\frac{k}{2} \langle \rho^{k-1} \rangle \,.$$

When these terms are combined with Equation (B.56), we have another expression for I, such that

$$I = \frac{1}{4} \langle \rho^{k-1} \rangle - \frac{n(k-1)}{k} \langle \rho^{k-2} \rangle + \frac{(k-2)\ell(\ell+1)}{k} \langle \rho^{k-3} \rangle \,. \tag{B.57}$$

Combining this result with Equation (B.55) to eliminate I, we find the recursion formula,

$$k \langle \rho^{k-1} \rangle = 2n(2k-1) \langle \rho^{k-2} \rangle + [k(k-1)(k-2) - 4(k-1)\ell(\ell+1)] \langle \rho^{k-3} \rangle \,. \tag{B.58}$$

For the mean values of r^n, we need $\rho = 2Zr/na_0'$ so $\langle \rho^k \rangle = (2Z/na_0')^k \langle r^k \rangle$. This leads to the results

(a) $k = 0$

$$\langle \rho^{-2} \rangle = \frac{2\ell(\ell+1)}{n} \langle \rho^{-3} \rangle \tag{B.59}$$

or

$$\left\langle \frac{1}{r^3} \right\rangle = \frac{Z}{a_0' \ell(\ell+1)} \left\langle \frac{1}{r^2} \right\rangle . \tag{B.60}$$

(b) $k = 1$

$$\langle \rho^{-1} \rangle = \frac{1}{2n} \tag{B.61}$$

or

$$\left\langle \frac{1}{r} \right\rangle = \frac{Z}{n^2 a_0'} . \tag{B.62}$$

(c) $k = 2$

$$\langle \rho \rangle = 3n - 2\ell(\ell+1)\langle \rho^{-1} \rangle = 3n - \frac{\ell(\ell+1)}{n} \tag{B.63}$$

using Equation (B.61), so that

$$\langle r \rangle = \frac{a_0'}{2Z}[3n^2 - \ell(\ell+1)] . \tag{B.64}$$

(d) $k = 3$

$$3\langle \rho^2 \rangle = 10n\langle \rho \rangle + 6 - 8\ell(\ell+1) = 30n^2 + 6 - 18\ell(\ell+1) \tag{B.65}$$

using Equation (B.63), or

$$\langle r^2 \rangle = \frac{n^2 a_0'^2}{2Z^2}[5n^2 + 1 - 3\ell(\ell+1)] . \tag{B.66}$$

where normalization assures us that $\langle \rho^0 \rangle = 1$. This process can be extended to obtain $\langle r^k \rangle$ for $k > 2$, but there is a gap for the negative powers since we cannot obtain $\langle r^{-2} \rangle$ by this method, and without this value, we cannot find $\langle r^{-3} \rangle$. For this expression, we must use another method. The result for $\langle r^{-2} \rangle$ is

$$\left\langle \frac{1}{r^2} \right\rangle = \frac{2Z^2}{n^3 a_0'^2 (2\ell+1)} , \tag{B.67}$$

and using this result, we find $\langle r^{-3} \rangle$ from Equation (B.60) to be

$$\left\langle \frac{1}{r^3} \right\rangle = \frac{Z^3}{n^3 a_0'^3 \ell(\ell+\frac{1}{2})(\ell+1)} . \tag{B.68}$$

C

Answers to Selected Problems

C.1 The Foundations of Quantum Physics

Problem 1.2 Defining $u \equiv hc/\lambda\kappa T$, then $u_m = 4.965$ so that $\lambda_{\text{max}} = .002898$ m-K or $\lambda_{\text{max}} = .2497\,\mu$-eV, where the wavelength is in microns and the temperature is expressed in eV. For example, in Fig. 2.1, the middle curve corresponds to 2400 K, so $T = 2400/11604 = .2068$ eV and $\lambda_{\text{max}} = .2497/.2068 = 1.21\ \mu$, as indicated in the figure.

Problem 1.10 *Mean values.*

(a) $|A| = \sqrt{3/2a}$.

(b) $\langle x \rangle = 0$.

(c) $\langle x^2 \rangle = a^2/10$.

(d) $\langle p \rangle = 0$.

(e) $\langle k^2 \rangle = 3/a^2$.

(f) $(\Delta x)(\Delta p) = \sqrt{3/10}$ which is 9.5% above the minimum.

Problem 1.18 $|A| = 1/(\pi a_0^3)^{1/2}$, $\langle x \rangle = 0$, $\langle x^2 \rangle = a_0^2$, so $(\Delta x)^2 = a_0^2$. $\langle 1/r \rangle = 1/a_0$.

Problem 1.23 $[\hat{L}_x, \hat{L}_y] = \mathrm{i}\hbar \hat{L}_z$.

C.2 The Schrödinger Equation in One–Dimension

Problem 2.4 *Partial Ans.* The meaning of $|C/A| > 1$ is that the probability/m is higher on the right side since the beam *slows down* and particles are *closer together*. Nevertheless, the *current* on the right is less than on the left since $T = 1 - R < 1$.

Problem 2.6 $V_i = E - \sqrt{E(E-V)}$, and $a = (n+\frac{1}{2})\pi\hbar/[2m\sqrt{E(E-V)}]^{1/2}$ where n is an integer.

Problem 2.18 *Nonclassical behavior.*

(a) $A = \sqrt{\hbar\omega/k}$.

(b) $P_{|x|>A} = \mathrm{erfc}(kA) = \mathrm{erfc}(1) = 0.1573$.

(c) $P_{|x|>2A} = \mathrm{erfc}(2) = 0.00468$.

Problem 2.21 *Mixed states.*

(a)

$$\Psi(x,t) = \sqrt{\frac{2}{7}}\Psi_0(x,0)\mathrm{e}^{-\mathrm{i}\omega t/2} + \frac{\mathrm{i}}{\sqrt{7}}\Psi_1(x,0)\mathrm{e}^{-3\mathrm{i}\omega t/2} + \frac{2}{\sqrt{7}}\Psi_2(x,0)\mathrm{e}^{-5\mathrm{i}\omega t/2}$$

(b) $\langle E\rangle = \frac{25}{14}\hbar\omega$.

Problem 2.22 *Expectation values from the raising and lowering operators.*

(a) $\langle\psi_n|x|\psi_{n+1}\rangle = \sqrt{n+1}/\sqrt{2}\alpha$. $\langle\psi_n|x|\psi_{n-1}\rangle = \sqrt{n}\sqrt{2}\alpha$.

(b)

$$\begin{aligned}
\langle\psi_n|x^2|\psi_n\rangle &= \frac{n+\frac{1}{2}}{\alpha^2}, \\
\langle\psi_n|x^2|\psi_{n+2}\rangle &= \frac{\sqrt{(n+1)(n+2)}}{2\alpha^2}, \\
\langle\psi_n|x^2|\psi_{n-2}\rangle &= \frac{\sqrt{n(n-1)}}{2\alpha^2}.
\end{aligned}$$

All other values vanish.

(d)

$$\begin{aligned}
\langle\psi_m|q^4|\psi_n\rangle = \frac{1}{4}[&\sqrt{(n+1)(n+2)(n+3)(n+4)}\delta_{m,n+4} \\
&+(4n+6)\sqrt{(n+1)(n+2)}\delta_{m,n+2} \\
&+3(2n^2+2n+1)\delta_{m,n} + (4n-2)\sqrt{n(n-1)}\delta_{m,n-2} \\
&+\sqrt{n(n-1)(n-2)(n-3)}\delta_{m,n-4}]\,. \qquad \text{(C.1)}
\end{aligned}$$

C.3 The Schrödinger Equation in Three Dimensions

Problem 3.8 *Uncertainty in angular momentum.*

(a)

$$L_x = \tfrac{1}{2}(L_+ + L_-)\,, \qquad L_y = \tfrac{1}{2i}(L_+ - L_-)\,.$$

(b)

$$\begin{aligned}
\langle L_x \rangle &= \langle L_x \rangle = 0\,, \\
\langle L_z \rangle &= m\hbar\,, \\
\langle L_x^2 \rangle &= \tfrac{\hbar^2}{2}[\ell(\ell+1) - m^2]\,, \\
\langle L_y^2 \rangle &= \tfrac{\hbar^2}{2}[\ell(\ell+1) - m^2]\,, \\
\langle L_z^2 \rangle &= m^2\hbar^2\,.
\end{aligned}$$

Hence the uncertainties are

$$\begin{aligned}
(\Delta L_x) &= \hbar\sqrt{\ell(\ell+1) - m^2}/\sqrt{2}\,, \\
(\Delta L_y) &= \hbar\sqrt{\ell(\ell+1) - m^2}/\sqrt{2}\,, \\
(\Delta L_z) &= 0\,.
\end{aligned}$$

(c) Since $\langle L^2 \rangle = \langle L_x^2 \rangle + \langle L_y^2 \rangle + \langle L_z^2 \rangle$, this means that $\langle L_x^2 \rangle + \langle L_y^2 \rangle = 0$, and since both terms are nonnegative, each must individually vanish. Since $\langle L_x \rangle = \langle L_y \rangle = 0$ also, this implies that $(\Delta L_x) = (\Delta L_y) = 0$.

Problem 3.15 $r_{\max} = 2n/\alpha_n = n^2 a_0/Z$.

Problem 3.22 *Neutron potential well depth.* $V_0 = 19.8$ MeV.

Problem 3.25 *Charmonium energy levels.*

(a)

$$\begin{aligned}
M_1c^2 &= 2.42 + (.3)2.3381 = 3.121 \text{ GeV}\,, \\
M_2c^2 &= 2.42 + (.3)4.0879 = 3.646 \text{ GeV}\,, \\
M_3c^2 &= 2.42 + (.3)5.5206 = 4.076 \text{ GeV}\,, \\
M_4c^2 &= 2.42 + (.3)6.7867 = 4.456 \text{ GeV}\,,
\end{aligned}$$

$$\begin{aligned}
3.121/3.097 &= 1.0079 \quad \text{so} \quad +0.79\%\,, \\
3.646/3.686 &= 0.9892 \quad \text{so} \quad -1.08\%\,, \\
4.076/4.100 &= 0.9942 \quad \text{so} \quad -0.58\%\,, \\
4.456/4.414 &= 1.0095 \quad \text{so} \quad +0.95\%\,.
\end{aligned}$$

(b) $V_0 = 1.26$ GeV, $g = 1.13$ GeV$/f$.

(c) $r_m = (1.26 - .583)/1.13 = 0.594$ f.

(d) $\Delta E = 2(.197/.297)^2/1.84 = 0.478$ GeV. The lowest state energy is then $E = E_1 + .478 = 3.575$ GeV which is 2.4% above 3.49 GeV, the average of the three $\ell = 1$ levels.

Problem 3.26 *Partial Ans.* (c) $A = r_0^{12}V_0 = 2.59 \times 10^{-118}$ eV-m^{12}, $B = 2r_0^6 V_0 = 4.56 \times 10^{-59}$ eV-m^6, and $\frac{1}{2}\hbar\omega = 4.1 \times 10^{-20}$ J or 0.257 eV.

C.4 Total Angular Momentum

Problem 4.1 *Eigenfunctions for $\hat{J}_y$.*

(a)

$$\begin{aligned}
\chi_0 &= \tfrac{1}{\sqrt{2}}(|1,1\rangle + |1,-1\rangle)\,.\\
\chi_1 &= -\tfrac{i}{2}|1,1\rangle + \tfrac{1}{\sqrt{2}}|1,0\rangle + \tfrac{i}{2}|1,-1\rangle\,.\\
\chi_{-1} &= \tfrac{i}{2}|1,1\rangle + \tfrac{1}{\sqrt{2}}|1,0\rangle - \tfrac{i}{2}|1,-1\rangle\,.
\end{aligned}$$

(b)

$$\begin{aligned}
\chi_{\frac{3}{2}} &= -\tfrac{i}{2\sqrt{2}}|\tfrac{3}{2},\tfrac{3}{2}\rangle + \tfrac{\sqrt{3}}{2\sqrt{2}}|\tfrac{3}{2},\tfrac{1}{2}\rangle + \tfrac{i\sqrt{3}}{2\sqrt{2}}|\tfrac{3}{2},-\tfrac{1}{2}\rangle - \tfrac{1}{2\sqrt{2}}|\tfrac{3}{2},-\tfrac{3}{2}\rangle\,.\\
\chi_{\frac{1}{2}} &= -\tfrac{i\sqrt{3}}{2\sqrt{2}}|\tfrac{3}{2},\tfrac{3}{2}\rangle + \tfrac{1}{2\sqrt{2}}|\tfrac{3}{2},\tfrac{1}{2}\rangle - \tfrac{i}{2\sqrt{2}}|\tfrac{3}{2},-\tfrac{1}{2}\rangle + \tfrac{\sqrt{3}}{2\sqrt{2}}|\tfrac{3}{2},-\tfrac{3}{2}\rangle\,.\\
\chi_{-\frac{1}{2}} &= \tfrac{i\sqrt{3}}{2\sqrt{2}}|\tfrac{3}{2},\tfrac{3}{2}\rangle + \tfrac{1}{2\sqrt{2}}|\tfrac{3}{2},\tfrac{1}{2}\rangle + \tfrac{i}{2\sqrt{2}}|\tfrac{3}{2},-\tfrac{1}{2}\rangle + \tfrac{\sqrt{3}}{2\sqrt{2}}|\tfrac{3}{2},-\tfrac{3}{2}\rangle\,.\\
\chi_{-\frac{3}{2}} &= \tfrac{i}{2\sqrt{2}}|\tfrac{3}{2},\tfrac{3}{2}\rangle + \tfrac{\sqrt{3}}{2\sqrt{2}}|\tfrac{3}{2},\tfrac{1}{2}\rangle - \tfrac{i\sqrt{3}}{2\sqrt{2}}|\tfrac{3}{2},-\tfrac{1}{2}\rangle - \tfrac{1}{2\sqrt{2}}|\tfrac{3}{2},-\tfrac{3}{2}\rangle\,.
\end{aligned}$$

Problem 4.4 *Eigenfuctions of $\hat{S}_x$ and $\hat{S}_y$.* First, we have for S_x,

$$\chi_{\frac{1}{2}} = |\rightarrow\rangle = \frac{1}{\sqrt{2}}\begin{pmatrix}1\\1\end{pmatrix}, \qquad \chi_{-\frac{1}{2}} = |\leftarrow\rangle = \frac{1}{\sqrt{2}}\begin{pmatrix}1\\-1\end{pmatrix}.$$

Then for S_y, we have

$$\chi_{\frac{1}{2}} = |\nearrow\rangle = \frac{1}{\sqrt{2}}\begin{pmatrix}1\\i\end{pmatrix}, \qquad \chi_{-\frac{1}{2}} = |\swarrow\rangle = \frac{1}{\sqrt{2}}\begin{pmatrix}1\\-i\end{pmatrix}.$$

C.5 Approximation Methods

Problem 5.3 *First order with $\hat{\mathcal{H}}^{(1)} = cx^4$.*

(a)

$$E_n^{(1)} = \frac{3c}{4\alpha^4}(2n^2 + 2n + 1)\,. \tag{C.2}$$

(b)

$$\psi_n^{(1)} = \sum_{j \neq n} a_{nj}\psi_n^{(0)}\,,$$

where the nonzero coefficients are

$$\begin{aligned}
a_{n,n+4} &= -\frac{c\sqrt{(n+1)(n+2)(n+3)(n+4)}}{16\alpha^4\hbar\omega} \\
a_{n,n+2} &= -\frac{c(2n+3)\sqrt{(n+1)(n+2)}}{4\alpha^4\hbar\omega} \\
a_{n,n-2} &= \frac{c(2n-1)\sqrt{n(n-1)}}{4\alpha^4\hbar\omega} \\
a_{n,n-4} &= \frac{c\sqrt{n(n-1)(n-2)(n-3)}}{16\alpha^4\hbar\omega}\,.
\end{aligned}$$

Problem 5.4 *Second order with $\hat{\mathcal{H}}^{(1)} = cx^4$.*

$$E_n^{(2)} = -\frac{c^2}{8\alpha^8\hbar\omega}(34n^3 + 51n^2 + 59n + 21)\,. \tag{C.3}$$

The total energy through second order is then

$$E_n = \hbar\omega\left(n + \frac{1}{2}\right) + \frac{3c}{4\alpha^4}\left[2n^2 + 2n + 1 - \frac{c}{6\alpha^4\hbar\omega}(34n^3 + 51n^2 + 59n + 21)\right] \tag{C.4}$$

Problem 5.8 *Two-dimensional harmonic oscillator.*

(a)

$$\begin{aligned}
\psi_{00} &= \psi_0(x)\psi_0(y)\,, & E_{00} &= \hbar\omega\,, \\
\psi_{10} &= \psi_1(x)\psi_0(y)\,, & E_{10} &= 2\hbar\omega\,, \\
\psi_{01} &= \psi_0(x)\psi_1(y)\,, & E_{01} &= 2\hbar\omega\,.
\end{aligned}$$

(b)

$$E_{00}^{(2)} = -\frac{21\hbar\omega b^2}{64\alpha^4}\,.$$

The two zero-order degenerate eigenstates are ψ_{10} and ψ_{01}, labeled

$$\psi_n = \frac{1}{\sqrt{2}}(\psi_{10} + \psi_{01}),$$
$$\psi_\ell = \frac{1}{\sqrt{2}}(\psi_{10} - \psi_{01}).$$

The eigenvalues are $E_n = \hbar\omega(2 + 3b/4\alpha^2)$, and $E_\ell = \hbar\omega(2 - 3b/4\alpha^2)$

Problem 5.15 *Spring break.* $P_1 = 0.956$, $P_3 = 3.2 \times 10^{-4}$.

Problem 5.16 *Tritium decay.* $P_1 = 0.702$

Problem 5.19

$$E = \left(\frac{3}{2}\right)^{5/3}\left(\frac{\hbar^2 C^2}{m}\right)^{1/3}.$$

Problem 5.25 *Transmission coefficient.* $\eta = 0.188$, $E = 98.8$ eV.

C.6 Atomic Spectroscopy

Problem 6.3 *Particle exchange operator.*

(a) $P_{12}\psi^*_{ab}\psi_{ab} = \psi^*_{ba}\psi_{ba}$ which is not the same. In a similar fashion, $P_{12}\psi^*_{ba}\psi_{ba} = \psi^*_{ab}\psi_{ab}$, which again is not the same.

(b)

$$P_{12}\psi^*_S\psi_S = \tfrac{1}{2}(|\psi_{ba}|^2 + |\psi_{ab}|^2 + \psi^*_{ba}\psi_{ab} + \psi^*_{ab}\psi_{ba}),$$

which is unchanged. For the antisymmetric case,

$$P_{12}\psi^*_A\psi_A = \tfrac{1}{2}(|\psi_{ba}|^2 + |\psi_{ab}|^2 - \psi^*_{ba}\psi_{ab} - \psi^*_{ab}\psi_{ba}),$$

which is again unchanged.

(c) Yes, because eigenfunctions of the interchange operator (ψ_S and ψ_A) are invariant in time (since P_{12} commutes with the Hamiltonian) while states described by the individual pieces are not invariant.

Problem 6.6 *Two equivalent d electrons.* The appropriate figure is given in Figure C.1 and the corresponding table is given in Table C.1.

Problem 6.7 This figure is similar to the figure for the previous problem except that the exclusion principle plays no role, and the L levels combine a d level ($\ell = 2$) with an f level ($\ell = 3$), so the L levels are $L = 1, 2, 3, 4, 5$ (P, D, F, G, H) instead of $L = 0, 1, 2, 3, 4$. The levels are sketched in Figure C.2.

TABLE C.1
Values of m_ℓ and m_s for two equivalent d electrons.

$m_{\ell 1}$	$m_{\ell 2}$	m_{s1}	m_{s2}	label	$m_{\ell 1}$	$m_{\ell 2}$	m_{s1}	m_{s2}	label	$m_{\ell 1}$	$m_{\ell 2}$	m_{s1}	m_{s2}	label
2	2	+	+	out	0	2	+	+	6	−2	2	+	+	14
2	2	+	−	1	0	2	+	−	8	−2	2	+	−	16
2	2	−	+	1	0	2	−	+	7	−2	2	−	+	15
2	2	−	−	out	0	2	−	−	9	−2	2	−	−	17
2	1	+	+	2	0	1	+	+	19	−2	1	+	+	27
2	1	+	−	3	0	1	+	−	21	−2	1	+	−	29
2	1	−	+	4	0	1	−	+	20	−2	1	−	+	28
2	1	−	−	5	0	1	−	−	22	−2	1	−	−	30
2	0	+	+	6	0	0	+	+	out	−2	0	+	+	36
2	0	+	−	7	0	0	+	−	31	−2	0	+	−	38
2	0	−	+	8	0	0	−	+	31	−2	0	−	+	37
2	0	−	−	9	0	0	−	−	out	−2	0	−	−	39
2	−1	+	+	10	0	−1	+	+	32	−2	−1	+	+	41
2	−1	+	−	11	0	−1	+	−	33	−2	−1	+	−	43
2	−1	−	+	12	0	−1	−	+	34	−2	−1	−	+	42
2	−1	−	−	13	0	−1	−	−	35	−2	−1	−	−	44
2	−2	+	+	14	0	−2	+	+	36	−2	−2	+	+	out
2	−2	+	−	15	0	−2	+	−	37	−2	−2	+	−	45
2	−2	−	+	16	0	−2	−	+	38	−2	−2	−	+	45
2	−2	−	−	17	0	−2	−	−	39	−2	−2	−	−	out
1	2	+	+	2	−1	2	+	+	10					
1	2	+	−	4	−1	2	+	−	12					
1	2	−	+	3	−1	2	−	+	11					
1	2	−	−	5	−1	2	−	−	13					
1	1	+	+	out	−1	1	+	+	23					
1	1	+	−	18	−1	1	+	−	25					
1	1	−	+	18	−1	1	−	+	24					
1	1	−	−	out	−1	1	−	−	26					
1	0	+	+	19	−1	0	+	+	32					
1	0	+	−	20	−1	0	+	−	34					
1	0	−	+	21	−1	0	−	+	33					
1	0	−	−	22	−1	0	−	−	35					
1	−1	+	+	23	−1	−1	+	+	out					
1	−1	+	−	24	−1	−1	+	−	40					
1	−1	−	+	25	−1	−1	−	+	40					
1	−1	−	−	26	−1	−1	−	−	out					
1	−2	+	+	27	−1	−2	+	+	41					
1	−2	+	−	28	−1	−2	+	−	42					
1	−2	−	+	29	−1	−2	−	+	43					
1	−2	−	−	30	−1	−2	−	−	44					

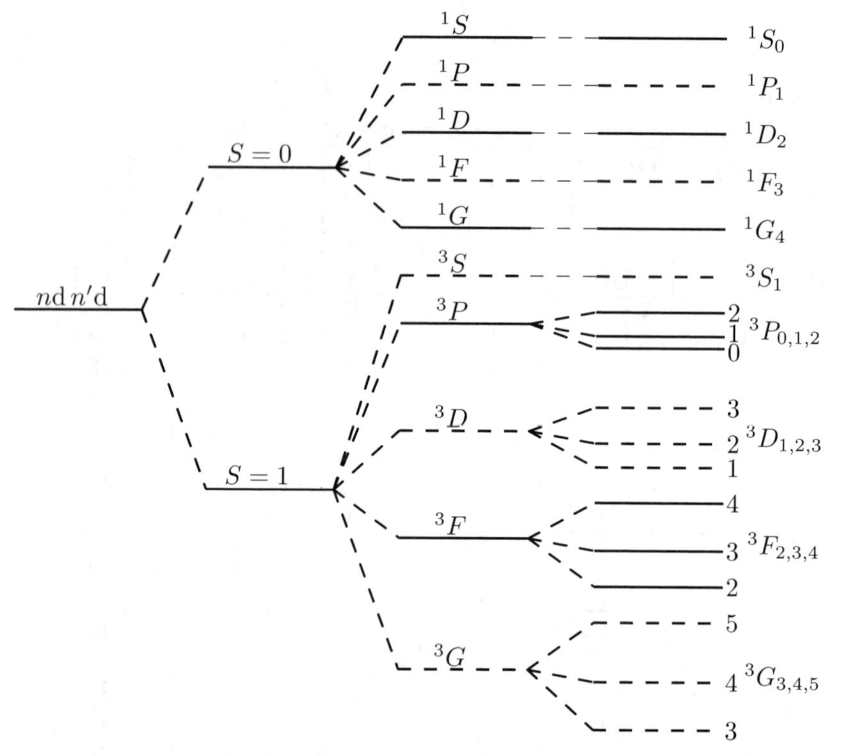

FIGURE C.1
LS coupling diagram for two d electrons. If $n = n'$, then the exclusion principle excludes states with dashed lines.

Problem 6.9 For the upper $^4\mathrm{F}_{3/2}$ state, $g = 2/5$ so the energy levels are $\Delta E = \mu_B B(\pm 1, \pm 3)/5$. For the lower $^4\mathrm{D}_{5/2}$ state, $g = 48/35$ so the energy levels are $\Delta E = 24\mu_B B(\pm 1, \pm 3, \pm 5)/35$. The splitting is into the 12 levels $\Delta W = \mu_B B(\pm 3, \pm 17, \pm 31, \pm 51, \pm 65, \pm 99)/35$.

Problem 6.10 *Zeeman width.* (a) The maximum spread is from $-\frac{41}{15}\Delta$ to $\frac{41}{15}\Delta$, or $\frac{82}{15}\Delta$. The classical case is with $g = 1$, $M_J = \pm 1$, so the classical spread is 2Δ, and the ratio is $41/15$.

C.7 Quantum Statistics

Problem 7.3 *The Stirling approximation to the factorial function. Partial Answer*: (c) See Table C.2.

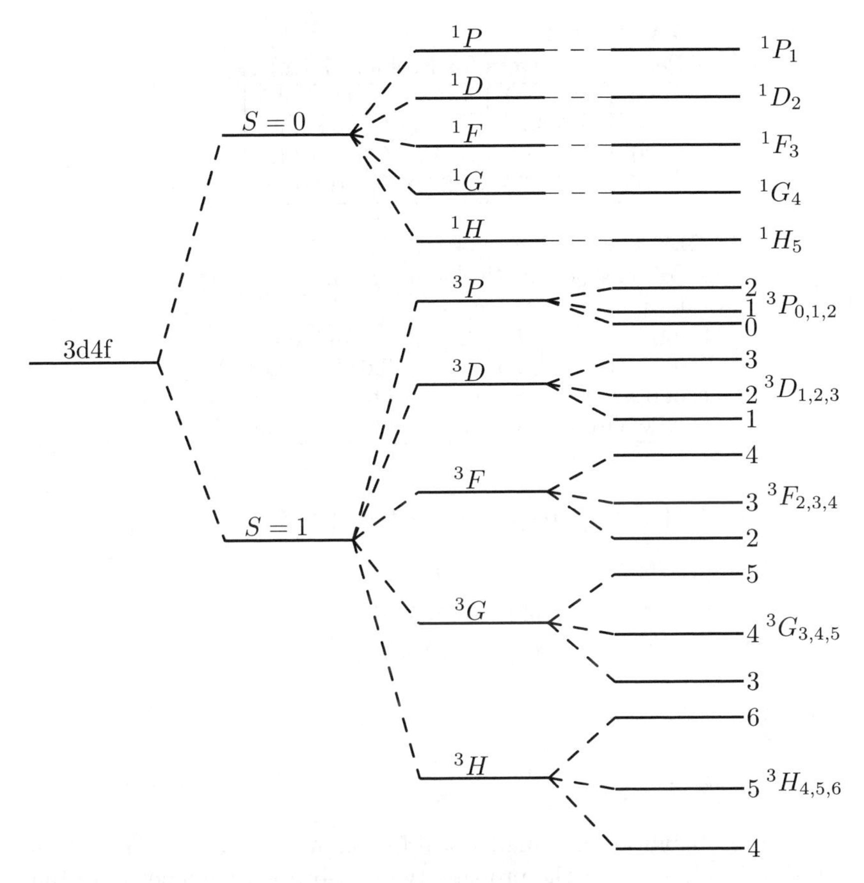

FIGURE C.2
LS coupling diagram for a 3d and a 4f electron.

Problem 7.7

$$\frac{1}{\beta} = \frac{2E}{3N} = \frac{2}{3}\langle \epsilon \rangle ,$$
$$\alpha = \frac{3}{2}\ln\left(\frac{4\pi m E}{3Nh^2}\right) - \ln\frac{N}{V} .$$

Problem 7.8 *Ionization of hydrogen.*

(a)

$$n_{21}(\epsilon) = \frac{d_2}{d_1}\mathrm{e}^{(\epsilon_1-\epsilon_2)/\kappa T}$$

(b) Table of energies

TABLE C.2

Percentage errors for Problem 7.1(c).

	$\delta n = \pm 1$	$\delta n = \pm 2$	$\delta n = \pm 5$
5	5.36%	4.39%	-6.55%
20	0.81%	0.77%	0.49%
400	.02083%	.02078%	.02042%

TABLE C.3

Table of energies for Problem 7.8(b).

State	n	d_n	ϵ_n
Ground state	1	2	-13.6 eV
First excited	2	8	$-13.6/4 = -3.4$ eV
Second excited	3	18	$-13.6/9 = -1.51$ eV
Third excited	4	32	$-13.6/16 = -.85$ eV

Then since $5,000^\circ$ K corresponds to $\kappa T = .43085$ eV,

$$n_{21} = 2.09 \times 10^{-10}$$
$$n_{31} = 5.87 \times 10^{-12}$$
$$n_{41} = 2.25 \times 10^{-12}$$

(c) For the $3 \rightarrow 2$ Balmer line,

$$n_{32} = \frac{d_3}{d_2} e^{(\epsilon_2 - \epsilon_3)/\kappa T} = \frac{18}{8} e^{(-3.4+1.51)/.43085} = .028$$

and the probabilities are smaller still for the higher Balmer transitions ($4 \rightarrow 2, 5 \rightarrow 2$, etc.) so the probability of absorption is higher than the probability of emission (since more electrons in the lower state). Hence, the black-body radiation from the sun is preferentially absorbed at those wavelengths, producing an absorption spectrum.

Problem 7.14 The Fermi energy relative to the valence band edge is given by Equation (7.42), so for silicon, we have

$$\epsilon_F = 0.538 \text{ eV}.$$

For InSb, we have

$$\epsilon_F = 0.152 \text{ eV}.$$

For Si, the intrinsic density is

$$n_i = 1.03 \times 10^{16}$$

which is within 3% of the listed value, while for InSb,

$$n_i = 2.68 \times 10^{22}$$

which is about a factor of two higher than the listed value.

Problem 7.18 *Degenerate stars.*

(b) $\langle E\rangle_e/m_ec^2 = 0.23$ and $\langle E\rangle_n/m_nc^2 = 1.7 \times 10^{-7}$.

Problem 7.23 *The λ point.* $\sum_{p=1}^{\infty} \frac{1}{p^{3/2}} \simeq 2.61$. The number of atoms is given by the total mass divided by the mass per atom, so the density is given by

$$\frac{N}{V} \equiv n = \frac{M_{tot}}{V}\frac{1}{M_a} = 2.26 \times 10^{28}\,.$$

Then

$$T_c = \left(\frac{nh^3}{2.61}\right)^{2/3} \frac{1}{2\pi m\kappa} = 3.2^\circ\,.$$

C.8 Band Theory of Solids

Problem 8.6 $\langle p\rangle = \langle p_k\rangle + \hbar k$ where

$$\langle p_k\rangle = \int_D^{D+a} \pi_k^* \frac{\hbar}{\mathrm{i}}\frac{\mathrm{d}\pi_k}{\mathrm{d}x}\,\mathrm{d}x\,.$$

Problem 8.7 $\langle T\rangle = \frac{\hbar^2k^2}{2m} + \frac{\hbar k}{m}\langle p_k\rangle + \langle T_k\rangle$, where

$$\langle T_k\rangle = -\frac{\hbar^2}{2m}\int_D^{D+a} \pi_k^* \frac{\mathrm{d}^2\pi_k}{\mathrm{d}x^2}\,\mathrm{d}x\,.$$

Problem 8.9 *Hall coefficient.*

(a) The Hall coefficient is given by $R = 1/ne$, where $n =$(mass/volume)/(mass/atom)$= 8.32\times10^3/(62.9\cdot1.66\times10^{-27}) = 7.97\times10^{28}$ /m^3. Therefore $R = 7.84 \times 10^{-11}$.

(b) $R = 1/(10^{19}\cdot 1.6\times10^{-19}) = .625$.

(c) The Hall voltage is $V_T = \mathcal{E}_T w$ where w is the width across which the voltage is measured, and where $\mathcal{E}_T = R_H JB$, with $I = Jwh$ where h is the other cross sectional dimension. Therefore, $V = R_H IB/h$. For copper, the current is $I = hV/R_HB = 10^{-3}\times10^{-3}/(7.84\times10^{-11}\cdot1) = 1.27\times10^4$ A. For the doped silicon, $I = 10^{-3}\times10^{-3}/(.625\cdot1) = 1.6\times10^{-6}$ A.

Problem 8.12 *Junction diode.*

$$V = \frac{\kappa T}{e}\ln(10001) = 0.238 \text{ V}.$$

C.9 Emission, Absorption, and Lasers

Problem 9.1 *Spontaneous decay rate.* Partial ans.

$$A = \left(\frac{2}{3}\right)^8 \frac{c\alpha^4}{a_0'}$$

Problem 9.2 *Balmer-α transition.* Partial ans., $\ell = 2 \to \ell = 1$.

$$\begin{aligned}\langle 322|d_x|211\rangle &= 2.123ea_0' \equiv d_0\\ \langle 320|d_x|210\rangle &= 2d_0/sqrt3\end{aligned}$$

Partial ans., $\ell = 1 \to \ell = 0$.

$$\begin{aligned}\langle 311|d_x|200\rangle &= 1.251ea_0' \equiv d_1\\ \langle 310|d_x|200\rangle &= \sqrt{2}d_1\end{aligned}$$

C.10 Scattering Theory

Problem 10.5

$$r_{\min} = \frac{K}{v_0^2}(1 + \csc\theta_c/2)\,.$$

Problem 10.9

$$\sigma = \frac{16\pi\mu^2\beta^2}{\mu_n^2\hbar^2(\mu_n^2\hbar^2 + 8\mu E)}\,.$$

C.11 Relativistic Quantum Mechanics and Particle Theory

Problem 11.3 The various operators are

$$\begin{aligned}\hat{\mathcal{O}}_{11} &= \hat{\mathcal{O}}_{22} = \frac{\hbar}{\mathrm{i}}\frac{\partial}{\partial t} + e\phi + mc\\ \hat{\mathcal{O}}_{12} &= \hat{\mathcal{O}}_{21} = \hat{\mathcal{O}}_{34} = \hat{\mathcal{O}}_{43} = 0\\ \hat{\mathcal{O}}_{13} &= -\hat{\mathcal{O}}_{24} = \hat{\mathcal{O}}_{31} = -\hat{\mathcal{O}}_{42} = c\left(\frac{\hbar}{\mathrm{i}}\frac{\partial}{\partial z} - eA_z\right)\end{aligned}$$

$$\hat{\mathcal{O}}_{14} = \hat{\mathcal{O}}_{32} = c\left[\left(\frac{\hbar}{\mathrm{i}}\frac{\partial}{\partial x} - eA_x\right) - \mathrm{i}\left(\frac{\hbar}{\mathrm{i}}\frac{\partial}{\partial y} - eA_y\right)\right]$$

$$\hat{\mathcal{O}}_{33} = \hat{\mathcal{O}}_{44} = \frac{\hbar}{\mathrm{i}}\frac{\partial}{\partial t} + e\phi - mc$$

$$\hat{\mathcal{O}}_{23} = \hat{\mathcal{O}}_{41} = c\left[\left(\frac{\hbar}{\mathrm{i}}\frac{\partial}{\partial x} - eA_x\right) + \mathrm{i}\left(\frac{\hbar}{\mathrm{i}}\frac{\partial}{\partial y} - eA_y\right)\right].$$

Problem 11.8

$$\rho = ef^* f(|A|^2 + |B|^2)\,,$$

$$J_x = 0\,,$$

$$J_y = 0\,,$$

$$J_z = ecf^* f(A^* B + B^* A)\,.$$

D

The Fundamental Physical Constants, 1986

These values are taken from *Physics Today*, August 1995, and are listed as the "1986 recommended values of the fundamental physical constants." The original source is found in Rev. Mod. Phys. **57**, 1121 (1987). The most current values may be found at http:\\pdg.lbl.gov. Only SI units are given, whose basic units are:

Quantity	**Units**	
	Name	**Symbol**
length	meter	m
mass	kilogram	kg
time	second	s
electric curent	ampere	A
thermodynamic temperature	kelvin	K
amount of substance	mole	mol
luminous intensity	candela	cd

In the following table, digits in parenthesis indicate the standard deviation uncertainty in the last digits of the given value.

Useful equivalence

$$1 \text{ eV} \leftrightarrow 11,604 \text{ K}.$$

Quantity	Symbol	Value	Units
Universal Constants			
Speed of light in vacuum	c	299792458	m/s
Permeability of vacuum	μ_0	$4\pi \times 10^{-7}$	N/A^2
Permittivity of vacuum	ϵ_0	$1/\mu_0 c^2$	
		$= 8.854187817$	10^{-12} F/m
Constant of Gravitation	G	6.67259(85)	$10^{-11}\ m^3/kg\text{-}s^2$
Planck Constant	h	6.6260755(40)	10^{-34} J-s
$h/2\pi$	$\hbar$	1.05457266(63)	10^{-34} J-s
Electromagnetic Constants			
Elementary charge	e	1.60217733(49)	10^{-19} C
Magnetic flux quantum, $h/2e$	$\mathbf{\Phi}_0$	2.06783461(61)	10^{-15}Wb
Bohr Magneton, $e\hbar/2m_e$	μ_B	9.2740154(31)	10^{-24} J/T
Nuclear magneton, $e\hbar/2m_p$	μ_N	5.0507866(17)	10^{-27} J/T
Atomic Constants			
Fine-structure constant	α	7.29735308(33)	10^{-3}
inverse fine-structure constant	α^{-1}	137.0359895(61)	
Rydberg constant, $m_e c\alpha^2/2h$	R_∞	10973731.534(13)	m^{-1}
Bohr radius, $\alpha/4\pi R_\infty$	a_0	0.529177249(24)	10^{-10} m
Electron			
Mass	m_e	9.1093897(54)	10^{-31} kg
in electron-volts, $m_e c^2/\{e\}$		0.51099906(15)	MeV
Compton wavelength, $h/m_e c$	λ_C	2.42631058(22)	10^{-12} m
Classical radius, $\alpha^2 a_0$	r_e	2.81794092(38)	10^{-15} m
Magnetic moment	μ_e	928.47701(10)	10^{-26} J/T
in Bohr magnetons	μ_e/μ_B	1.001159652193(10)	
g-factor	g_e	2.002319304386(20)	
Proton			
Mass	m_p	1.6726231(10)	10^{-27} kg
in electron-volts, $m_p c^2/\{e\}$		938.27231(28)	MeV
Proton-electron mass ratio	m_p/m_e	1836.152701(37)	
Magnetic moment	μ_p	1.41057138(47)	10^{-26} J/T
Gyromagnetic ratio	γ_p	26752.2128(81)	$10^4\ s^{-1}T^{-1}$
Neutron			
Mass	m_n	1.6749286(10)	10^{-27} kg
in electron-volts, $m_p c^2/\{e\}$		939.56563(28)	MeV
Gyromagnetic ratio	γ_n	26752.2128(81)	$10^4\ s^{-1}T^{-1}$
Magnetic moment	μ_n	0.96623707(40)	10^{-26} J/T
Other Constants			
Avogadro constant	N_A	6.0221367(36)	$10^{23}\ mol^{-1}$
Molar gas constant	R	8.314510(70)	J/mol-K
Boltzmann constant, R/N_A	k	1.380658(12)	10^{-23} J/K
Stefan-Boltzmann constant	σ	5.67051(19)	$10^{-8}W/m^2\text{-}K^4$
Atomic mass unit (unified)	u	1.6605402(10)	10^{-27} kg

Bibliography

[1] Abramowitz, M. and Stegun, I., *Handbook of Mathematical Functions with Formulas, Graphs, and Mathematical Tables*, National Bureau of Standards, U.S. Government Printing Office, 1964. (Also available from Dover.)

[2] Jeffries, H. *Proc. London Math. Soc.* **23**, 428 (1923).

[3] Wentzel, G., *Zeitschrift für Physik* **38**, 518 (1926).

[4] Kramers, H.A., *Zeitschrift für Physik* **39**, 828 (1926).

[5] Brillouin, L., *Journal de Physique* **7**, 353 (1926).

[6] Leighton, R.B., *Principles of Modern Physics*, McGraw-Hill, 1959.

Supplementary Resources

- Anderson, E.E., *Modern Physics and Quantum Mechanics*, W. B. Saunders, 1971.
- Griffiths, D.J., *Inroduction to Quantum Mechanics — Second Edition*, Pearson Prentice Hall, Inc., 2005.
- Scherrer, R., *Quantum Mechanics: An Accessible Introduction*, Addison Wesley, 2006.
- Winter, R.G., *Quantum Physics*, Faculty Publishing, Inc., Davis, California 1986.
- Zettili, N., *Quantum Mechanics — Concepts and Applications*, John Wiley & Sons, Inc., 2001.

Index